Risk Analysis and Management of Natural and Man-Made Hazards

Proceedings of the third conference sponsored by the Engineering Foundation

and co-sponsored by the
National Science Foundation
U.S. Army Corps of Engineers, Institute for Water Resources
Universities Council on Water Resources
Task Committee on Risk Analysis and Management
   of the Committee on Water Resources Planning
   of the ASCE Water Resources Planning and Development Division

Approved for publication by the Water Resources
Planning and Development Division of the
American Society of Civil Engineers

Santa Barbara, California
November 8-13, 1987

Edited by Yacov Y. Haimes and Eugene Z. Stakhiv

Published by the
American Society of Civil Engineers
345 East 47th Street
New York, New York 10017-2398

ABSTRACT

The third Engineering Foundation Conference on Risk-Based Decisionmaking titled "Risk Analysis and Management of Natural and Man-Made Hazards" was held in Santa Barbara, California on November 8-13, 1987. Its proceedings papers focus on the use and application of risk analysis to the management of natural and man-made hazards. In particular, recent advances in the theory and methodology of risk assessment and management are presented and their efficacy are evaluated. Engineers, natural scientists, as well as, social and behavioral scientists share their experiences. The following aspects of risk analysis are covered: risk analysis perspectives; risk communication decisionmaking; environmental health hazards; and risk management strategies for natural hazards. Finally, a summary of the participants' responses to a questionnaire is also included.

Library of Congress Cataloging-in-Publication Data

Risk analysis and management of natural and man-made hazards: proceedings of a conference, Santa Barbara, California, November 8-13, 1987/sponsored by the Engineering Foundation and co-sponsored by the National Science Foundation . . . [et al.]: edited by Yacov Y. Haimes and Eugene Z. Stakhiv.
    p.   cm.
  Includes indexes.
  ISBN 0-87262-688-1
  1. Risk assessment—Congresses. I. Haimes, Yacov Y. II. Stakhiv, Eugene Z. III. Engineering Foundation (U.S.) IV. National Science Foundation (U.S.)
T174.5.R553   1989   89-366
363.1—dc19   CIP

PREFACE

In this decade, our society has been preoccupied with man-made hazards and disasters: dam failures, nuclear power plant failures, ocean dumping of waste, groundwater contamination, and hazards associated with nuclear and toxic waste disposal. Such natural hazards as earthquakes, landslides, tsunamis, windstorms, floods, volcanoes, and wildfires have recently been in the forefront of our consciousness, along with the less visible but potentially adverse effects of global warming and acid rain. The coming decade has been designated as an International Decade for Natural Hazard Reduction by the U.S. National Academy of Sciences. This program will attempt to heighten the awareness of the public and policy makers of the constant threat of natural hazards to society, making clear that these threats will intensify if mitigative and adaptive measures are not integrated into the overall strategies for growth and development.

Regardless of whether the hazards are man-made, natural, or a combination of both, and whether the primary risks are to humans or the environment, fundamental and difficult choices must be made about the most effective and efficient adaptive strategies. Budgetary constraints and fears of adverse economic consequences make the right choice even more difficult to make. Added to these difficulties is the reality that hazard mitigation strategies normally reflect a combination of structural, nonstructural, and regulatory measures whose combined performance has not yet been found reliable. For example, flood warning and evacuation systems are currently proposed as the sole method for flood hazard reduction in many communities across the United States. There are many uncertainties associated with the implementation of such systems, including weather forecasts, streamflow forecasts, issuance of warnings, dissemination of information, and response characteristics of the populace. Structural measures such as levees and reservoirs are more reliable, but the risks of safety are transposed to the low-probability high-consequence failure regime should those structural systems also fail.

The issues of how analysts and policy makers make and convey choices about risk and reliability are the subject of these conference proceedings. There have been many other conferences and workshops on this general theme conducted by virtually every professional society that deals with risk and uncertainty. This is the third such conference sponsored by the Engineering Foundation. Although there were slight variations in the titles of these conferences—in 1980 at Asilmar, California on "Risk/Benefit Analysis in Water Resource Planning and Management"; in 1985 at Santa Barbara on "Risk-Based Decisionmaking in Water Resources"; and this one, in 1987 on "Risk Analysis and Management of Natural and Man-Made Hazards"—all three conferences address the same subject: the assessment and management of risk; namely, the identification, quantification, evaluation and management of risk.

The key difference between the series of Engineering Foundation Conferences and all the other excellent specialized conferences is that here we have attempted to mold the viewpoints of academic researchers with those of regulators and practicing engineers. It is regulators who are ultimately responsible for how policies control the social practices that create or exacerbate hazards; it is engineers who design measures to prevent or mitigate the consequences of these hazards. The practitioners need analytical methods that aid the evaluaion and decisionmaking processes—such methods typically spelled out in a series of increasingly constrained, complex, yet conveniently vague legislative statutes, court decisions, and agency regulations. Elegant abstract theories and solutions that simply do not reflect situations in the real world need to evolve further, so that decisionmaking practices match the realities of the human world, which is often influenced by unstated objectives and political uncertainties.

What has changed during the past decade to warrant yet another conference? Each conference has had roughly the same goals:

- familiarization with advances in risk assessment methods

- exchange of information and perspectives

- identification of current unresolved and emerging risk issues

- attainment of consensus on future courses of action regarding methods and research needs, and

- influence throughout all interested professions, resulting in the promotion of risk assessment applications

What has changed is that we, in this new professional niche, have matured. Those of us involved in risk-based decisionmaking are experiencing the same evolutionary process that systems analysts and systems engineers went through a decade or two ago and maybe are still going through. That is, we are realizing and appreciating both the efficacy and also the limitations of mathematical tools and systemic analysis. In fact, there are many who simply see risk analysis as a specialized extension of the body of knowledge and evaluation perspectives that have come to be associated with systems analysis. Today there is a strong public awareness of the subject of risk: environmental risk, technological and natural risks, and human health and safety risks. The professional community is responding much more forcefully and knowledgeably as well and, in many instances, leading what has ultimately come to be a political debate. We are more critical of the tools that we have developed, because we recognize their ultimate importance and usefulness in the resolution of critical societal problems. We are more willing to accept the premise that a truly effective risk analysis study must, in most cases, be cross-disciplinary, relying on social and behavioral scientists, engineers, regulators, and lawyers.

We are able to acknowledge the fact that the ultimate utility of decision analysis, including risk-based decisionmaking, is not necessarily to articulate the *best* policy option, but rather to avoid the worst, the disastrous policies—those actions in which the cure is worse than the disease.

We are convinced that risk assessment and management must be a integral part of the decisionmaking process, rather than a gratuitous add-on technical analysis. We are realizing that viable risk management must be done within a multiobjective framework, where trade-offs can be explicitly evaluated to reflect social risk preferences, professional or expert opinion, and of course, the economic consequences of alternative courses of action. Some of us are becoming more and more convinced of the grave limitations of the traditional and commonly used expected value concept, and complementing and supplementing this concept with conditional expectation, where decisions about extreme and catastrophic events are not averaged out with more commonly occurring high-frequency/low-consequence events. These are some of the trends that distinguish the conference of 1987 from those of 1980 and 1985.

We are grateful for the wonderful response to the meeting. We are grateful to all participants for attending and sharing their experiences, knowledge, perceptions, and prejudices with each other. We need not be embarrassed by acknowledging the lack of our expertise, because only when we do so will we be able to improve, expand, and advance our knowledge base. We need not be inhibited from exposing our biases, because only through a dialogue will we be able to free ourselves from them. Let us not be rigid in holding on to our perceptions, because their modifications might be a good sign of growth. And let us share our experiences so that we can learn from each other's successes and mistakes. To realized these goals, we have conducted our meetings in the true spirit of the Engineering Foundation Conference, by minimizing lecturing and maximizing discussion.

Several organizations and individuals were instrumental in making this conference possible. Dr. Eleonora Sabadell, program manager of the National Science Foundation research program on "Natural and Man-Made Hazards," was especially helpful in insuring that the conference objectives would be realized by providing generous financial support. Mr. J. Randall Hanchey, Director of the U.S. Army Institute for Water Resources, was equally generous in providing financial support for the conference, as was Mr. Harold Commeror of the Engineering Foundation. We thank, also, the ASCE Task Committee on Risk-Based Decisionmaking and the Universities Council on Water Resources for co-sponsoring this conference and providing the valuable resource base for a large number of eminent speakers and conference participants. Ultimately, the value of a conference such as this resides in the intangible nature of the ideas that were presented and debated and the influence that the discourse had on each of the participants' perspectives. We hope that these perspectives are reflected in the proceedings of this conference.

All papers have been reviewed, edited, and accepted for publication in these proceedings by the editors. The papers are eligible for discussion in the Journal of Water Resources Planning and Management and are also eligible for ASCE Awards.

Yacov Y. Haimes
Charlottesville, Virginia
Eugene Z. Stakhiv
Fort Belvoir, Virginia

CONTENTS

RISK ANALYSIS PERSPECTIVES

RISK COMMUNICATION DECISIONMAKING

ENVIRONMENTAL RISK ANALYSIS

ENVIRONMENTAL HEALTH HAZARDS

GLOBAL WARMING AND CLIMATE CHANGE

RISK MANAGEMENT STRATEGIES FOR NATURAL HAZARDS

RISK MANAGEMENT STRATEGIES FOR TECHNOLOGICAL HAZARDS

ALTERNATIVE RISK EVALUATION PARADIGMS

W. D. Rowe

ABSTRACT

Increased public and regulatory concern with risks imposed by technological undertakings has focused attention on risk analysis as a tool for aiding in risk based decisions. Promulgating health, safety and environmental regulations; addressing product and environmental impairment liability in the light of recent adverse court decisions; qualifying new chemicals, pesticides and technological facilities to meet regulations are but a few of the many areas that require new tools and approaches to identify and sort out the many issues involved. Risk analysis, encompassing the assessment and management of risk, has been promoted as one such approach.

In some areas risk analysis has been used quite effectively in both the assessment and management of risk. In these cases, the adequacy of the data base and the decision criteria encompass the inherent uncertainty in the risk analysis process in a robust manner. However, the situations where adequate data is available and uncertainty ranges are narrow are rare; resulting in severe limitations to the use of risk analysis in a technical or scientific sense. These limitations to risk analysis are real, and can result in misspent resources and costly mistakes in lives and credibility.

This paper addresses risk analysis from the policy making point of view as means of identifying the opportunities for and limitations of risk analysis; and how to avoid the limitations. A fundamental precept in the use of risk assessment as a policy tool is: finding the method to solve the problem, rather than finding the problem to fit the method.

1. ESTABLISHING REFERENCE POINTS FOR COMMUNICATION ABOUT RISK: DEFINITIONS

The late world futurist, Herman Kahn, in opening his many lectures often asked his audience to name the three or four books which they considered as most important to them: for example, the Judeo-Christian Bible, Pilgrim's Progress, etc. This establishes the cultural basis for communication, and provides a reference point from which understanding and misunderstanding derive. In the same manner when we talk about risk, each of us ought to provide his or her definition of risk. I don't mean that we should argue the validity of a definition, but rather provide a reference for our risk paradigms.

My definition of risk is "the downside of a gamble." A gamble implies a probability of outcome, and the gamble may be involuntary or voluntary, avoidable or unavoidable, controllable or uncontrollable.

W.D. Rowe, Ph.D., P.E., is with Rowe Research and Engineering Associates, Inc., Alexandria, Virginia.

1

Table 1

Some Alternative Risk Paradigms

Part A. Paradigm Basis

Use Basis

>   Spectrum of Uses
>   Regulatory Analyses
>   Management Analyses
>   Public Awareness
>   Hierarchy of the Analysis
>   Generic
>   Specific
>   Target Audience
>   Public
>   Stakeholders
>   Technocracy
>   Risk Analysts and Peers
>   Decisionmakers

Approach Basis

>   Top-down vs. Bottom-up Risk Analysis Approaches
>   Quantitative vs. Qualitative Health-Based Risk Analysis
>   Probabilistic vs. Consequential Approaches
>   Absolute vs. Relative Risk Assessments
>   Normative vs. Descriptive Analyses

Part B. Types of Uncertainties in Risk Analyses

>   Random Processes
>   Measurement Uncertainty
>   Interpretation of Measurements
>   Model Choice Uncertainty
>   Identification and Aggregation of Margins of Safety

The total gamble in which the risk is imbedded must be addressed if the risk is to be analyzed, both the upside (benefits) and downside. Further, I define risk assessment to mean the estimation of risk, and risk management to mean the reduction or control of risk to an "acceptable" level, whether or not the level can be explicitly set. In reality these two processes are not separable since the uncertainty in one affects the judgments we make about the other and vice versa. They may be separated in practice for convenience, but the uncertainties in each area may be the dominant factors in any analysis of risk.

This leads to my definition of risk analysis. Risk analysis is a policy analysis tool that uses a knowledge base consisting of scientific and science policy information to aid in resolving decisions. Risk analysis is thus a subset of decision theory, and its importance and utility derive from its applications and how well the decisions involved were resolved.

TABLE 2a

Regulatory Analyses

A. KINDS OF ANALYSES CONDUCTED BY REGULATORY AGENCIES

1). Screening Analyses - To determine if a risk exists and is high enough to be considered for regulatory control.

2). Regulatory Impact Analysis - To justify regulatory actions and satisfy administrative law requirements.

3). Compliance Analyses - To demonstrate regulatory violations.

4). Responding Analyses - In response to judicial and legislative challenges.

B. ANALYSES MADE BY OTHERS IN RESPONSE TO EXISTING REGULATIONS

1). Environmental Impact Statements
2). Permitting Requirements
3. Compliance Monitoring

C. ANALYSES MADE BY OTHERS TO DEFEND AGAINST UNWARRANTED REGULATORY ACTION

1). Response To Requests For Comments By Regulators-Industry response to agency actions above.

2). Support of Judicial Actions

a. Response to improper agency actions
b. Defense against enforcement proceedings

---

2. ALTERNATIVE RISK PARADIGMS

Two different ways of looking at risk paradigms derive from the use to be made of the analysis and the approach to be used to carry out the analysis. Given limited space, only the paradigms listed in Table 1 will be addressed. Though not necessarily exhaustive, these particular paradigms seem to be the most divisive ones at present.

The Spectrum Of Uses Of Risk Analysis

There is no such thing as the risk analysis. There is a whole spectrum of risk analyses based, at least in part, on the use to be made of the risk analysis. Tables 2a, 2b, and 2c provide a spectrum of different uses of risk analysis which is neither collectively exhaustive nor mutually exclusive. Table 2a lists some regulatory uses of risk analysis, Table 2b management uses, and Table 2c analyses used to affect public awareness. The point is that each use requires a different type

TABLE 2b

Management Support Analyses

A. MARKETING

1). Absolute Risk - Demonstrate that a product or process is safe or harmless on an absolute risk basis,--that is, the risk on an absolute basis is below some standard or regulation implying an acceptable level of risk.

2). Relative Risk - Demonstrate that a product or process is relatively safer and less harmful than alternative and competitive products or processes.

B. PLANNING

1). Research and Development

    a. Risk Reduction - Identify areas of high risk (or relatively high risk) in particular products or processes to:

          1. Forestall the need for regulation.
          2. Reduce exposure to future liability claims.
          3. Develop defensive strategies to bound risk liability.
          4. Identify new markets for risk control technology.

    b. Improved Analysis Capability

2). Cost-Effective Use of Resources - Focus resources on the most risk reduction for the dollar.

3). Evaluation of Alternative - Systems or processes

C. RISK MANAGEMENT

1). Prevent Risks from Occurring - by anticipating and controlling them.

2). Reducing Exposure - for health and safety and financial risks for a given, existing process or product. Conduct analyses for:

    a. System Safety - Reduce risk within a system.
    b. Product Safety and Liability. - Reduce exposure to legal proceedings.
    c. Third Party Assumption of Risk

          1. Insurance - As a means to hedge against risks
          2. Malpractice - Laws to limit liability

TABLE 2c

Public Education

A. INCREASE PUBLIC AWARENESS

1). Seek Rational Public Responses - A knowledgeable public will hopefully act on information rather than preset beliefs.

2). Fulfill Regulatory Requirements for Public Disclosure - A good, simplified and accurate disclosure can also be a useful educational tool.

B. ANXIETY FACTORS

1). Bring Perceived Risks More Closely into Alignment with Objective Risks - Anxiety reduction; may also be a defensive strategy.

2). Frighten People into Action or Agreement - An offensive strategy attempting to stir fear and anxiety

---

TABLE 3

Classification of Risk Analyses by Scope of the Application

MICRO TO MACRO CLASSIFICATION

SITE-SPECIFIC STUDIES

UTILITY PLANNING STUDIES

POWER GRID PLANNING STUDIES

NATIONAL ENERGY SUPPLY PLANNING

GLOBAL PLANNING

INTERNATIONAL ENERGY PLANNING

SPECIAL PURPOSE APPLICATIONS

ENERGY SUBSYSTEM INVESTMENT

EVALUATION OF POTENTIAL PROBLEMS IN NEW ENERGY SOURCES

TO SUPPORT OR REJECT AN ENERGY OPTION

TABLE 4

Definitions

TOP-DOWN RISK ANALYSIS

THE PROCESS WHEREBY THE RISK ANALYSIS METHODOLOGY IS TAILORED TO THE POLICY NEEDS AND ITS FEASIBILITY DETERMINED.

BOTTOM-UP RISK ANALYSIS

TAKING EACH EVENT THAT CAN OCCUR IN A SYSTEM, ANALYZING THE PATHWAYS LEADING TO THE RANGE OF POSSIBLE CONSEQUENCES, AND AGGREGATING THESE OVER THE TOTAL SPECTRUM OF EVENTS AND THEIR ASSOCIATED PROBABILITIES.

JOINT RISK ANALYSIS

THE LIMITED BOTTOM-UP RISK ANALYSIS CARRIED OUT AS A RESULT OF TOP-DOWN RISK ANALYSIS, AND SUBSEQUENTLY MERGED WITH THE TOP-DOWN ANALYSIS TO FORM THE FINAL POLICY ANALYSIS.

---

of risk analysis, and the different analyses are not interchangeable. Appendix A provides a detailed explanation of each type of analysis and its requirements.

A second aspect of use is the level of specificity of the analysis. Table 3 provides an illustration of the range of analyses from the specific to the general, in this case specific to a risk analysis of alternative energy systems. Each level requires a different kind of analysis.

A third aspect is the particular target audience or audiences. The public and lay stakeholders require a presentation that is different from the technical audience, including risk analyst peers, and from the decisionmakers (who may or may not be technically oriented). If the presentation is different, the analyses themselves for each audience may be different.

Conclusively, the use determines the method, not the other way around. Picking the method to resolve the problem defined by the use is more important than developing universal methods.

3.   SOME DIFFERENT APPROACHES TO RISK ANALYSIS

The policy analysis aspect of risk suggests that the problem drives the methodology of the analysis. This implies the need for an approach that starts with the problem and works downward to the methods of detailed analysis. This is called the top-down approach. Table 4 provides definitions of top-down, bottom-up, and joint risk analysis approaches.

Top-down and Bottom-up Risk Analysis

a.   Top-down Risk Analysis

    Top-down risk analysis is a paradigm for determining the most
appropriate risk analysis for a given decision situation, for making
visible the key decision parameters and value judgments involved, for
identifying viable alternative strategies for resolution of issues, for
identifying scientific and other information critical to the decision
process as well as the needed precision of such information, and for
communicating the decision process to those affected. The top-down
analysis tailors the risk analysis to resolve the issues at hand, and
this aspect of the analysis does not necessarily analyze the risks. The
top-down approach shows whether it is possible to resolve the policy
issues by a subsequent risk analysis; and, if not, identifies the value
conflicts that prevent issue resolution by other than political means.

b.   Bottom-up Risk Analysis

    In contrast, bottom-up risk analysis starts from the basic
scientific information and attempts to use this information for policy
analysis by way of prescriptive methodologies. In nearly all cases,
problems arise from large uncertainties in the scientific information
base. These problems are usually addressed by retaining and aggregating
the ranges of uncertainty, most often in a semi-qualitative manner, or
by use of the value judgments of experts or groups of experts. In the
first case, the uncertainties are often too large for the analysis to
result in meaningful conclusions; or if forced to conclusions,
assumptions may mask the uncertainty. The second case substitutes
science policy for science. This does not imply that all bottom-up
analyses are not useful, but often the resources entailed in making such
analyses are very large; and the results are often inconclusive,
especially when such an analysis purports to serve all policy purposes.

c.   Joint Risk Analysis

    The risk analysis that is made as a result of a top-down risk
analysis need only use information necessary to resolve the decision (if
it is resolvable by other than political means) and the information used
must only be as precise as is necessary for resolution. The combination
of a top-down risk assessment with the needed (and only the needed)
bottom-up analyses are called the joint top-down, bottom-up approach.
It is this paradigm which is preferred, since it joins the best of the
two approaches. Appendix B lays out and discusses the steps in one
approach to joint risk analysis. This approach has been used
effectively in a number of different areas, but there is no reason to
believe it is any better than other well planned approaches. It is
provided to illustrate the different perspective involved.

Quantitative Versus Qualitative Health Analysis

a.   Qualitative Health Risk Analysis

    The National Academy of Science, EPA, OSHA, and a number of other
federal and state agencies subscribe to a particular paradigm for
estimating the risks of exposure to toxic chemicals. I have termed this

approach--extrapolating results of bioassays at high doses in laboratory animals to relative potencies at low doses in humans--the quantitative approach to health risk assessment. There is no empirical evidence supporting the determination of relative cancer-inducing potencies of two or more chemicals from laboratory animal data without human epidemiological data. Therefore, the methodology is based upon a number of biological and stochastic models, which can only be validated at high dose levels in animals and at the limits of epidemiological studies in human victims of historic or accidental exposure. Pharmacokinetic studies of the response of a chemical and its metabolites in the various species under test, including man when ethical, can help in adjusting the biological models, but still do not allow actual validation of the models.

b. Uncertainties

    There are several types of uncertainties involved:

    Random Fluctuation   - Inherent random processes

    Measurement Uncertainty   - Precision of the measurement
                                   system
                                Accuracy of measurement

    Interpretation of Measurements
                                - Difference in judgment
                                - Biases of experts

    Model Choice Uncertainty
                                - Unverifiable uncertainty in the
                                   selection of alternative models
                                    Biological models
                                    Extrapolation models
                                    Statistical models

    Margins of Safety   - Use of margins of safety to account for
                                   uncertainty
                                - Aggregation of margins of safety

    The first two types are well established, and there are established methods for analyzing these uncertainties. The resulting uncertainty when experts disagree about the interpretation of measurements is not as well understood, and often not even recognized as a major source of uncertainty. Moreover, differences in interpretation can be the result of differences in training, methodology used, and, unfortunately, biases based upon other than scientific motivation.

    Model choice uncertainty is illustrated in Table 5. Seven different types of models are shown that are required to make an estimate of environmental risk, specifically in this case of a toxic organic compound, radio-iodine and nitrogen dioxide. The range of uncertainty due to the choice of plausible, available alternative models is shown. The bottom of the range represents situations with low variability, e.g. a diffusion model for a flat, even terrain. The higher values are for cases with high variability, e.g. a diffusion model operating on a hilly terrain with shear wind strata. When these

TABLE 5

Model Uncertainty in the Bottom-up Risk Analysis
of Illustrative Releases to the Environment

STEP	MODEL	UNCERTAINTY FACTOR RANGE		
		Toxic Organic	Radio Iodine	Nitrogen Dioxide
1.	Source Term.			
	Averaging in Space	1.1 to 3	1.1 to 2	1.1 to 10
	Averaging in Time	1.1 to 3	1.1 to 3	1.1 to 5
2.	Pathways.			
	Diffusion Models	2 to 10	1.1 to 2	1.1 to 3
3.	Metabolic Pathways and Fate.			
	Organ Intake Models	2 to 10	1.1 to 3	1.1 to 3
	Distribution Models	2 to 4	1.1 to 2	1.1 to 2
	Retention Models	2 to 4	1.1 to 1.5	1.1 to 1.5
4.	Dose Estimate.			
	Exposure Time Profile	2 to 10	1.1 to 10	2 to 10
	Maximum vs. Average Individual	2 to 10	1.1 to 5	2 to 10
5.	Dose-Effect Relationship.			
	Extrapolation From Animal to Man			
	Choice of Scaling Model	40	2 to 10	20
	Metabolic Differences	2 to 100	1.1 to 1.5	2 to 5
	Extrapolation From High to Low Dose			
	Choice of Model*	1000	3	100
	Margins of Safety*	10 to 1000	2	2 to 10
6.	Individual Risk Estimate.			
	Real vs Hypothetical Individual	4 to 20	4 to 20	4 to 20
7.	Population Risk Estimate.			
	Integration vs Averaging Models	2 to 10	2 to 5	2 to 10
	Multiplicative Ranges			
	Low	5×10^7	2×10^2	5×10^4
	High	1×10^{15}	4×10^6	3×10^{10}

*Use either, but not both. The first is for non-threshold dose-effect relationships, the latter for threshold types.

ranges are aggregated, they do so multiplicatively; and, since they are uniform distributions, their endpoints must be used. As can be seen from the multiplicative ranges in the table, the ranges vary from about two orders of magnitude (factors of ten) to fifteen orders of magnitude. For a toxic organic, that range goes from seven orders of magnitude to fifteen. With this kind of uncertainty, risk analysis cannot be very robust unless the means of reducing risk is so effective as to overcome these margins of safety. A good example is a hazardous waste facility with a destruction removal efficiency of 99.9999 percent (for PCBs and TCDDs) and a diffusion factor that results in dilution to one part in ten million. This results in a total risk reduction of thirteen orders of magnitude, enough to overcome most of the uncertainty due to model choice. Conversely a hazardous waste landfill cannot account for more than three orders of magnitude of risk reduction and cannot overcome the effect of the aggregated margins of safety.

c.   Qualitative Health Risk Analysis

Qualitative approaches attempt to establish "no-observed-effect levels" (NOELS) in test animals and provide margins of safety--for example, a factor of 100--for acceptable human exposure levels. The inherent models are that the test animal and man are metabolically similar, that a no-effect threshold exists, and that an adequate margin of safety below the threshold will account for uncertainties.

In either case the uncertainties should be displayed, not masked. If the total range of uncertainties is displayed, quantitative risk analysis may be a useful means for displaying what is known and what is not known about risk. However, it may not be a useful technique for regulating risk in the manner recommended by the National Academy of Science, EPA, OSHA, etc. A case-by-case examination of the problem is probably required rather than a procedural process minimizing the need for judgment by those using the procedures. Generally, this has occurred, in spite of the lip service given to quantitative risk analysis.

Probabilistic vs. Consequential Risk Analysis

a.   Probabilistic Risk Analysis

Since risk involves a gamble, probabilistic outcomes are inherent. However, the measurement of probability is based upon belief in the meaning of probability. Objectivists base probability estimates on either the frequency of outcomes of previous trials (frequentist approach) or predetermined models (a priori approach). Subjectivists believe that they can make judgments about probability, and they use these judgments in conjunction with data to improve their estimates (Bayes approach). These fundamentally different approaches are not reconcilable, and these different views spill over into risk analysis.

Probabilistic risk assessment stresses the measurement of probabilities. If the events occur frequently enough, probabilities of future events can be inferred from past experience. Even, fault, and decision trees and statistical decision theory provide useful techniques for estimating the absolute risk of alternative outcomes, absolute risk being the measurement of risk in terms of probabilities and consequences for the gamble. However, when events are rare--that is, they occur so infrequently that actual data cannot be obtained--the use of probabilistic analysis to define risk on an absolute basis breaks down. The amount of information inferred from sparse events depends upon one's belief about probability, specifically whether one is an objectivist or a subjectivist. The objectivists believe that the measured data contain all the information available. The subjectivists believe that additional data can be obtained. The estimates of absolute risk are thus based upon belief structures, and are unverifiable.

Several approaches have been used to get around this fundamental problem:

System Structures - Measurements on frequent events in subsystems are used to project overall system behavior.

Data Interpolation - Measurements in other systems thought to have identical or similar behavior are used to expand the data base to learn about a system for which little data is available.

These approaches have major limitations in application, e.g., common mode failure in system structuring and invalid pooling of data in data interpolation. As a result the absolute application of probabilistic risk assessment for rare events is not meaningful. It depends upon the belief about probability of the analysts, which is a doubtful process upon which to base either scientific or public policy.

On the other hand, relative risk analysis of rare events can be useful. Relative risk is the estimation of the risks of alternative gambles in relation to each other. It generates less error than absolute risk analysis since many of the errors in the absolute case are identical for all alternatives. So they wash out in a relative analysis. Probabilistic risk analysis can be used to evaluate the relative risk of alternative systems, to identify vulnerabilities to failure in alternatives or component systems, and to estimate the relative cost-effectiveness of risk reduction among alternatives. Having made a relative risk analysis, a handle on absolute risk may be obtained by "pegging" one alternative to an absolute measure. Pegging involves making an absolute risk assessment for the most well-known alternative. While there will be high uncertainty in the absolute estimate of risk, stakeholders who have experienced a well-known gamble can substitute their own perception of the absolute risk for the peg. The other alternatives are scaled upward or downward as the peg varies. Even though the pegging is "soft" it can be related to the perception of stakeholders as opposed to a belief structure of analysts.

b.   Consequential Risk Analysis

When absolute or relative risk estimates cannot be made effectively for low probability (rare) events, other approaches are in order. Consequential risk analysis assumes that an event occurs and attempts to prevent a particular outcome from occurring or to mitigate the impact of the outcome. Some of the techniques used involve analytical processes such as sensitivity analysis, optimization and cost-effectiveness of risk reduction, and contingency processes that mitigate consequences such as emergency response planning, early warning systems, and evacuation operations.

Normative vs. Descriptive Risk Analyses

Normative risk analyses are undertaken for a specific purpose such as establishing standards, meeting regulatory requirements, etc. Descriptive risk analyses are undertaken to better understand risks and risk processes, to evaluate alternate systems and means for reducing risk, or simply to present risks and uncertainties to interested parties. Many normative risk analyses are used to support standard setting. Descriptive risk analyses are usually research activities or after-the-fact evaluations.

Normative analyses have tended to mask uncertainties by replacing ranges of uncertainties with procedural surrogates. Margins of safety are used to account for uncertainty, as in quantitative health risk analysis. The cost of these margins of safety and the uncertainties involved should be made explicitly visible to all audiences. Often the uncertainty in a risk analysis is so great that it cannot be used normatively, but it may be used descriptively to identify the critical uncertainties, to better understand the problem, and to set research and information acquisition agendas for effective reduction of uncertainty. Normative analyses often require simplicity while descriptive analyses often tend to be complex.

4. CONCLUSION

There are many other classifications or paradigms for risk analysis; those presented here are particularly critical. The policy requirements and the problem content (the actual data) determine the methods needed, not the structure of risk analysis. What works should be used, what doesn't should be discarded or at least put away until better data is available.

Structuring the problem using a variety of available approaches before selecting solution methodologies is a requirement for realizing useful risk analyses. The top-down approach illustrated here is one such process, particularly when addressing policy decisions. Pursuit of method independent of application assumes that the methods will have wide application. This is seldom true. They have limited application for a specific purpose or area; therefore, one must address the application and find the methods most useful. Use what works; discard what doesn't.

APPENDIX A

USES OF RISK ANALYSIS

The uses of risk analysis as outlined in Tables 2a, 2b, and 2c are explained here more fully. In addition the sponsor and the primary target users are indicated, in terms of risk analysis of alternate energy systems.

REGULATORY ANALYSES

Regulatory analyses are conducted by regulatory agencies and their contractors for a variety of reasons, primarily to establish and enforce regulations. Another set of analyses is used to adhere to existing regulations, both those set for environmental impact analysis (controlled in the United States by the National Environmental Policy Act [NEPA, 1970] & the Code of Federal Regulations, Section 1502) and those set by industry to meet existing regulatory requirements.

A. ANALYSES CONDUCTED BY REGULATORY AGENCIES

These analyses are sponsored by regulatory agencies and are usually directed at risk managers and the technical community. Many countries require public disclosure as well, and the presentation of technical material to the public presents special problems. For energy systems, the primary issues are environmental issues such as sitting, water resource and land use, pollution, and health and safety issues.

1.) Screening Analyses to determine if a risk exists and is high enough to be considered for regulatory control

Risk to individuals and to exposed populations is estimated using crude estimates, and the results are compared against formal or informal deminimis levels to determine if the risks are high enough to warrant further action. Either the risk level (individual and population risks) or the cost-effectiveness of risk reduction or both can be used as criteria for decisionmaking.

2.) Regulatory Impact Analysis to justify regulatory actions and satisfy administrative law requirements.

Every regulation establishing standards for health, safety, and the environment must go through a formal administrative process laid out by the laws of the country involved. In the United States this is governed by a plethora of regulation. Those for environmental regulations have many procedural steps, which include advance notice of proposed rule-making, a draft, and a final regulation, all involving public input and comment. The analysis used to support these regulatory steps generally has very large uncertainties in cause-effect relationships and exposure estimates and uses margins of safety in the direction of increased protection of human health to address these uncertainties. The margins of safety, used at each step, are aggregated throughout the total process leading to large overstatements of risk. Both individual and populations risks should be addressed, but this is not always done in practice.

Additional margins of safety are often added to provide increased agency credibility in the eyes of the agency's constituency. This process is to be avoided. The large amount of systematic error introduced is often hard to reverse once the process is established. Arguments to keep estimates within realistic ranges must take place at every step of the administrative process. Strategies aimed at providing early input into the analysis process and involvement at every step of the process are necessary and can be effective in keeping estimates of risk realistic. Industry has had a great deal of influence on the process by this approach, although the cost, including industry risk analyses to offset the agency analyses, is expensive.

3.) Compliance Analyses to demonstrate regulatory violations

When an agency takes an enforcement action, it makes an analysis of the violation of the standard with a "chain of evidence" to support the violation. Risk analyses are required when the regulation is based upon risk. Once again improper use of margins of safety used as means to address uncertainties must be kept within reasonable limits.

4.) Analyses in response to judicial and legislative challenges

When regulations are challenged in the courts or by legislative bodies (Congress and state legislatures in the United States), risk analyses are often made by the agencies responsible for the regulation to defend their actions. These analyses are often extensive and biased to support the action taken.

B. ANALYSES MADE BY OTHERS IN RESPONSE TO EXISTING REGULATIONS

Environmental and safety regulations require utilities or industrial organizations, public or private, to make risk analyses of proposed or existing facilities. The utility or industrial organization becomes the sponsor of the analysis. The regulating authority is the primary target user, but the analysis must undergo technical scrutiny and, in many countries, review by the general public.

1.) Environmental Impact Statements

The mechanism for carrying out these analyses in the United States is the National Environmental Policy Act (NEPA). It has been in force since 1970, and has been a pacesetter in establishing needs for formal environmental analysis. The process was fraught with difficulties in the beginning, but has resulted in better understanding of the risk/cost/benefit process. Some countries in Europe are adapting some of the more advantageous aspects of the NEPA process, such as carrying out the analysis of alternatives without the overly legalistic framework of the American system.

This process basically requires a relative risk analysis among alternatives, one of which is usually a "no-action" alternative. The process does not require margins of safety, except to account for uncertainties. Essentially only a best estimate is required, but the analysis must assure that an underestimate of risk does not take place.

Margins of safety for protection, as in regulations, are not required; but when information cannot be acquired or costs of acquisition are exorbitant, a worst-case analysis is required. The sponsors of the analysis and the decisionmakers must learn to "let it all hang out," state the risks as they are, put risks into perspective, and let the decisionmaker make his decision among the alternatives. In the United States the law implies that if the NEPA procedure is met, the value decision is up to the sponsoring agency, and the courts will not interfere. The analysis must be impartial, and those conducting the analyses should not be advocates of any alternative prior to the decision. After the analysis is completed, the sponsor may select a preferred alternative, and it may be presented with the other alternatives.

2.) Permitting Requirements

Specific regulations requiring permits, such as for the Resource Conservation and Recovery Act (RCRA) in the United States, require some form of risk analysis in the development of the justification for the permit. These analyses are usually conducted by the organization requesting the permit and must demonstrate that designated criteria are met.

3.) Compliance Monitoring

Enforcement actions by national, state, or local governments under existing regulations often require analyses, made by the regulating agency, to justify the compliance action. These are generally offense risk analyses.

C. ANALYSES MADE BY OTHERS TO DEFEND AGAINST UNWARRANTED REGULATORY ACTION

1.) Response to requests for comments by regulators

Utilities and industry respond to agency actions as described in Paragraph A2 above.

2.) Support of judicial actions

a. Response to improper agency actions

Utilities and industry respond to agency actions as described in Paragraphs A2 and A4 above.

b. Defense against enforcement proceedings

Enforcement actions by national, state, or local governments under existing regulations often require analyses made by the accused party for defensive purposes after the compliance action takes place.

MANAGEMENT SUPPORT ANALYSES

Utilities and industrial organizations, public and private, make analyses to assure adequate safety of operations, increased productivity, and cost-effective operation. These are sponsor-originated studies aimed at policy makers and risk managers at the strategic level, and may be aimed at technical people at the operational level.

A. MARKETING

These analyses demonstrate that a product or process is safe or harmless on an absolute basis, or that it is relatively more safe and less harmful than alternative and competitive products or processes. Often they are aimed at the general public as the ultimate consumers. There is, of course, no such thing as zero risk, but some try to sell it.

B. PLANNING

1.) Research and Development

a. Risk Reduction Safety Analysis

The purpose is to identify areas of high risk (or relatively high risk) in particular products or processes and to seek means to provide cost-effective solutions to reduce these risks. This is normally done for the following reasons:

> 1. Ensure a safe operation, which makes good business sense. Outages because of failure can cause loss of production.
>
> 2. Forestall the need for regulation.
>
> 3. Reduce exposure to future liability claims.
>
> 4. Develop defensive strategies to bound risk liability.
>
> 5. Identify new markets for risk control technology.

b. Improved Analysis Capability

Identify areas of high uncertainty in risk analyses, and undertake programs to cost-effectively reduce these uncertainties. This is a particularly necessary strategy for combating overzealous use of margins of safety which have been used in the face of uncertainty. This is also true for new product or process areas, or at least for areas where risk analysis has not been used effectively in the past as a result of such uncertainties.

2.) Cost-Effective Use of Resources

These analyses seek the most risk reduction for the dollar, and can be used by both the private and public sectors. Utilities and industrial organizations can use this approach to best allocate resources addressed to safety. This approach can be used in the public sector, by individual agencies, across agencies, and within other organizations as a way to cost-effectively address risk reduction. The idea is to serve the public by spending tax dollars for risk reduction cost-effectively and fairly. The depth of analysis is less than that for regulatory purposes and uses relative risk estimates and the relative cost of reducing risks. This is an area where probabilistic risk assessment techniques can be very effective.

3.) Evaluation of alternative systems for conducting a process

Alternative systems can be evaluated on a relative risk basis to provide perspectives on the types and magnitude of risks for alternative systems for implementing a specific process or product.

C. RISK MANAGEMENT

Risk management analyses involve preventing risks from occurring by anticipating and controlling them. This is accomplished by reducing exposure, health, safety, and financial risks for a given existing process or product. One can make analyses for:

1.) System safety

These analyses consider the system for points of possible failure and provide technological "fixes" for weak points in the system. The use of probabalistic risk analysis, using fault trees and event trees, has met some success in this area. Evaluation of margins of safety in the system is another approach that may be more fruitful.

2.) Product safety and liability

The courts become involved when parties are believed to have suffered physical and mental damage or stress from a product that either fails to operate as expected or causes harm. In the United States this has become a major problem, especially in the cases of large jury awards when the courts base liability on ability to pay rather than relating cause and effect.

3.) Third party assumption of risk

Third party assumption of risk occurs primarily through insurance. The objective is to spread the risk more equitably among subscribers. Of course, insurance companies sell this service with a profit motive in mind. As a result, the insurance industry has controls applied to it and is regulated at the state level to varying degrees.

a.   Insurance

Insurance companies can spread risk by pooling risks among subscribers for all types of coverage. Insurance companies operate on a profit basis for both stock and mutual underwriters. Large compensation awards in the United States courts have led to withdrawal from underwriting in many areas by most insurance companies. For example, at this writing it is impossible to obtain environmental liability insurance for hazardous waste disposal or insurance for vaccine producers of side-effect risks. The latter may soon be underwritten through government action, since many pharmaceutical companies have ceased to make certain critical vaccines.

b.   Malpractice

Large liability claims and awards for medical malpractice have caused a large proportion of effort to be directed at preventing malpractice suits rather than directly reducing risks. Moreover, the cost of malpractice insurance for individual practitioners has become exorbitant. Essentially more attention is being given to the third party risks of operating any business with definable risk consequences, in some instances more than the attention being given to the risks themselves.

PUBLIC EDUCATION

The public desire for adequate supplies of environmentally safe energy sources at low prices makes it incumbent on public officials to display energy options and their advantages, disadvantages, and problems to their constituents in a manner that promotes better understanding of the issues. However, if such information is biased toward specific applications or energy sources, it may become suspect and its intentions may be misinterpreted. It is necessary to provide information and commentary in an open manner, leaving the reader to draw his own conclusions, if such bias is to be avoided. This does not mean that summaries and commentaries should be avoided, only that they not be slanted.

A.   PUBLIC AWARENESS

1)   Seek rational public responses

We hope that a knowledgeable public will act on information rather than preset beliefs. The expectation is that such presentations will be without bias, if they are sponsored and carried out by public or private organizations which are not stakeholders.

2)   Fulfill regulatory requirements for public disclosure

Even though public disclosure may be required by law, a good, simplified, and accurate disclosure can also be a useful educational tool, whether or not the sponsor is a stakeholder.

B.   ANXIETY FACTORS

1)   Bring perceived risks more closely into alignment with objective risks

Public anxiety can be allayed to some extent by changing the perception of risks by making the objective risk estimates meaningful to people. While anxiety reduction is a worthwhile end in itself, it may also be used as a defensive strategy since an informed public can often be expected to act in a rational manner.

2)   Frighten people into action or agreement

For those seeking attention, a premature statement or an overstatement of risks to the public can arouse anxiety. This is an offensive strategy, attempting to stir fear and anxiety.

APPENDIX B

GENERIC STEPS IN A JOINT RISK ANALYSIS

I.   TOP-DOWN RISK ANALYSIS (Steps 1-13)

Part A.   Develop a framework and identify key value issues and
          conflicts

Part B.   Specify the framework of the analysis

II.   BOTTOM-UP RISK ANALYSIS (Steps 14-16)

Part C.   Acquire data

   Step 14.   Conduct required studies to obtain required information

   Step 15.   Acquire the data

Part D.   Implement the analysis

   Step 16.   Conduct the analysis

III.   IMPLEMENT THE DECISION FRAMEWORK (Steps 17-19)

Part E.   Merge the results of the bottom-up analysis into the framework

   Step. 17.   Reduce the conclusions to the implications of alternative
               policy options based upon the analysis

Part F.   Present the results of the analysis

   Step 18.   Present executive summary and policy analysis document

   Step 19.   Present technical backup documents

GENERIC STEPS IN A TOP-DOWN RISK ANALYSIS

Part A. Develop a framework and identify key value issues and conflicts

Step 1. Determine the use for which the analysis is to be made

Step 2. Identify a minimum set of critical variables

Step 3. Generate a set of combinational scenarios for the intersection of the variable conditions (states of nature)

Step 4. Develop a set of alternative solution strategies

Step 5. Develop a decision model problem structure

Step 6. Identify the critical decisionmakers

Step 7. Have each (or group of) decisionmaker determine his choice of alternatives for each scenario or identify the information needed to make such a choice

Step 8. Classify each scenario into one of three classes

Step 9. Find means to resolve value conflicts, if possible; if not, stop

Part B. Specify the framework of the analysis

Step 10. Finalize the structure of the analysis

Step 11. Develop the data requirements for the bottom-up analysis

Step 12. Identify the limitations due to uncertainties and estimate the robustness of the joint analysis

Step 13. Provide a report on the top-down analysis, providing specifications, limitations, and recommendations

---

ENGINEERING APPROACHES TO RISK AND RELIABILITY ANALYSIS

Ben Chie Yen

ABSTRACT

Engineering projects inevitably involve uncertainties in their design and operation. In this paper, reliability is first defined mathematically, providing a basis for quantitative reliability evaluation. The limitations of the traditional approach, using the return period of geophysical phenomena and an arbitrarily assigned safety factor, are discussed. Recently developed procedures for evaluating engineering reliability and methods for combining uncertainties of various individual contributing factors for a total reliability determination are introduced. The fault tree is a useful tool to identify different failure modes and to divide the work according to teams of different specialties. The evaluation of the reliability of a culvert design is given as an example. The existing reliability evaluation methods are not perfect, and refinement can be made through future research. Nonetheless, the new techniques are undoubtedly a great improvement over the past techniques, which considered only the return period and safety factors. The probable maximum precipitation (PMP) is also discussed in the context of reliability analysis. In view of the nonstationary dynamic global environment, it is inappropriate to designate a frequency for PMP.

1. INTRODUCTION

Reliability of engineering projects designed to achieve desired objectives and avoid failure has long been a technical interest as well as a public concern. Throughout the years, methods have been developed for the piecemeal reliability evaluation of some aspects of specific components of an engineering system. A typical example is frequency analysis, which deals with the uncertainty associated with natural randomness in hydrology, wind, and seismic engineering. Gradually in recent decades, methods have been developed in different engineering disciplines to account for uncertainties in individual components and to integrate them to evaluate total engineering systems reliability, notably advanced systems in nuclear, space, and communication engineering.

There are several reasons for the relatively slow development in comprehensive engineering reliability analysis. First, it requires the engineer to possess an overall view and comprehensive perception of the entire engineering system. Second, the analytical tools suitable for considering and combining component uncertainties and the computers suitable for the necessary computations were not sufficiently advanced

---

Ben Chie Yen is Professor of Civil Engineering, University of Illinois at Urbana-Champaign, Urbana, Illinois.

until the past quarter of a century. Third, the significant achievements of hydrologists in frequency analysis and probability distribution functions have obscured their development in more comprehensive risk analysis. Last but not least is the confusion of terminology used in different disciplines, especially the term "risk analysis."

Authoritative dictionaries give as a major definition of "risk" the chance of damage, injury, or loss. Thus, in various engineering fields such as structures, seismic, and nuclear engineering, risk analysis is referred to as the analysis of the probability of failure, where failure is defined as not achieving the intended performance level of the objective. Conversely, in economic and related decision sciences, and recently in health and environmental sciences, risk analysis has usually been referred to as the analysis and assessment of the consequences of failure, particularly its cost and/or social consequences. The majority of the participants in this conference are in decisionmaking or regulatory fields. Therefore, in order to minimize confusion in this presentation, let us agree that the term "engineering reliability analysis" refers to the quantitative evaluation of the probability of failure, whereas the term "risk analysis" will be avoided.

The purpose of this presentation is to introduce briefly some of the existing methods useful for total reliability evaluation of engineering systems. Reliability, determined quantitatively using these methods, would serve as the most basic information necessary for any scientific risk cost analysis and decisionmaking. However, discussion on risk cost analysis, failure consequence assessment, and decision analysis is beyond the scope of this presentation. In the following, first reliability is defined in a rigorous mathematical manner. The limitations of the traditional approach, using the return period of geophysical phenomena and an arbitrarily assigned safety factor, are discussed. Next the procedure for reliability evaluation and the methods to express and combine uncertainties of various contributing factors to produce total reliability are introduced. Finally, the probable maximum precipitation is examined in the context of engineering reliability.

2.   DEFINITION OF ENGINEERING SYSTEM RELIABILITY

The failure of an engineering system can be defined as the external causes or loading on the system, L, exceeding the target performance level or resistance of the system, R, such that the system fails to accomplish its intended objective. The resistance is the actual flow capacity of the pipe, while the loading is the approaching flow from upstream. In the example of a culvert or sewer, a failure may be considered to exist when the pipe is unable to deliver the flood flow. For a water supply project, failure may be defined as the demand (loading) exceeding the supply (resistance) in discharge, pressure, or both. For river flood forecasting, failure may be regarded as flood level exceeding a specific river stage, where the loading is the approaching flood and resistance is the given river stage.

It is possible and not unusual that for a given system, different failure levels are specified corresponding to different objectives. For example, for a dam overtopping, failure occurs when the water stage (loading) exceeds the dam crest elevation (resistance). At a different time, the same dam may fail to maintain a certain storage (corresponding to a certain stage of resistance) for certain reservoir inflow (loading), whereas the collapse (structural failure) of the dam occurs under a different set of loading and resistance criteria. In other words, "failure" is defined here in a rather flexible way, adaptable to the objective of the system. In water resources engineering, most failures belong to one of the following two types:

(a) Structural failure, which involves damage or alteration of the structure, resulting in diminishment or termination of the system's ability to function, or
(b) Performance failure for which a performance threshold or target level is exceeded, usually resulting in an undesirable consequence (e.g., flooding due to inadequate sewer capacity) without altering the structure and ability of the system.

Failure events are usually the joint occurrence of excessive loads or demand together with weak resistance or inadequate capacity of the system. Generally, engineering systems are dynamic, both resistance and loading changing with time. For instance, the strength of a dam varies with its maintenance operation, whereas its loading varies with the hydrometeorological conditions. The system may change with time due to deterioration, maintenance and operational procedures, changes due to expansion or modification, changes in demand, or changes in climate or environment.

The reliability of a system can be defined as the probability that at any time the load does not exceed the resistance, i.e.,

$$\text{Reliability} = P(L \leq R) \tag{1}$$

whereas the probability of failure, P_f, is

$$P_f = 1 - \text{Reliability} = P(L > R) \tag{2}$$

From Equation 2, P_f can be expressed in terms of the joint probability density function (pdf) of R and L, $f_{R,L}(r,\ell)$ as

$$P_f = \int_a^b \int_c^\ell f_{R,L}(r,\ell) \; dr \; d\ell \tag{3}$$

in which c is the low bound of R, and a and b are the lower and upper bounds of L, respectively. Usually a=c=0 and b=∞. The loading, L, and resistance, R, are of course each a function of many influential factors, X_i and Y_j, i.e.,

$$L = G_1(X_1, X_2, \ldots, X_n) \tag{4}$$
$$R = G_2(Y_1, Y_2, \ldots, Y_k) \tag{5}$$

If L and R are statistically independent and their probability distributions are $f_L(\ell)$ and $f_R(r)$, respectively, then Equation 3 can be written as

$$P_f = \int_a^b \int_c^\ell f_L(\ell) \; f_R(r) \; dr \; d\ell \qquad (6)$$

The risk relationship given in Equation 6 can be shown graphically, as in Figure 1. For a given loading L = A, the cumulative probability of R < A is the shaded area to the left of A and under $f_R(r)$.

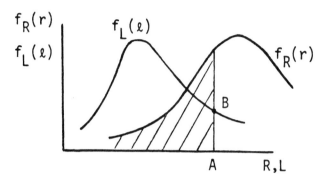

Figure 1. Probability density functions of load and resistance

3. UNCERTAINTIES, SAFETY FACTOR, AND RETURN PERIOD

Engineering systems are inevitably subject to various kinds of uncertainties in their planning, design, and operation. The uncertainty most frequently dealt with and best known to water resources engineers is the uncertainty that future geophysical events such as rainfall and floods will occur. There are other uncertainties that are significant but often ignored, however: each of the factors X_i and Y_j in Equations 4 and 5 is a potential contributor to uncertainties. The sources of uncertainties include the following:

(a) Geophysical uncertainties associated with random natural processes.
(b) Demand uncertainties, which arise from the variability of future demands on the engineering system or project.
(c) Operational uncertainties associated with operational and maintenance procedures, construction, manufacture, or deterioration of facilities, and other human operational factors.
(d) Data uncertainties, which include (i) measurement inaccuracy and errors, (ii) inadequacy of data sample, (iii) data handling and transcription errors.

(e) Model structural uncertainty, which reflects the inability of a simulation model or design method to represent exactly the real physical behavior or process of the system.
(f) Model parameter uncertainties, which arise from the variability in the determination of parameter values to be used in the model.

The first type of uncertainties, the randomness of natural geophysical events, has been traditionally expressed in terms of the average recurrence interval or return period. The return period, T_r, in years, is the reciprocal of the probability of an event X exceeding a given magnitude x, $p(X > x)$, on an annual basis. This probability is conventionally determined from frequency analysis and order statistics or through time series analysis, using available limited sample data of an unknown population. For a time-invariant (stationary) engineering system corresponding to the present global geophysical environment, once the value of T_r for a given threshold magnitude is known, the probability that this threshold, $x(T_r)$, will be exceeded in an n-year period is

$$P_n(X > x) = 1 - (1 - \frac{1}{T_r})^n \qquad (7)$$

For large T_r, approximately

$$P_n(X > x) = 1 - \exp(-\frac{n}{T_r}) \qquad (8)$$

The probability for the T_r-year event to occur fewer than m times in n years can also be calculated (Yen, 1970).

Since the return period accounts for only one of the aforementioned different sources of uncertainties, a safety factor has traditionally been used in engineering projects as a means to counter our ignorance and omission of other uncertainties. Yen (1979) pointed out that there are actually several different safety factors (Table 1). Those usually used by engineers are the characteristic safety factor and the arbitrarily assigned safety factor. With recent developments in reliability analysis, the safety factors can now be determined on a more scientific and rational basis.

4.   PROCEDURE FOR ENGINEERING RELIABILITY EVALUATION

Because of uncertainties, the performance of an engineering system can never be predicted exactly. There are two basic probabilistic approaches to evaluating the reliability of an engineering system: One is an assessment of historical failure data; the other approach is by simulation, considering the performance of each contributing factor and combining probabilistically the contributions of these factors to yield the system reliability. The former approach, failure data statistical analysis, is a lumped system approach requiring no knowledge of the constitution and working of the components of the system. This approach by deduction from past experience often cannot be carried out because of the inadequacy or lack of data. Moreover, this approach usually requires the background of the system to be time-invariant (stationary), which is usually not the case of the system over its service period.

Table 1. Different Types of Safety Factor (after Yen, 1979)

Type of safety factor	Definition
Pre-assigned	Assigned number
Central	μ_R/μ_L
Mean	$\bar{R}/\bar{L}$
Characteristic	R_o/L_o
Partial	$1/\gamma = N_L/N_R$ where $$P_f = P(L > \gamma R) = P(N_L \hat{L} > N_R \hat{R})$$
Reduced	ν where $$P_f = P(L > \frac{R}{\nu})\ P(N > \nu)$$

For example, data available from most countries on failure of dams (incidents involving structural and functional damages that could result in serious downstream consequences) yield failure statistics between 10^{-3} to 10^{-4} per dam-year. Similar statistics are now available on nuclear power plants. These statistics provide a global view of the dams or plants collectively, but they do not provide useful information on the safety of a particular dam or plant. Even if the data is sufficient to allow statistical refinement based on the type (e.g., concrete, gravity, earth, arch) and age of a dam, refined statistics are still at best a rough indication of the safety of the dam, which actually changes with time.

The second approach is a distributed system approach. In order to evaluate the total system reliability, the procedure should account for all the contributing factors and their uncertainties. Such a procedure is described as follows.

Determining engineering reliability involves the evaluation of the probability of failure, P_f, in Equation 3. Generally this evaluation involves the following procedures:

(a) Identify the engineering system, specify the objective, and define failure accordingly.

(b) Identify the different possible modes of failure to facilitate division of teamwork to be assigned to experts of different disciplines. For example, evaluation of the possibility of an earth embankment failure due to piping would be assigned to geotechnical engineers. The possibility of failure due to flood overtopping would be the task of hydraulic engineers. In this phase a fault tree analysis can help to identify the different levels of component failure modes and to indicate their combination for total reliability. A simple fault tree showing the failure of upstream flooding due to an inadequate culvert, unable to deliver the water through, is shown in Figure 2 as an example.

(c) Identify the factors that contribute to the behavior of the system. These factors are represented as X_i and Y_j in Equations 4 and 5.

Also, formulate or adapt a simulation model that represents the behavior of the system in transforming input to produce output. In other words, identify the functions G_1 and G_2 in Equations 4 and 5 or similar model functions. The simulation models may consist of submodels of different levels describing different failure modes to be worked out by experts in different disciplines. If an identified factor does not appear in the simulation model, it implies that the model is not complete and only an approximation. Either an improvement must be made to include this factor in the model, or the factor may be omitted and the omission recognized as part of the modeling error.

(d) Assess the uncertainty in terms of variability for each of the contributing factors. Different factors are assessed by different experts according to failure modes and disciplines. The variability may be expressed as a probability distribution or in terms of statistical quantities such as mean, variance, and skewness.

(e) Combine the component uncertainties contributed by each of the factors to yield the total probability of failure. This is essentially the integration of Equation 3. The technique of integration depends on the reliability evaluation method used, which will be discussed further. The combination of uncertainties may be performed in divided levels using different methods. Again the fault tree is helpful in this step.

(f) Identify the relative significance of the factors, if required. It might also be possible to identify the weakest factors, improvement of whose uncertainties could significantly enhance the reliability.

The evaluation of the reliability of an engineering system or its complement, the failure probability given in Equation 3, is not a simple task. Both resistance R and loading L consist of many influential factors (Equations 4 and 5) from very diversified sources, and they belong to different disciplines requiring experts of different specialties to work together as a team. Without the collective contribution of experts from different disciplines, factors may be omitted, producing only a partial reliability evaluation. Conversely, it is very likely that even with great care some factors may still be omitted. Nevertheless, good teamwork will usually account for the major factors. Thus, it is not inconceivable that in general the system reliability determined, while not the true total value, is a close approximation to it.

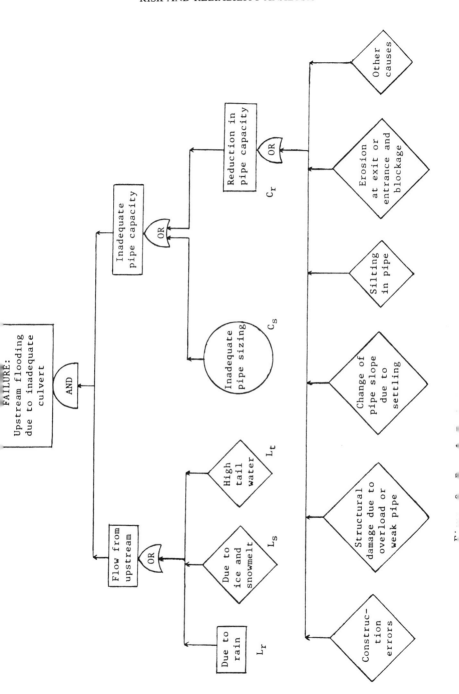

5.   METHODS FOR RELIABILITY CALCULATION

The most important parts of reliability evaluation are steps (d) and (e) in the procedure just outlined. Basically, they involve the integration of the probability density function $f_{R,L}$ in Equation 3, or f_R and F_L in Equation 6. Since the resistance R and loading L are each a function of many variables, $f_{R,L}$, f_R, and f_L are actually weighted probability distributions reflecting the joint effects of the probability distributions of the individual basic variables X_i and Y_j in Equations 4 and 5. The true pdfs of the basic variables are usually unknown, although sometimes they may be assumed based on limited information. In fact, hydrologists have made significant contributions concerning the distribution of precipitation and flood. Even if the pdfs of the basic variables are given, their combination to obtain the joint pdf, $f_{R,L}$ in Equation 3 or f_R and f_L in Equation 6, is not a simple task, if possible at all. Therefore, various techniques have been developed to combine the individual variable uncertainties and estimate the total probability of failure, P_f. Table 2 compares briefly some of the major properties of these methods.

In performing the evaluation, often it is more convenient to deal with a performance variable Z(R,L) instead of R and L separately. Two useful forms of Z are

$$Z = \ln(R/L) \tag{9}$$

and

$$Z = R - L \tag{10}$$

Accordingly, from Equations 1 and 2, the reliability of the system is

$$\text{Reliability} = P(Z \geq 0) \tag{11}$$

and

$$P_f = P(Z < 0) = \int_{-\infty}^{0} f_Z(z) \ dz \tag{12}$$

where Z is a function of the individual variables contributing to L and R (Equations 4 and 5), i.e.,

$$Z = g(X_1, X_2, \ldots, X_m) \tag{13}$$

If Z is normally distributed, integration of Equation 12 yields

$$P_f = \Phi(-\beta) = 1 - \Phi(\beta) \tag{14}$$

where β is called the "reliability index" and is defined as

$$\beta = \frac{\mu_Z}{\sqrt{\text{Var}(Z)}} \tag{15}$$

Table 2. Brief Comparison of Methods for Engineering Reliability Calculation (after Yen, 1987)

Method	Direct Integration	Monte Carlo	Reliability Index	MFOSM	AFOSM	Second Order Method
Capability to account for different factors	limited	yes	yes	yes	yes	yes
Information needed on probability distribution of factors	extensive	moderate	first two statistical moments	only the combined distribution; for factors the first two statistical moments suffice	only the combined distribution; for factors the first two statistical moments suffice	moderate
Complexity in application	complicated	moderately complicated	moderate	moderate	moderate	moderate
Amount of computations	moderate to extensive	extensive	simple to moderate	simple to moderate	moderate	moderate to extensive
Capability to estimate total system reliability	difficult	extensive computations	no	yes	yes	yes
Result adaptability for risk cost analysis	yes	yes	no	yes	yes	yes

in which μ_Z denotes the mean value of Z, Var(Z) is the variance of Z, and $\Phi(\beta)$ in Equation 14 denotes the cumulative standard normal distribution evaluated at the given value of β. The value of Φ can be found in tables in standard statistics and probability reference books; it increases as β increases. Note that β is simply the inverse of the coefficient of variation of Z. Other formulas for P_f derived from Equation 12 for selected probability distributions of the performance variable Z are listed in Table 3. Among the different distributions of Z, the most popular is the normal distribution because the combined distribution of linearly related normally distributed contributing basic variables X_i is also normally distributed.

Detailed descriptions of these methods can be found in Yen (1987), Yen et al. (1986), Ang and Tang (1984), Tung and Mays (1980), Plate and Duckstein (1987) and, for the second order method, Madsen et al. (1986).

Among the reliability calculation methods listed in Table 2, the method of direct integration of Equations 3, 6, or 12 is the most accurate but also the least practical one. It is accurate if the pdf of Z is precisely known. It is impractical except for very simple problems because in engineering problems Z is a function of many basic variables X_i (Equation 13) whose pdfs can at best be estimated. Their individual evaluation and combination for the joint pdf, f_Z in Equation 12, is a rather difficult task. Some of the commonly used probability distributions for floods, heavy rain, high wind, and earthquakes are discussed in Ang et al. (1985).

The Monte Carlo method is a "brute force" technique requiring a large amount of calculation. It also requires knowledge of the probability distributions of the basic variables, which are often unavailable. The probability of failure is calculated from the statistics of the large amount results. The major drawback of this method is the expense in computations in order to achieve a desirable level of accuracy, especially when the number of variables is large and the failure probability is low. However, with recent advancements in computer technology and reduced computational costs, this method is becoming increasingly practical.

The reliability index β (Equation 15) does not evaluate the reliability itself. Rather, it offers a comparative reliability evaluation of different systems or different alternatives of a system. Therefore it cannot be used for risk cost assessment.

The first-order second-moment (FOSM) method is an approximate analysis performed by truncating the second-order and higher order terms of the Taylor series expansion of the performance parameter Z, which is a function of the basic random variables (Equation 13). A second-moment analysis utilizes only the first two statistical moments, i.e., the expected value and variance (or coefficient of variation) of the random variables, evaluated at the point of Taylor's expansion. In actual engineering practice the distributions of the constituent variables, X_i, are usually not well defined; often, information on these variables is limited to means and variances. Values of higher statistical moments are either unreliable or unavailable. In such cases, approximate methods consistent with the type and quality of available information indeed are more sensible, and hence the first-order second-moment method may be an appropriate approach.

In the mean-value first-order second-moment (MFOSM) method, the first-order Taylor series for Z is expanded about the mean values $\bar{x}_i$ of the variables X_i:

$$Z = g(\bar{x}_i) + \sum_{i=1}^{m} (X_i - \bar{x}_i) \frac{\partial g}{\partial X_i} \tag{16}$$

where the derivatives are evaluated at $\bar{x}_i = (\bar{x}_1, \bar{x}_2, \ldots, \bar{x}_m)$. Taking the first and second moments of Z in Equation 16 and neglecting terms higher than second-order, we have

$$E(Z) = \bar{Z} = g(\bar{x}_i) \tag{17}$$

$$Var(Z) = \sum_{i=1}^{m} C_i^2 \, Var \, (X_i) \tag{18}$$

where C_i's are the values of the partial derivatives $\partial g/\partial X_i$, evaluated at $\bar{x}_1, \bar{x}_2, \ldots, \bar{x}_m$. The variables X_i's are assumed to be statistically independent. Accordingly,

$$\sigma_Z = \left[\sum_{i=1}^{m} (C_i \sigma_i)^2 \right]^{1/2} \tag{19}$$

in which σ_Z and σ_i are the standard deviations of Z and X_i, respectively. The reliability index β is calculated from Equation 15 and P_f is evaluated from Equation 14. The coefficient of variation (cov) of Z, Ω_Z, can be evaluated in a first-order sense from the cov of the basic variables as

$$\Omega_Z^2 = \frac{1}{\bar{g}^2} \sum_{i=1}^{m} C_i^2 \, \bar{X}_i^2 \, \Omega_{xi}^2 \tag{20}$$

In the advanced first-order second-moment (AFOSM) method, the Taylor series is expanded at a likely point on the failure surface described by the mathematical function $Z = g(X_i) = 0$. Detailed procedures of this method can be found in Yen et al. (1986). Like MFOSM, only the first two statistical moments (mean and variance) of the individual basic variables are needed to describe their uncertainties. The MFOSM method may incur significant errors if the functional relationship between the mean and failure surface is highly nonlinear. In this case the AFOSM method offers a more accurate alternative. However, the location of the failure point is initially unknown. It must be found through iterations, resulting in more computations for AFOSM than MFOSM. Theoretically, the basic variables X_i can assume any probability distributions. But this will make the AFOSM computation rather complicated. The problem is much simplified if the distributions of X_i are transformed into an equivalent normal distribution in a manner similar to that proposed by Rackwitz (1976). And f_z in Equation 15 is normally distributed accordingly. Moreoever, the development of the FOSM method requires the basic variables X_i to be statistically independent. Therefore, for correlated original variables it is necessary to perform a transformation to independent (at least very weakly dependent) new variables.

The second-order method is based on the concept of retaining the second-order terms in Taylor's expansion. Obviously the method is far more involved than the FOSM method. Different techniques have been proposed. The method is still in its development stage, but practical procedures are expected soon.

6.   EXAMPLE: RELIABILITY OF CULVERT DESIGN

The engineering approach to reliability analysis is best demonstrated by using a simple example: determining the upstream flooding probability of a culvert design. The MFOSM method is adopted here because of its practicality and relative simplicity suitable for illustration, despite the fact that AFOSM is probably the better method for this particular problem. It should be emphasized that this example is presented merely as an illustration, not for its detailed completeness and precision in the analysis. A real problem should be dealt with in a refined manner.

By referring to the procedure given previously, in step (a) the system of culvert is identified. The highway culvert should drain the run-off from a 1104-acre (4.47 km^2) watershed of farm land. The culvert is 100 ft (30 m) in length, consisting of four 25-ft(7.6 m)-long circular concrete pipes on a 0.005 slope. The entrance to the culvert is a vertical head wall set flush with the pipe and symmetric 45° wingwalls. The pipe invert at the entrance is at elevation 695 ft (211.8 m). The outlet has a flushed end wall with a short apron to protect the downstream from erosion. Under the design condition, the maximum allowable headwater elevation upstream from the culvert is 705 ft (214.9 m) and upstream storage is negligible. The expected tailwater elevation is 698 ft (212.8 m), although it may vary from 697 to 700 ft (212.4 to 213.4 m).

The formula relating rainfall intensity i (in/hr) to return period T_r (years) and duration t_d (min) is

$$i = \frac{a\, T_r^m}{t_d^k + b} \tag{21}$$

in which a = 30, b = -1, m = 0.16, and k = 0.75 for this location for t_d = 1 to 12 hours and T_r = 3 to 200 years. The design culvert service period is 50 years. The design rainfall duration is assumed equal to the time of concentration, which is estimated by using Kirpich's formula with the basin length L = 8500 ft (2590 m) and average slope S = 0.0012. Accordingly, t_d = 0.0078 $(L/\sqrt{S})^{0.77}$ = 110 min. Using the design culvert service period of 50 years as T_r and t_d = 110 min, the design rainfall intensity as given by Equation 21 is 1.70 in/hr (43.2 mm/hr).

Failure is defined here as the flow from upstream (loading), Q_L, exceeding the culvert capacity (resistance), Q_c, i.e., $P_f = P(Q_L > Q_c)$. The performance variable is defined here as $Z = \ln(R/L)$.

In step (b), the example fault tree is created as shown in Figure 2. Different elements in the fault tree may be assigned to experts in different disciplines. For example, the analysis pertinent to structural damage of the culvert pipe during the service period is assigned to a structural engineer. In the fault tree, an "AND" gate (or junction) corresponds to intersection in probability theory, whereas an "OR" gate corresponds to union. Thus, upstream flooding occurs when there is inflow from upstream due to rain, L_r, or ice and snow melt, L_s, or high tailwater back-up from downstream of the culvert, L_t, and at the same time the capacity of the culvert (which may change with time) is inadequate (events C_s and C_r in Figure 2) due to various possible reasons. Thus mathematically,

$$P_f = P\left[(L_r \cup L_s \cup L_t) \cap (C_s \cup C_r)\right] \tag{22}$$

in which C_r can be further decomposed according to the fault tree.

In step (c) the factors contributing to each of the elements in the fault tree are to be identified and the mathematical simulation models are set up for each of the elements. For brevity, assume that there is not an ice and snow melting problem at this location and only inadequate capacity due to pipe sizing considered in the following example. The contribution of reduced pipe capacity due to various causes (C_r in Figure 2) can be evaluated similarly and the individual results from different teams can be combined according to the probabilistic relationship given in Equation 22, which can now be rewritten as follows for the conditions just specified:

$$P_f = P[(L_r \cup L_t) \cap C_s] + P[(L_r \cup L_t) \cap C_r]$$
$$- P[(L_r \cup L_t) \cap C_s] - P[(L_r \cup L_t) \cap C_r] \tag{23}$$

As to the simulation models to describe the system behavior, two models are set up in this example, one for the run-off (loading) and the other for the pipe capacity (resistance). A number of models exist to estimate the run-off from a drainage basin. The rational formula is used here merely for the sake of simplicity, not as an endorsement. With a run-off coefficient of 0.20, the expected peak run-off rate is

$$Q_o = \overline{C}i\overline{A} = 0.20 \times 1.70 \times 1104 = 375 \text{ cfs } (10.6 \text{ m}^3/\text{s})$$

Since the upstream storage of the culvert is negligible, Q_o can be used directly as the design discharge.

Flow design through a culvert can be either open channel or pressurized conduit and the control may be upstream or downstream. Hence it is often difficult to identify precisely a culvert's critical design condition. Bodhaine (1968) classified culvert flow into six types. Yen (1986) considered the entrance, pipe flow, and exit conditions and identified a refined classification of 27 cases. Bodhaine's Types 4 to 6 (submerged entrance) or Yen's cases IV-10-D and II-2-B are most likely the design conditions. For these flow conditions, the Bernoulli head relationship between the approaching and tailwater flows is

$$[y_u + (V_u^2/2g)] = [y_d + (V_d^2/2g)] + h_f + h_L \tag{24}$$

in which y is the water surface elevation of flow; V is velocity; the subscripts u and d represent upstream and downstream, respectively; h_f is the head loss due to friction in the pipe, which can be estimated by using Manning's formula

$$h_f = \frac{4^{10/3} n^2 L\ Q^2}{2.21\ \pi^2 D^{16/3}} \tag{25}$$

and h_L includes the entrance, exit, and other losses, if any, i.e.,

$$h_L = (K_{ent} + K_{exit})\ 8Q^2/\pi^2 gD^4 \tag{26}$$

in which K_{ent} and K_{exit} are the entrance and exit loss coefficients, respectively. Substituting Equations 25 and 26 into Equation 24 and neglecting the velocity head of the flow in the approaching and exit channels, one obtains the culvert capacity model,

$$Q = \pi\sqrt{y_u - y_d}\ \left[\frac{46.0\ n^2 L}{D^{16/3}} + (K_{ent} + K_{exit})\ \frac{8}{gD^4}\right]^{-1/2} \tag{27}$$

which can be solved for the pipe diameter D.

For the present culvert design, $n = 0.013$ with a possible range from 0.011 to 0.15 and $K_{ent} = 0.35$ with a range of 0.15 to 0.45. For Bodhaine's Type 6 Flow, $K_{exit} = 1$ and y_d is the crown elevation of the pipe at the exit if it is higher than the tailwater elevation. Thus, with $Q = 375$ cfs, $L = 100$ ft, and $y_u - y_d = 705 - (D + 695 - 100 \times 0.005)$, the required diameter as computed from Equation 24 is 5.99 ft. Hence 6-ft pipes are used.

In step (d), the uncertainties of the variables are assessed in terms of the coefficient of variation (second statistical moment) as follows.

For the uncertainties on resistance:

The resistance is the actual discharge capacity of the culvert. Uncertainties exist as to whether the critical flow condition of maximum discharge is open channel flow or pressurized conduit flow, and if the control is upstream or downstream. There are further uncertainties as to the hydraulic characteristics of the entrance and exit, the geometry, slope, and roughness of the conduit, and errors in modeling the flow (e.g., using a steady flow formula to approximate unsteady flow). Assuming the maximum discharge occurs with submerged entrance (Bodhaine's Types 4 to 6 [1968] or Yen's cases IV-10-D or III-2-B [1986]), applying Equation 20 to Equation 27, the cov for the capacity part of Z for the culvert is

$$\Omega_c = \left[\Omega_{\lambda c}^2 + 0.25\Omega_y^2 + \alpha_\ell \Omega_\ell^2 + \alpha_n \Omega_n^2 + \alpha_D \Omega_D^2 + \alpha_{Kent} \Omega_{Kent}^2 + \alpha_{kexit} \Omega_{Kexit}^2 \right]^{1/2} \tag{28}$$

in which the subscript y represents the head difference $y_d - y_d$, ℓ represents the culvert length L, λ_c is the capacity modeling error correction factor, and

$$\alpha_\ell = \left[\frac{23.0 \, \bar{L} \, \bar{n}^{-2}}{MD^{-16/3}} \right]^2 \qquad \alpha_n = \left[\frac{46.0 \, \bar{n}^{-2} \bar{L}}{MD^{-16/3}} \right]^2$$

$$\alpha = \left\{ \frac{1}{M} \left[\frac{122.7 \, \bar{L} \, \bar{n}^{-2}}{D^{-16/3}} + (\bar{K}_{ent} + \bar{K}_{exit}) \frac{16}{g\bar{D}^{-4}} \right] \right\}^2$$

$$\alpha_{Kent} = \left[\frac{4\bar{K}_{ent}}{Mg\bar{D}^{-4}} \right]^2 \qquad \alpha_{Kexit} = \left[\frac{4\bar{K}_{exit}}{Mg\bar{D}^{-4}} \right]^2$$

where

$$M = \frac{46.0 \, \bar{n}^2 \bar{L}}{\bar{D}^{16/3}} + (\bar{K}_{ent} + \bar{K}_{exit}) \frac{8}{g \bar{D}^4}$$

For the design condition that the distribution mode values of the variables are $\bar{D}$ = 6.0 ft, $\bar{y}_u - \bar{y}_d$ = 4.5 ft, $\bar{n}$ = 0.013, $\bar{K}_{ent}$ = 0.35, $\bar{K}_{exit}$ = 1.0, and $\bar{L}$ = 25 ft (for each pipe section), M = 0.0003138. The corresponding values of the coefficient $\alpha = (\partial g / \partial X_i)^2 \, \bar{X}_i^2 / \bar{g}^2$ for the Ω^2s of the resistance parameters are also given in Table 4.

In computing the values of cov, the ranges of variation of the variables are as follows. The water surface y_u ranges from 704 to 705 ft and y_d 700.5 ± 0.05 ft. Hence the range of $y_u - y_d$ is from 3 to 5 ft. The value of K_{exit} is 1.0±0.05 and K_{ent} from 0.15 to 0.45. The range of n is 0.011 to 0.015. According to ASTM Standard C76-74, for a 6-ft concrete pipe the diameter variation is ±1%, and for a 25-ft-long pipe the allowable length error is ±0.5 in., or ±0.167%, which yields the same percentage error for the entire 100-ft length. For all these variables, the distributions are assumed triangular, with the mode at the bar-value used in the design. In a real situation the assumed distributions of the parameters should be replaced by the statistical means and variance obtained from the sample data for each parameter.

For the uncertainties on loading:

There are four parameters contributing to the uncertainty of the estimated loading Q_o given by the rational formula: the rain intensity i, run-off coefficient C, basin size A, and modeling error correction factor λ_L. The intensity is obtained from the intensity-duration-frequency relationship represented by Equation 21.

For the reliability of rainfall intensity i, applying first-order analysis to Equation 21 yields the contribution of T_r on uncertainty of i as measured by cov as (Yen et al., 1976; Yen and Tang, 1976)

$$\Omega_{iT} = \left[(\frac{1}{m} - 1) \sqrt{1 - 2m} \; T_r^m + 0.45 \right]^{-1} \tag{29}$$

Similarly, uncertainty of the duration t_d on i due to the cov of t_d is

$$\Omega_{id} = \frac{k}{(1 - m)} \frac{a}{(t_d^k + b)} \, (1 - \frac{b}{t_d^k + b}) \frac{1}{\bar{i}_T} \Omega_{td} \tag{30}$$

where

$$\bar{i}_T = \frac{a}{t_d^k + b} \left[T_r^m + \frac{m}{1 - m} \frac{0.45}{\sqrt{1 - 2m}} \right] \qquad (31)$$

The effect of limited rainfall record of N-year observation on i is

$$\Omega_{ir} = \frac{1}{\sqrt{N}} \frac{m}{\sqrt{1 - 2m}} \frac{1}{(1 - m)} \frac{a}{(t_d^k + b)} \frac{1}{\bar{i}_T} \qquad (32)$$

Since average rainfall intensity decreases with increasing watershed size, and the constants in Equation 21 are obtained from point rainfall data, consequently watershed size contributes to the uncertainty of i. With no better information available, this uncertainty Ω_{ia} is arbitrarily assumed equal to 0.001. Likewise, for errors in instrumentation and measurement of rainfall data, in data handling and interpolation, the uncertainties are assumed to be 0.03 and 0.05, respectively. Finally, modeling errors arise in the derivation of the rainfall intensity formula and assumption on distributions. For this modeling, error $\Omega_{i\lambda}$ is assumed equal to 0.05. A summary of these uncertainties is given in Table 5 in terms of the values of cov. The combined uncertainty of rainfall intensity, Ω_i^2, is computed according to Equation 20 as the sum of component Ω^2s. The expected rain intensity in a 50-year period, $\bar{i}_T = 1.80$ in/hr, is computed from Equation 31, and this value is greater than i = 1.70 in/hr, given by Equation 21.

For the other three parameters, C, A, and λ_L, the mean and cov of the run-off coefficient C are obtained from the following values, given independently by 22 trained students: 0.23, 0.20, 0.20, 0.10, 0.22, 0.15, 0.20, 0.20, 0.15, 0.25, 0.23, 0.10, 0.30, 0.15, 0.20, 0.25, 0.20, 0.20, 0.20, 0.30, 0.17, and 0.24. The mean and cov for the basin area A consists of two parts: that due to map error (assume $\Omega = 0.0001$) and that due to measurement error. The latter is provided by a statistical analysis of different measurements from the map independently by different persons. Past statistics indicate that for areas about the size 1 to 10 sq mi (3 to 25 km^2), measured from a USCS 7.5- or 15-min map, the value of cov is around 0.05. A separate study with the rational formula revealed that the discharge is overestimated as the basin size becomes larger if the value of the run-off coefficient C is selected from standard tables such as that proposed in the ASCE Manual No. 37. For drainage basins around 1000 acres, the correction factor is about 0.9 and $\Omega_{\lambda L}$ is 0.10. The uncertainties for loading are summarized in Table 6. The value of Ω_i is obtained from Table 5. The combined uncertainty of loading is computed according to Equation 20 as

$$\Omega_L^2 = \Omega_{\lambda L}^2 + \Omega_C^2 + \Omega_i^2 + \Omega_A^2 \qquad (33)$$

and $Q_L = \bar{\lambda}_L \bar{C} \bar{i}_r \bar{A}$.

Table 4. Uncertainty of Resistance (Culvert Capacity)

Parameter	Mean	Coefficient of Variation Ω	α
$y_u - y_d$	4.17 ft	0.102	0.250
n	0.013	0.063	0.031
L	100 ft	0.001	0.008
D	6.0 ft	0.004	4.477
K_{ent}	0.32	0.197	0.011
K_{exit}	1.00	0.02	0.093
λ_c	1.0	0.05	1.0
Q_c	366 cfs	0.076	

Table 5. Uncertainty in Rainfall Intensity Prediction

Factor		Coefficient of Variation
Design period (50 yr)	Ω_{iT}	0.117
Duration (110 min)	Ω_{id}	0.047
Basin area (1104 ac)	Ω_{ia}	0.001
Limit record (40 yr)	Ω_{ir}	0.019
Instrument and measurement errors	Ω_{im}	0.03
Data handling and interpolation errors	Ω_n	0.05
Modeling	$\Omega_{i\lambda}$	0.05
Intensity ($\bar{i}_T$ = 1.80 in./hr)	Ω_i	0.149

Table 6. Uncertainty of Loading

Factor	Parameter	Mean	Coefficient of Variation Ω
Run-off	C	0.20	0.253
Intensity	i_T	1.80 in./hr	0.149
Basin area	A	1104 ac	0.05
Modeling error	λ_L	0.9	0.10
Flood (Loading)	Q_L	397 cfs	0.314

In step (e) the component uncertainties are combined and the failure probability P_f is computed according to Equation 14. For this example,

$$\bar{Z} = \ln(\bar{R}/\bar{L}) = \ln(366/397) = -0.0813$$

$$\text{Var}(Z) = \Omega_R^2 + \Omega_L^2 = 0.076^2 + 0.314^2 = 0.104$$

Hence, from Equation 15, $\beta = -0.252$. The corresponding failure probability, which is $P[(L_r \cup L_t) \cap C_s]$ in Equation 23, is

$$P[(L_r \cup L_t) \cap C_s] = \phi(-\beta) - \phi(0.252) = 0.60$$

The other failure probability, $P[(L_r \cup L_t) \cap C_r]$ in Equation 23, can be evaluated similarly. Assuming its value is 0.20, then the total failure probability can be compared from Equation 23 as

$$P_f = 0.60 + 0.20 - 0.60 \times 0.20 = 0.68$$

7.   RELATIVE CONTRIBUTION OF FACTORS TO UNCERTAINTY

A useful by-product of the reliability analysis is a quantitative comparison of the significance of the influential factors in their contribution to the total reliability. The uncertainty contribution of the X_i parameter is $\alpha_{xi} \Omega_{xi}^2$ where

$$\alpha_{xi} = \left[\frac{\partial g}{\partial X_i}\right]_0^2 \ \bar{X}_i^2 / \bar{g}^2 \qquad (34)$$

Table 7 illustrates the uncertainty contributions for the factors involved in the culvert reliability example. It can be seen that in this case much improvement can be made on the culvert reliability if the run-off coefficient and rainfall determination are improved.

8.   APPLICATIONS OF FIRST-ORDER METHODS TO WATER RESOURCES PROBLEMS

The mean value and advanced first-order methods have been popular for reliability analysis in structural, seismic, and nuclear engineering. Their application to water resources problems has also been increasing. Tang and Yen (1972) applied the MFOSM method to the design of sewers. Later Yen et al. (1976) refined the technique for risk-based least cost sewer system design. Garen and Burges (1981) employed the MFOSM analysis using a modified Stanford Watershed Model to derive catchment run-off prediction error bounds due to assumed parameter uncertainties and compared with Monte Carlo simulation results. The problem was further investigated by McBean et al. (1984). Tung and Mays (1980, 1981) applied a combination of MFOSM and direct integration to culvert and levee problems. Yen and Tang (1977) formulated a MFOSM procedure for flood routing in channels.

Cheng et al. (1982) applied the AFOSM method to investigate the overtopping probability of an earth dam. Mol et al. (1983) used AFOSM for the design of rubble mound backwater. Vrijling (1987) applied AFOSM

Table 7. Relative Contribution to Uncertainties

Parameter	Cost of Variation Ω	α	$\partial\Omega^2/\Omega_Z^2$ %
	Resistance		
$y_u\text{-}y_d$	0.102	0.250	2.49
n	0.063	0.031	0.12
L	0.001	0.008	0.01
D	0.004	4.477	0.07
K_{ent}	0.197	0.011	0.41
K_{exit}	0.02	0.093	0.04
λ_c	0.05	1.0	2.40
Q_c	0.076	1.0	5.53
	Loading		
C	0.253	1.0	61.3
i_T	0.149	1.0	21.3
A	0.05	1.0	2.4
λ_L	0.10	1.0	9.6
Q_L	0.314	1.0	94.5
Total	0.323		100

to the Oosterschelde storm surge barrier (Delta project) in the Nether-lands. Melching and Yen (1986) applied AFOSM to study the slope effect on sewer design. Melching (1987) formulated an AFOSM procedure for catchment flood real-time forecasting. Sitar et al. (1987) employed AFOSM to a groundwater solute transport problem and compared the result with a Monte Carlo simulation. Cheng et al. (1986) used AFOSM to investigate wind effects on overtopping of embankments and the relia-bility implication of free board.

The aforementioned examples demonstrate the various types of water resources engineering problems to which FOSM methods are applicable. The methods can be used for reliability-based engineering design and operation, for forecasting, for data worth assessment and network design, and for providing basic probabilistic information for risk cost analysis and decisionmaking.

For example, in traditional real-time catchment flood forecasting, as when a heavy rain is falling or predicted, a warning as to flood stage or discharge may be issued, but not with the reliability of this new prediction method. It is not unusual that serious flood warnings have not been issued because the reliability of predictions was uncertain. FOSM methods can add useful information. Figure 3 shows the

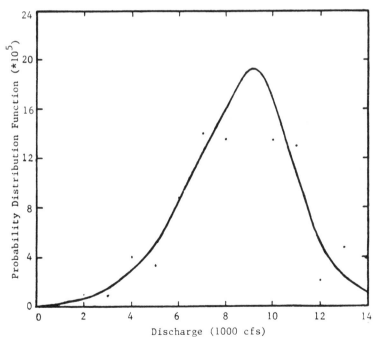

Figure 3. Probability Distribution of Real-Time Forecasted Peak Dis-charge of a Rainstorm-Produced Flood (Melching, 1987)

forecast of a flood event at Pontiac, Illinois, made by Melching (1987). The prediction was made for the May 5, 1965, rainstorm over the 579-mi^2 catchment upstream of Pontiac. The rainfall-run-off model used was HEC-1 and the reliability analysis method used was AFOSM. The figure shows that there is a near-bell-shaped probability distribution of the forecasted peak discharge produced by the rainstorm. Under such circumstances, the flood warning officer may then issue a warning at the mode, mean, or any other value of the flood according to the rules, together with the confidence interval or other reliability measure of the issued flood magnitude. Thus, the quantitative reliability values will be very useful to the national flood-plain delinearization and management program.

Similarly, the reliability methods will provide vital information for dam safety evaluation. For instance, in the national dam safety program discussed in Moser and Stakhiv (1987), one basic type of information needed is dam failure probability under different conditions.

9. PROBABLE MAXIMUM PRECIPITATION

The concept of probable maximum precipitation (PMP) and the associated probable maximum flood (PMF) has been used extensively in water resources engineering as a safeguard against unacceptable catastrophic losses, particularly when loss of lives is involved. It may be worthwhile here to re-examine this concept within the context of engineering reliability.

The concept of PMP evolved in the 1930s when frequency analysis was just about to take firm root in hydrology. It was then obvious that current frequency analysis techniques and the data available were unable to handle the extremal problems demanded by engineers. In response to this demand, PMP was proposed by meteorologists as an "upper ceiling" and "envelope" value. This concept certainly is compatible with the idea that no monetary value can be assigned to human life and hence it cannot be subject to cost-benefit analysis.

In the American Meteorological Society (AMS) "Glossary of Meteorology," PMP is defined as the "theoretically greatest depth of precipitation for a given duration that is physically possible over a particular drainage at a certain time of the year." Meteorological experts in the U.S. Weather Bureau made significant contributions to the development of the concept and proposed estimated PMP values to be used as standard for the United States. As recently as 1982, in the U.S. National Weather Service's Hydrometeorological Report No. 52 entitled "Application of Probable Maximum Precipitation Estimates--United States East of the 105th Meridian," the above definition was essentially accepted, with slight modification replacing the words "over a particular drainage" by "over a given size storm area at a particular geographical location." Before AMS adopted the term PMP into its glossary, the term "maximum possible precipitation" was also frequently used (Wang, 1984).

The concept of PMP is fine and admirable, but problems arise in the estimation of its value. No satisfactory method has been developed that is generally acceptable. Estimations have been made from historical information, precipitable moisture analysis, storm transposition and maximization, and frequency analysis (which was originally used only as a tool to provide consistent values, with no intention to imply a method to determine PMP). However, in recent years, several attempts have been made to assign a frequency to PMP.

If PMP is indeed the theoretically _largest_ rainfall that conceivably could happen under the most severe conditions at a location, the only acceptable way to determine it is through rigorous physical principles governing the distribution, transport, and transformation of water and energy. In view of our weather forecasting abilities today, it will be a long time before this ideal approach becomes attainable and practical. Meanwhile we will continue to struggle to find an acceptable method that can provide generally acceptable PMPs on a comparative basis.

If the PMP is the theoretically largest rainfall, it should be the upper ceiling and no rainfall can exceed it at that location. If a probability must be assigned to it, the only logical nonexceedance probability that can be designated is one, and accordingly its return period is infinity. However, this probability and return period of PMP is meaningless, dangerous, and should not be used. The key issue is the time-frame being considered. If the global environment were time-invariant, "stationary" in hydrologic jargon, the unity nonexceedance probability and infinite return period would make sense. But the global environment is a dynamic, nonstationary system. Some million years later the particular location may become the bottom of an ocean. Customarily, engineering projects and hydrologic analyses are viewed within a human time-frame, not a geological time-frame. Within the human time-frame, the environment is often assumed stationary (which is usually acceptable but not always valid: e.g., the effects of deforest-ation or large urban development on precipitation, and also recent discussions on the global greenhouse and nuclear winter effects). The PMP based on the human time-frame bears no direct relation to the precipitation probability of a nonstationary environment over geological time. Should one desire to know the probability of the value cor-responding to the stationary system PMP determined using a human time-frame, one should investigate this value in the nonstationary framework. Conversely, if the true nonstationary dynamic system PMP is determined, there is no sense to assign it a probability corresponding to the stationary system frequency in human time-frame. In cost-benefit analysis, optimization, and decisionmaking in engineering, it is more scientific and logical to deal with reliability analysis than to assign PMP a probability.

PMF should be viewed similarly. In fact, the problem with PMF is more serious than PMP because run-off is strongly affected by the land surface condition, which may change rapidly.

In this connection it should be remarked that in frequency analysis the probability distribution used is no more than a human-assigned working tool, hopefully approximating well the process that the

distribution is intended to fit. Most of the distributions used in
hydrology have problems at their extremes (upper or lower tails), which
the determination of extreme reliability relies upon. Most of these
distributions have an infinite upper tail, implying an infinitely large
flood or precipitation. In reality the true dynamic PMP or PMF, if it
could be determined, is the upper limit of the correct distribution.
The total amount of global water, ocean and all, one trillion km^3, is a
large but still a finite number. Furthermore, a three-parameter distribution, such as the Pearson Type 3, may fit a given set of data better
than a two-parameter distribution such as the Gumbel distribution,
because the former has one more degree of freedom in fitting. But it
does not imply that the former is more reliable than the latter in
extending the distribution for prediction.

10. CONCLUSION

     The reliability of the design and operation of an engineering
project depends on many factors. In water resources engineering,
traditional evaluations of failure probability due to naturally random
floods or rainfall followed a frequency analysis, wherein the uncertainties and ignorance of other factors were protected through an arbitrarily assigned safety factor. Recent advances in the field of engineering reliability provide several quantitative and systematic methods
to evaluate the reliability of an engineering system, accounting for all
the identifiable contributing factors. The best known methods are
summarized in Table 2. They vary in their degrees of complexity and
practicality. An example of how the uncertainties of the contributing
factors can be accounted for is illustrated through a first-order,
second-moment evaluation of the reliability of typical culvert design.
These new methods are a definite improvement over the arbitrary safety
factor scheme.

     The reliability determined by using these methods serves as the
basic information for any risk/cost-benefit or risk-based decisionmaking
analysis. Without estimating the probability of failure, the risk cost
cannot be scientifically estimated. The conventional manner of using
only the frequency of flood or precipitation accounts for only part of
the risk. Any risk analysis performed without considering how other
factors contribute to reliability is incomplete.

     Undoubtedly, at this stage of development our knowledge of relia-
bility analysis is incomplete and the existing methods are imperfect.
Future improvements and refinements are expected. Nevertheless, the
existing methods make a significant advance over the past situation of
being incapable of evaluating quantitatively the reliability of engineering systems. Even if some of these methods cannot provide the exact
true value of the reliability, they can provide the estimations needed
in many engineering problems. They may be used to compare the relative
reliability of an engineering system under different situations, of
different engineering designs or operational alternatives, or of different systems. Moreover, it is usually better to consider as many
factors contributing to system reliability as possible--preferably all,
even in a rudimentary manner--than to consider only one or a few selected
factors, ignoring others simply because of our incomplete knowledge or

lack of precise techniques to handle them. The imperfections in existing methods should not hinder our use of them. They represent the stage of progress we have currently achieved, and they are indeed advancements and improvements over the past methods.

REFERENCES

Ang, A. H.-S., and W. H. Tang. 1984. Probability Concepts in Engineering Planning and Design. Vol. II: Decision, risk, and reliability. New York: John Wiley & Sons.

Ang, A. H.-S., W. H. Tang, Y. K. Wen, and B. C. Yen. 1985. Methods for engineering hazard analysis. Proceedings, PRC-US-Japan Trilateral Symposium/Workshop on Multiple Natural Hazards Mitigation, Beijing.

Bodhaine, G. L. 1968. Measurement of peak discharge at culvert by indirect methods. In Techniques of water-resources investigation, book 3, Washington, D.C.: U.S. Geological Survey.

Bowles, D. G. 1987. A comparison of methods for risk assessment of dams. In L. Duckstein and E. J. Plate (eds.), Engineering Reliability and Risk in Water Resources. NATO Advanced Study Institute, Tucson, Arizona. Dordrecht, The Netherlands: Martinus Nijhoff.

Cheng, S.-T., B. C. Yen, and W. H. Tang. 1982. Overtopping risk for an existing dam. Civil Engineering Studies: Hydraulic Engineering Series no. 37, University of Illinois at Urbana-Champaign, Illinois.

Cheng, S.-T., B. C. Yen, and W. H. Tang. 1986. Wind induced overtopping risk of dams. In B. C. Yen (ed.), Stochastic and Risk Analysis in Hydraulic Engineering. Littleton, Colorado: Water Resources Publications.

Garen, D. C., and S. J. Burges. 1981. Approximate error bounds for simulated hydrographs. Journal of Hydraulics Division, ASCE 107(HY11): 1519-1534.

Madsen, H. O., S. Krenk, and N. C. Lind. 1986. Methods of Structural Safety. Englewood Cliffs, New Jersey: Prentice-Hall.

McBean, E. A., J. Penel, and K.-L. Siu. 1984. Uncertainty analysis of a delineated floodplain. Canadian Journal of Civil Engineers 11:387-395.

Melching, C. S. 1987. A reliability analysis on flood event forecasting with uncertainties. Ph.D. thesis, Department of Civil Engineering, University of Illinois at Urbana-Champaign, Illinois.

Melching, C. S., and B. C. Yen. 1986. Slope influence on storm sewer risk. In B. C. Yen (ed.), Stochastic and Risk Analysis in Hydraulic Engineering. Littleton, Colorado: Water Resources Publications.

Mol, A., H. Ligteringen, and A. Paape. 1983. Risk analysis in breakwater design. In Proceedings of Conference on Breakwater Design Construction, London, May 1983. London: Telford Ltd.

Moser, D. A., and E. Z. Stakhiv. 1987. Risk analysis considerations for dam safety. In L. Duckstein and E. J. Plate (eds.), Engineering Reliability and Risk in Water Resources. NATO Advanced Study Institute, Tucson, Arizona. Dordrecht, The Netherlands: Martinus Nijhoff.

Plate, E. J., and L. Duckstein. 1987. Reliability in hydraulic design. In L. Duckstein and E. J. Plate (eds.), Engineering Reliability and Risk in Water Resources. NATO Advanced Study Institute, Tucson, Arizona. Dordrecht, The Netherlands: Martinus Nijhoff.

Rackwitz, R. 1976. Practical probabilities approach to design. Bulletin 112, Comite European du Beton, Paris.

Sitar, N., J. D. Cawfield, and A. Der Kiureghian. 1987. First-order reliability approach to stochastic analysis of subsurface flow and contaminant transport. Water Resources Research 23(5):794-804.

Tang, W. H., and B. C. Yen. 1972. Hydrology and hydraulic design under uncertainties. In Proceedings of International Symposium on Uncertainties in Hydrologic and Water Resources Systems, Tucson, Arizona.

Tung, Y.-K., and L. W. Mays. 1980. Risk analysis for hydraulic design. Journal of Hydraulics Division, ASCE 106(HY5):893-913.

Tung, Y.-K., and L. W. Mays. 1981. Risk models for flood levee design. Water Resources Research 17(4):833-841.

Vrijling, J. K. 1987. Probability design of water retaining structures. In L. Duckstein and E. J. Plate (eds.), Engineering Reliability and Risk in Water Resources. NATO Advanced Study Institute, Tucson, Arizona. Dordrecht, The Netherlands: Martinus Nijhoff.

Wang, B.-H. 1984. Estimation of probable maximum precipitation: case studies. Journal of Hydraulics Division, ASCE 110(HY10):1457-1472.

Yen, B. C. 1970. Risk in hydrologic design of engineering projects. Journal of Hydraulics Division, ASCE 96(HY4):959-966.

Yen, B. C. 1970. Safety factors in hydrologic and hydraulic engineering design. In E. A. McBean, K. W. Hipel, and T. E. Unny (eds.), Reliability in Water Resources Management. Littleton, Colorado: Water Resources Publications.

Yen, B. C. 1986. Hydraulics of sewers. In B. C. Yen (ed.), Advances in Hydroscience, vol. 14. New York: Academic Press.

Yen, B. C. 1987. Reliability of hydraulic structures possessing random loading and resistance. In L. Duckstein and E. J. Plate (eds.), Engineering Reliability and Risk in Water Resources. NATO Advanced Study Institute, Tucson, Arizona. Dordrecht, The Netherlands: Martinus Nijhoff.

Yen, B. C., and W. H. Tang. 1976. Risk-safety factors relation for storm sewer design. Journal of Environmental Engineering Division, ASCE 102(EE2):509-516.

Yen, B. C., and W. H. Tang. 1977. Reliability of flood warning. In Stochastic Processes in Water Resources Engineering, Proceedings of Second International Symposium on Stochastic Hydraulics, Lund, Sweden, 1976. Littleton, Colorado: Water Resources Publications.

Yen, B. C., S. T. Cheng, and C. S. Melching. 1986. First order relia-bility analysis. In B. C. Yen (ed.), Stochastic and Risk Analysis in Hydraulic Engineering. Littleton, Colorado: Water Resources Publica-tions.

Yen, B. C., H. G. Wenzel, Jr., L. W. Mays, and W. H. Tang. 1976. Advanced metholodies for design of storm sewer systems. Research report 112, Water Resources Center, University of Illinois at Urbana-Champaign, Illinois.

AN OVERVIEW OF HEALTH RISK ANALYSIS IN THE ENVIRONMENTAL PROTECTION AGENCY

Timothy M. Barry

ABSTRACT

Risk analysis consists of two interdependent steps: risk assessment and risk management. The Environmental Protection Agency (EPA) uses risk analysis as a tool to support environmental decisionmaking. This paper reviews some of the methods used by EPA to assess risks from exposure to carcinogens and systemic toxicants and discusses the general regulatory context in which EPA risk management decisions must be made.

1. INTRODUCTION

The Environmental Protection Agency (EPA) frequently finds itself faced with regulatory and policy choices in which there is uncertain information, incomplete data, or a slate of interdependent options all competing for the same scarce resources. To strengthen its decisionmaking, EPA utilizes a number of quantitative methods for arraying, integrating, and interpreting information. A number of these methods may, in very general terms, be described as risk analysis.

In order to understand risk analysis in EPA, it is important to understand the complex regulatory environment in which decisions must be made and the fluid, dynamic nature of the underlying science. This paper presents a broad overview of the process and issues surrounding health risk assessment and risk management at EPA. It does not include a discussion of EPA's risk assessment guidelines for mutagenicity or suspect developmental toxicants.

2. RISK ASSESSMENT: THE EPA PARADIGM

The National Research Council of the National Academy of Sciences published a report in 1983 on risk assessment in the federal government (National Research Council, 1983). The Council's report has come to serve as an EPA reference or starting point for definitions of risk assessment principles and processes.

Timothy M. Barry is Chief of the Science-Policy Integration Branch, Office of Policy Analysis, U.S. Environmental Protection Agency, Washington, D.C.

Remarks in this paper represent the views of the author and are not necessarily the official policies of the U.S. Environmental Protection Agency.

Within the context of human health, "risk assessment" may be defined as the quantification of the likelihood of adverse health effects associated with exposure of particular individuals or populations to specific chemical substances. Risk assessment consists of four steps: (1) hazard identification, (2) dose-response assessment, (3) exposure assessment, and (4) risk characterization.

Hazard Identification. Hazard identification is the qualitative step of evaluating the nature and strength of available evidence and judging whether or not exposures to a particular chemical will likely cause an increased incidence of some defined health effect in humans.

Dose-Response Assessment. Dose-response assessment is the process of defining a quantitative relationship between the dose of a chemical and the incidence of a health effect in the exposed population.

Exposure Assessment. An exposure assessment quantifies contact of a substance with the affected population. Exposure is commonly expressed as the dose (e.g., mg/kg-day) and takes into account the magnitude, frequency, duration, and route of exposure (i.e., ingestion, inhalation, or dermal contact).

Risk Characterization. Risk characterization is the combination of the dose-response relationships with exposure data to estimate incidence of a particular health effect in the exposed population. Risk characterization includes a discussion of the nature and effects of the various uncertainties inherent in the analysis.

3. EPA GUIDELINES FOR RISK ASSESSMENT

In 1986, EPA published a series of technical guidelines (U.S. EPA, 1986) which set forth general principles and procedures to aid agency staff responsible for performing risk assessments. The overall objectives of the guidelines are to promote consistency, quality, and predictability in EPA risk assessments. Five separate guidelines were published: (1) Guidelines for Carcinogenic Risk Assessment, (2) Guidelines for Mutagenicity Risk Assessment, (3) Guidelines for the Health Assessment of Suspect Developmental Toxicants, (4) Guidelines for the Health Risk Assessment of Chemical Mixtures, and (5) Guidelines for Estimating Exposures.

Prior to being released as official agency policy, EPA's risk assessment guidelines were extensively reviewed both within and outside the agency. An initial review was performed by workgroups comprised of EPA scientists. The draft guidelines were then published in the Federal Register for public review and comment. Next, the guidelines were submitted to EPA's Science Advisory Board for peer review by members of the scientific community. This review was followed by a review by all EPA offices and their staff. Finally, the guidelines were submitted to the Office of Management and Budget for their evaluation. At each step in this process, EPA staff analyzed the issues, comments, and suggestions which were received to identify and incorporate improvements in the guidelines. Only after the guidelines had survived this lengthy review were they released as official agency policy. Literally hundreds of hours of staff time were devoted to this effort.

EPA's risk assessment guidelines suggest specific methods and conventions to be followed by agency staff in performing risk assessments. Methods and procedures recommended in the guidelines should be followed, except in those specific cases in which there is convincing evidence or data which suggests that an alternative approach might be more appropriate. This is a pragmatic and flexible approach which simply reflects our evolving understanding of exposure and disease and recognizes that science will periodically forge ahead of the guidelines.

One general concern regarding guidelines is their tendency to mechanize the risk assessment process; that is, over time, "institutional momentum" may build to the point where it could hinder incorporating new knowledge in a timely fashion. In direct contrast, another concern is that elements of the guidelines are so general that they are of limited practical use. EPA is well aware of these issues. Risk assessment is a dynamic, evolving process and EPA is committed to revising the guidelines at appropriate times to incorporate new information. In fact, the cancer, mutagenicity, and exposure guidelines published in 1986 already represent updated versions of earlier guidelines (1976, 1980, and 1983). EPA is currently preparing an additional set of risk assessment guidelines for male and female reproductive effects, systemic toxicity, and neurotoxicity.

4.   EPA GUIDELINES FOR CANCER RISK ASSESSMENT

Hazard Identification for Carcinogens. The cancer risk assessment guidelines discuss the kind of information that should be reviewed in a hazard assessment for a suspected carcinogen. Relevant information includes: (1) physical-chemical properties, routes and patterns of exposure, (2) structure-activity relationships, (3) metabolic and pharmacokinetic properties, (4) toxicologic effects, (5) short-term tests, (6) long-term animal studies, and (7) human studies.

To characterize the overall weight of evidence for carcinogenicity, EPA has developed a system for stratifying the available information. The EPA classification system for human and animal carcinogenicity data includes five major categories: (A) carcinogenic to humans, (B) probably carcinogenic to humans, (C) possibly carcinogenic to humans, (D) not classifiable as to human carcinogenicity, and (E) evidence of non-carcinogenicity to humans. Only carcinogens classified as A or B are regarded by EPA as suitable for quantitative risk assessment.

Dose-Response Assessment for Carcinogens. The cancer risk assessment guidelines establish a set of hierarchical ground rules which express preferences for the kind of data that are useful for developing dose-response relationships:

- human epidemiological data is preferred over animal data

- in the absence of human data, data from animal species that respond most like humans is preferred

- data from long-term animal studies showing the greatest sensitivity should be given the greatest emphasis

- data from experiments employing the same exposure route as the chemical under consideration are preferred

- animals with one or more tumor sites or types showing significantly elevated tumor incidence should be pooled and used for extrapolation

- benign tumors should be generally combined with malignant tumors unless the benign tumors do not have the potential to progress to malignancies of the same histogenic organ.

Adequate epidemiologic data is rarely available to determine with statistical rigor whether a contaminant is carcinogenic to humans and if it is, what risks it poses for a range of probable exposures. The majority of our knowledge of carcinogenic potential comes from animal studies. There is considerable controversy in the scientific community surrounding the use of data from high-dose animal experiments to extrapolate to low-dose human exposures. Critics of animal studies (Abelson, 1987; also see responses by McConnel, 1987; Meisler, 1987) cite the very high doses needed to cause statistically significant effects in small test groups, inappropriate test species (inbred strains, supersensitive species), and questionable test endpoints (e.g., mouse liver tumors). It is not very likely that this controversy will be resolved very soon. The fundamental problem is that mechanisms for carcinogenesis are largely unknown.

For those chemicals whose toxicity data are derived from animal bioassays, it is necessary to extrapolate from the animal data to estimate the equivalent human response. Two kinds of extrapolations are involved: (1) <u>low-dose extrapolation</u>, and (2) <u>interspecies extrapolation</u>.

Low-dose extrapolation is the process of estimating responses to doses which are considerably below the lowest dose for which test data is available. A decision rule or model is needed to estimate responses outside the range of doses used in the animal bioassay. EPA assumes, as do most other federal agencies, that carcinogenesis is a non-threshold process*--that is, there is no level of exposure above zero dose for a carcinogenic agent that does not pose some risk, no matter how small (i.e., risk > 0 for all doses > 0).

---

* EPA scientists are reviewing an internal report which suggests that threshold mechanisms may be appropriate for thyroid follicular cell tumors where there is evidence of thyroid-pituitary hormonal imbalance.

EPA recommends using the linearized multistage procedure (LMP) for low-dose extrapolation of animal bioassay data (see Appendix A). The core of the LMP is the multistage model of carcinogenesis. The multistage model assumes that cancer starts with changes in a single cell and proceeds through a number of stages, eventually producing a tumor. Mathematically, the multistage model may be expressed in a simplified form as an exponential polynomial in dose, that is

$$R = 1-\exp(-q_0-q_1d-q_2d^2\ldots) \qquad q_i \geq 0. \tag{1}$$

where d is the dose and q_0, q_1, . . . q_n are constants determined by the data. At sufficiently low doses defined by $q_1d \ll \sum_{i=2} q_id^i$, excess risks (i.e., risks above the background) will be approximately proportional to dose, $\Delta R = R(D)-R(0) \sim q_1d$ (the linearization process). The LMP uses the upper 95% confidence limit on the coefficient of the linear term as the measure of carcinogenic potency. This value, designated q_1^*, when multiplied by dose, $\Delta R \sim q_1^* \cdot d$, represents the "plausible upper limit to the risk that is consistent with some mechanisms of carcinogenesis."

Interspecies extrapolation is a procedure for estimating the equivalent human dose from animal bioassay data. Ideally, an interspecies extrapolation procedure would account for such variables as life span, body size, genetic variability, population homogeneity, concurrent disease, metabolism, excretion factors, exposure patterns, etc. The most frequently used method to extrapolate from animal dose to an equivalent human dose is through the use of scaling factors. Scaling factors may be based on body weight (either mg per kg body weight per day or mg per kg body weight per lifetime), dietary intake (ppm in food or water), or surface area (mg per m^2 body surface area per day).

Unless there is data to indicate that alternative scaling methods would be more appropriate, EPA recommends that surface area be used to scale animal doses to equivalent human doses for carcinogens. This recommendation is based largely on the observation that certain pharmacological effects apparently scale on a surface area basis (Federal Register, 1986).

5.   SYSTEMIC TOXICANTS: NON-CANCER HEALTH EFFECTS

EPA is currently preparing guidelines to assess the health risks of systemic toxicants. Systemic toxicity* is a loose categorization used by the agency to describe any adverse health effect other than cancer and gene mutation. In contrast to carcinogenesis, for which EPA assumes no dose threshold, systemic effects are assumed to be associated with a dose threshold. Below the threshold, risks are thought to be unlikely. Systemic toxicants are assumed to exhibit dose thresholds because organ systems typically have adaptive and compensating mechanisms which allow them to tolerate a range of stresses and still function normally.

---

* General toxicity is another term frequently used to describe these health effects.

Defining and then estimating the risks associated with exposure to systemic toxicants is complicated by the very nature of their effects. Systemic toxicity is associated with effects for which there are generally a range of responses in an exposed population. Systemic toxicants will frequently produce a number of different biological effects in several of the various organ systems of the body. For example, exposure to a chemical might result in changes in liver weight. Under identical exposure conditions, we would expect a range of liver weight changes in the exposed population. These differences are due to variability in each person's ability to adapt or compensate for the stress. Depending on the level of exposure, some of the expected changes in liver weight may be biologically insignificant.

With systemic toxicants, it is necessary to judge if (or when) an observed effect is biologically significant (i.e., adverse). Is a 1% change in liver weight significant? Is a 20% change significant? With a few exceptions, questions like these are answered by the professional judgment of agency scientists on a case-by-case basis.

As with carcinogens, much of our information on systemic toxicity is derived from animal data. To extrapolate systemic toxicity data from animal to man for dose-effect-response estimation requires a greater knowledge of disease mechanisms than we currently possess. The traditional approach taken by toxicologists is the "safety factor method." In the safety factor approach, an initial no-effect dose is estimated from available toxicity data. This dose is then adjusted by safety or uncertainty factors (usually divisors in multiples of ten) to account for uncertainty in the available data. This adjusted dose then becomes a benchmark in the sense that risks are considered negligible for exposures below the adjusted dose. Risks for exposure above the benchmark are not known but are expected to be greater than zero.

Reference Dose Approach (RfD's). Reference doses, or RfD's, are benchmark doses identified by EPA scientists for systemic toxicants. Below the RfD, the risk of adverse effects from chronic exposure is considered to be unlikely.

RfD's are developed by workgroups of EPA toxicologists and health scientists. The literature is reviewed to identify the critical toxic (adverse) effect and the highest dose at which the critical effect has not been observed. This dose is termed the NOAEL ("no observed adverse effect level"). The NOAEL is then divided by an uncertainty factor, UF (usually between one and 1,000) to adjust for man-to-animal extrapolation, sensitive populations, conversion of subchronic exposures to chronic exposures, less than lifetime exposures, etc. An additional divisor, called a modifying factor, MF ($1 \leq MF \leq 10$), may be used to adjust for uncertainties in the data set. Mathematically,

$$RfD = \frac{NOAEL}{UF \cdot MF} , \quad in \frac{mg}{kg \; body \; weight \cdot day} \quad (2)$$

Methods for using RfD's in a risk assessment-like manner include: (1) comparing ratios of the dose (associated with each regulatory alternative) divided by the RfD, (2) comparing the number of people exposed at levels above the RfD for each regulatory option, (3) comparing the products of methods 1 and 2. Technically, analyses based on these ratios or products do not constitute a quantitative risk assessment. Nevertheless, it has become fairly common usage at EPA to refer to RfD-based studies as risk assessments.

Detailed procedures to calculate RfD's have only been worked out for the ingestion route of exposure. Procedures to calculate RfD's for inhalation are currently under development by EPA scientists.

6.   RISK ASSESSMENT GUIDELINES FOR CHEMICAL MIXTURES

Chemical mixtures--coke oven emissions, for example--pose some very difficult problems for risk assessors. Without knowledge of disease mechanisms or test data, it is not possible to say whether risks are independent or dependent (and, if dependent, whether antagonistic or synergistic). Risk independence implies that for small risks, risk addition is appropriate ($r = r_1 + r_2$). Risk dependence means that risks can interact either antagonistically ($r < r_1 + r_2$) or synergistically ($r > r_1 + r_2$).

EPA's guidelines for chemical mixtures recommend risk addition unless there is information on sufficiently similar mixtures and risk can be estimated directly. For carcinogens, risk addition is simply

$$r = \sum_j (q_1^*)_j \cdot d_j \tag{3}$$

For non-cancer health effects (exposures through ingestion), the guidelines recommend calculating a Hazard Index (HI),

$$HI = \sum_j \frac{d_j}{RfD_j}. \tag{4}$$

If $HI \ll 1$, the chemical mixture is assumed to pose no significant health risk. For $HI \gg 1$, significant risks are expected. If $HI \approx 1$, those chemicals contributing most to the index should be examined.

7.   EPA GUIDELINES FOR EXPOSURE ASSESSMENT

EPA's exposure assessment guidelines provide a general framework for estimating the degree of chemical contact with an affected population. Generally, an exposure assessment consists of five steps: (1) Source Characterization, (2) Pathways and Fate Analysis, (3) Estimation of Environmental Concentrations, (4) Demographic Analysis, and (5) Integration.

Source Characterization. Source characterization is the process of identifying the relevant sources of contamination and then assembling the appropriate data necessary for the subsequent steps of analysis. The scope of the source characterization phase is frequently determined by the authority under which the action is being taken. Source characterization data might include environmental loadings (i.e., emission rates, kg/sec, mg/1, etc.), source performance data (e.g., stack diameter, gas temperature, and stack height), available control technologies and their efficiencies, etc.

Pathways and Fate Analysis. A pathways and fate analysis is an exercise in tracking a pollutant from its source as it moves through the environment to the exposed population. In this phase the analyst focuses on both intramedia processes (e.g., chemical transformation) and intermedia processes (e.g., deposition). The analyst considers direct exposure pathways (inhalation, ingestion, dermal contact) and indirect pathways such as food chain exposures. For example, a pollutant discharged into a river might pose a risk through the air if it volatilizes. It might also pose a risk if it contaminates drinking water sources. If the pollutant bioaccumulates, it could pose a risk by contaminating commercial fish catches.

Estimation of Environmental Concentrations. Exposure data or ambient monitoring data for most toxic pollutants are usually not available (see Table 2 for a sampling of EPA's monitoring programs). When these data are available, they frequently have been gathered for other purposes and are of limited use in estimating exposures for specific risk assessments. These shortcomings and data gaps necessitate the extensive use of mathematical dispersion modeling by EPA to estimate environmental concentrations.

Modeling of Pollutant Transport. Environmental monitoring data is preferred for health risk assessments when it is available. Unfortunately, monitoring data are often not available or are too costly to obtain. EPA also frequently estimates risks from future exposures. In these situations, EPA relies on mathematical models to estimate the fate and transport of pollutants (US EPA, 1982).

Within any medium, the transport of pollutants is governed by three intramedia processes: advection, diffusion, and transformation. Superimposed on these intramedia processes are intermedia processes, which transfer pollutants from one medium to another. Familiar examples of intermedia processes include deposition and evaporation.

One very common mathematical approach to modeling the transport of environmental pollutants is based on conservation of mass (see Appendix B). Conservation of mass is a bookkeeping requirement which simply requires that the mass of contaminant in a volume equal the difference between the mass entering and the mass leaving plus the contributions from any pollutant sources.

Most pollutant transport models based on conservation of mass are derivatives of the advective-diffusion equation. For example, the mass balance equation governing the flow of a solute through a porous medium (UA EPA, 1984) is given by

$$\frac{\partial (nC)}{\partial t} = \nabla \bullet (nD_h \bullet \nabla C) - \nabla \bullet uC + \dot{M} - \lambda nC - \frac{\partial (\beta C_s)}{\partial t} - \lambda \beta C_s \qquad (5)$$

where C is the concentration of the solute; C_s is the absorbed concentration in the solid; n is the porosity of the medium; D_h is the coefficient of hydrodynamic dispersion; $u = u(x,y,z,t)$ is the Darcy velocity vector; β is the bulk density of the medium; M is the source release rate; and λ is the degradation rate.

For many problems it is reasonable to model the fluid flow as a constant (i.e., $u = V_x e_x$) and treat the components of the dispersion tensor as constants $D_h = (D_{xx}, D_{yy}, D_{zz})$. If it is reasonable to assume that the aquifer characteristics and sorption are in a state of linear isothermal equilibrium, then Equation 5 may be simplified to

$$R_d \frac{\partial C}{\partial t} + V_x \frac{\partial C}{\partial x} - D_{xx} \frac{\partial^2 C}{\partial x^2} - D_{yy} \frac{\partial^2 C}{\partial y^2} - D_{zz} \frac{\partial^2 C}{\partial z^2} + \lambda R_d C = \frac{1}{n}\dot{M} \qquad (6)$$

where $R_d = 1 + \dfrac{\beta K_d}{n}$ is the retardation factor and K_d is the distribution coefficient. Equation 6 is the model for unidirectional advective transport with three-dimensional dispersion in a homogeneous saturated aquifer. This equation serves as the basis for several EPA groundwater models and is easily solved for a number of boundary and initial conditions using Green's functions, separation of variables methods, etc. (Hwang, 1986; US EPA, 1984; Bear, 1979; Carslaw and Jaeger, 1986).

Exposure Estimates. Once the spatial and temporal distributions of environmental concentrations have been determined, either through monitoring or dispersion modeling, it is necessary to estimate exposures. Generally, the nature of the toxic health effect will determine the time period over which exposures are calculated. For most chronic exposures (Falco and Moraski, 1987), the exposure, in mg/kg-day, is taken as the average daily lifetime exposure:

$$\begin{array}{l} \text{Average Daily} \\ \text{Lifetime Exposure} \end{array} = \frac{\text{Total Dose (mg)}}{\text{Body Weight (kg)} \bullet \text{Lifetime (days)}} \qquad (7)$$

Total dose is calculated as

$$\begin{array}{l} \text{Total} \\ \text{Dose} \end{array} = \begin{array}{l} \text{Environmental} \\ \text{Concentration} \end{array} \times \begin{array}{l} \text{Contact} \\ \text{Rate} \end{array} \times \begin{array}{l} \text{Exposure} \\ \text{Duration} \end{array} \times \begin{array}{l} \text{Fraction} \\ \text{Absorbed} \end{array} \qquad (8)$$

and depends on the environmental concentration (units of mass per volume), the rate of contact with the pollutant through ingestion, inhalation, or dermal contact (units of volume per day; e.g., 18 m^3 per day breathing rate for inhalation), the length of time people are in contact with the pollutant (days), and the fraction or effective proportion (unitless, $0 \le \text{fraction} \le 1$) of the pollutant crossing an exchange membrane (i.e., skin, gastrointestinal tract, or alveolar membrane).

8. RISK MANAGEMENT

Risk Management. Risk management is the process of identifying and evaluating plausible risk reduction strategies and selecting among them. Regulatory decisionmaking involves both risk assessment and risk management. Most risk management activities in EPA are conducted under various legislative authorities (e.g., the Clean Air Act). In risk management, the decisionmaker brings together the risk assessment information with other pertinent considerations (e.g., legislative requirements, technical feasibility, costs, socio-economic factors, political aspects, etc.) to reach a decision.

Ideally, risk assessments are carried out in a manner which is essentially independent of the consequences of regulatory actions. However, risk assessments require resources--in some cases, very significant resources. Budget and staffing decisions must be made to support each of the agency's regulatory activities. These programmatic decisions invariably define the scope, level of analytic detail, and the overall timing of risk assessments. The objective then is to carefully tailor each risk assessment to the decision at hand.

Legislative Authorities. EPA is charged with administering more than a dozen separate environmental laws. In very broad terms, each of these laws implicitly requires EPA to perform risk assessment and risk management decisions. Each of the acts is fundamentally a response on the part of the Congress to perceived risks for which public action is deemed necessary. Each piece of enabling legislation involves decisions as to how to proceed. Specific requirements of each law help shape or direct the manner in which EPA implements the various elements of risk assessment. Table 1 lists the major pieces of legislation for which EPA is responsible (US EPA, 1987b).

Yosie (1986), in a discussion of the risk assessment culture at EPA, points out that differences in requirements from statute to statute limit the factors which can be considered in regulatory decisionmaking. Using Yosie's example, the Clean Air Act requires the administrator, without consideration of costs, to set primary National Ambient Air Quality Standards (NAAQS) that are protective of the public health with an adequate margin of safety. In contrast, the amendments to the Safe Drinking Water Act require the administrator to set health-based maximum contaminant levels (MCL's), taking into consideration technical feasibility and costs. While both pieces of legislation require the agency to perform what are essentially similar risk assessments, each act is quite different in its risk management directives. Differences such as these define the bounds of EPA's risk assessment/risk management opportunities.

Regulatory Analyses. The objective of a regulatory analysis is to lay out and evaluate the major issues and plausible options for a regulation (Carter, 1978; US EPA, 1979). In its most broad application, a regulatory analysis examines and quantifies a regulation's health and environmental effects, economic impacts, energy impacts, technical feasibility, barriers to implementation, and alternatives to direct regulation. Regulatory analyses are generally necessary to comply with the requirements of various executive orders or to comply with legislative directives.

Regulatory analyses are usually only performed for significant regulations. EPA presumes that all new regulations are significant unless they fall into a special exclusion category. Regulations that are commonly excluded include administrative or procedural regulations, minor amendments to existing regulations, regulations mandated by Congress in which EPA has been given no discretion to evaluate alternatives, and regulations designated as insignificant by the originating EPA office.

Major Regulations. Significant regulations may be subdivided into major regulations and routine regulations. Major regulations are defined by Executive Order (E.O.) 12291 (Reagan, 1981) and require a Regulatory Impact Analysis (RIA). A major regulation is any regulation whose annual costs exceed $100 million or which is likely to result in major cost or price increases, adverse impacts on competition, employment, investment, productivity, innovation, or the international competitive position of U.S. firms. E.O. 12291 requires, to the extent permitted by law, that "regulatory action shall be chosen to maximize the net benefits to society" and that "regulatory action shall not be undertaken unless the potential benefits to society for the regulation outweigh the potential costs to society."

E.O. 12291 directs the risk manager to select that regulatory option whose benefit-cost ratio is greater than one and which maximizes net benefits. From a purely economic perspective, using this decision rule to select among options makes a great deal of sense. However, in order to assess the net benefits of an environmental regulation, it is necessary to convert human health information derived from a risk assessment as well as ecological or welfare effects to monetary terms. Of all the issues surrounding benefit-cost analysis, one of the most controversial is the valuation of human health. There is no universally accepted method for converting reductions in human health risks (morbidity and mortality) to monetary terms.

It is common practice to avoid valuation of human life and instead use measures of cost-effectiveness when dealing with health issues. That is, the relative cost-effectiveness of each regulatory option is derived by using estimates of disease incidence (from the risk assessment) to estimate such measures as cost per case avoided, cost per life saved, etc.

Between 1981 and 1986, EPA issued approximately 1000 regulations (US EPA, 1987b). Of the 1000 regulations, 18 were designated as major rules (either proposed rules or final rules). For these 18 major rules, 15 RIAs were prepared at an average cost of $675,000.

A major regulation can potentially affect many segments of society. Because of the wide range of cross-cutting issues and concerns involved in developing and promulgating a major regulation, EPA has published guidelines for performing RIAs (US EPA, 1983, 1987a). These guidelines are intended to promote consistency and quality in the agency's RIAs.

Routine Regulations. Routine regulations are regulations with an annual cost of less than $100 million and do not fall into the special exclusion category. Regulatory analyses for a routine regulation are less sophisticated than RIAs for a major rule and will vary quite a bit depending on the specific regulation and relevant legislative directives. A regulatory analysis developed in support of a routine regulation will generally attempt to examine costs and economic effects. EPA's guidelines call for the lead office to examine such economic consequences as total costs, price and revenue impacts, job losses and gains, regional impacts, and energy effects.

Cost-effectiveness measures are often used to guide the decisionmaker in selecting an option whenever a regulatory analysis is prepared for a routine regulation. For example, one common measure is cost per ton removed for air pollutants. Other measures include cost per case avoided, cost per site, etc.

9. AN EXAMPLE: THE CLEAN AIR ACT

Integrated Risk Information System (IRIS). IRIS (US EPA, 1986) is a computerized compendium of the EPA's risk assessment and risk management information for specific chemical substances. IRIS is maintained by the Office of Health and Environmental Assessment and was designed to serve federal, state, and local health agencies by providing a consistent, ready source for EPA's latest information on health assessments and rulemaking for specific chemicals.

The following discussion is taken in large part from Appendix D in IRIS. It is an attempt to highlight some of the more salient risk management features of the Clean Air Act.

National Ambient Air Quality Standards (NAAQS). The Clean Air Act requires the administrator to set NAAQS (primary standards) for ubiquitous air pollutants which may reasonably be expected to endanger the public health and welfare. Pollutants for which NAAQS have been developed are called air criteria pollutants and include carbon monoxide, sulphur dioxide, lead, particulate matter, nitrogen oxides, and ozone.

Primary standards are designed to protect public health with an adequate margin of safety. The intent of Congress is requiring an "adequate margin of safety" was to address inconclusive scientific evidence and to protect against hazards not yet identified. Primary standards are based strictly on health considerations. In fact, the courts have specifically ruled that EPA may not consider costs or technical feasibility in setting primary standards. For each of the primary standards, EPA considers the nature and severity of the health effects associated with each pollutant, the make-up and size of the sensitive population, and the uncertainties in the available data.

EPA has promulgated primary standards for each of the criteria pollutants. Except for the new PM_{10} standard (i.e., the primary standard for particulate matter less than ten microns in size), each area of the country was required to be in compliance by the end of 1987.

Compliance problems are most acute for ozone and carbon dioxide. Failure to comply could result in the withdrawal of federal aid highway funds or bans on construction.

Secondary Standards. Secondary standards are standards for criteria pollutants which are designed to protect against any known or anticipated welfare effects. Welfare effects include such considerations as visibility impacts, effects on wildlife and agriculture, materials damage, ecosystem damage, and effects on personal comfort and well-being.

In contrast to the primary standards for which a specific timetable for compliance was specified, each area must comply with secondary standards in a reasonable amount of time. "Reasonable" has not been clearly defined by either EPA or the courts. Compliance costs may potentially be considered by EPA in setting secondary standards, but the courts have not ruled on this issue.

Hazardous Air Pollutants. A hazardous air pollutant is any air pollutant not covered under a NAAQS and which poses risks of increases in mortality or serious irreversible or incapacitating reversible disease. Under section 112 of the Clean Air Act, the administrator is required to list as hazardous those air pollutants for which he intends to issue emission standards, and within 180 days of listing, propose regulations for new and existing sources. Final regulations are then to be issued within 180 days of proposal.

EPA's risk management strategy for hazardous air pollutants is called the Air Toxics Strategy (ATS). ATS is a tiered program to assess the risks of toxic air pollutants and to develop appropriate regulatory strategies. ATS consists of three main elements: (1) federal NESHAPS regulations, (2) state and local regulations, and (3) research "hot spots" and urban air quality.

National Emission Standards for Hazardous Air Pollutants (NESHAPs). In order to identify toxic air pollutants which pose significant national risks, EPA first goes through an initial screening. Each pollutant is evaluated under the Hazardous Air Pollutant Prioritization System (HAPPS) on the basis of the available data on health effects, production information, and chemical or physical properties.

A Health Assessment Summary (HAS) and preliminary source and exposure assessment is performed for chemicals selected under HAPPS. If risks appear to be significant, a more detailed Health Assessment Document (HAD) is prepared. Health Assessment Documents are commonly peer-reviewed by EPA's Science Advisory Board. If health risks from stationary sources appear to be significant and federal action is warranted, EPA lists* the pollutant as a hazardous air pollutant and proposes final emission standards within 180 days. Promulgation of these standards must follow in another 180 days.

---

*In 1985, under congressional pressure to be more aggressive in regulating hazardous air pollutants, EPA published a list of pollutants intended later to be listed as hazardous air pollutants under Section 112. EPA has not determined whether this policy of prior listing notification will be continued for other pollutants.

A recent decision by the District of Columbia Court of Appeals has ruled that EPA must follow what is essentially a two-step process in establishing NESHAP. Step one is strictly a health-based process in which the agency is required to determine environmental exposure levels for hazardous pollutants which are "safe." In step two, EPA is required to establish NESHAPs, which will result in exposure levels at or below the health-based criteria with an ample margin of safety. The margin of safety is determined by the administrator and may vary from one standard to another, depending on a wide range of risk management issues.

State and Local Regulations of Hazardous Air Pollutants. Many sources of toxic air pollutants are specific to particular locations, pose no significant national risks, and thus do not warrant federal regulation under the Clean Air Act. For these high-risk, highly localized sources, EPA encourages state and local governments to develop their own emission standards and regulations. EPA works with the state and local agencies and provides technical assistance and advice.

Research. The third element of ATS primarily focuses on toxic air issues on which more research is needed. For example, in highly urbanized areas it often is very difficult to identify sources of many of the toxic air pollutants. This complex chemical soup may contain hazardous air pollutants at relatively high concentrations, posing significant risks to the public. Yet despite our best efforts, we cannot identify the major sources of a number of the pollutants. To address issues such as this, EPA has pledged to undertake research necessary to develop needed risk assessment methods.

New Source Performance Standards (NSPS). NSPS are performance emission standards for new or modified stationary sources whose air emissions may endanger the public health and welfare. NSPS are standards that reflect the best demonstrated technology for achievable emissions reductions, taking cost and other impacts into consideration.

If a NSPS is issued for air pollutants not regulated under other provisions of the Clean Air Act, states are required to promulgate regulations for existing sources of the same type identified in the NSPS. To assist the states, EPA is required to provide guidance on appropriate control technologies and their costs. States must have a strong justification for adopting any emission standard that is less stringent than those specified in the performance standard.

10. THE FUTURE: REDUCING UNCERTAINTIES IN RISK ASSESSMENT

EPA risk assessments are sometimes criticized for being overly conservative. Among some groups, a perception exists that so many hidden safety factors or conservative assumptions are multiplied together that by the time the analysis reaches the risk manager, risks are grossly overstated and unrealistic. Some have suggested that risks may be overestimated by several orders of magnitude. While that view oversimplifies the complex interaction of errors and uncertainty, it is clear that the effects of conservative assumptions, uncertainty, and error propagation need to be better understood in risk assessment and risk management.

To deal with this issue, in October 1987 the agency initiated an effort to identify the explicit and implicit assumptions in cancer risk assessments. A second objective is to identify and evaluate plausible alternative assumptions in terms of their consequences for risk assessment and implications for agency cancer risk assessment/risk management policies. The eventual goal is to strip away the padding and provide the risk manager with unbiased, plausible estimates of risk and the various uncertainties involved.

REFERENCES

Abelson, Philip H. 1987. Cancer phobia. Science, July 31, 1987.

Bear, Jacob. 1979. Hydraulics of Groundwater. New York: McGraw-Hill.

Carslaw, H. S., and J. C. Jaeger. 1986. Conduction of Heat in Solids. Oxford: Clarendon Press.

Carter, Jimmy. 1978. Improving government regulations. Executive order 12044. The White House, Washington, D.C., March 23, 1978.

Falco, James W. and Richard V. Moraski. 1987. Methods used in the United States for the assessment and management of health risk due to chemicals. U.S. Environmental Protection Agency, Washington, D.C.

Federal Register 51:33992-34003 (September 26, 1986).

McConnell, Ernest E. 1987. Letter. Science, October 16, 1987, p. 259.

Meisler, Miriam. 1987. Letter. Science, October 16, 1987, p. 259.

National Research Council. 1983. Risk Assessment in the Federal Government: Managing the Process. Washington, D.C.: National Academy Press.

Reagan, Ronald. 1981. Federal regulation. Executive order 12291. The White House, Washington, D.C., February 17, 1981.

U.S. Environmental Protection Agency. 1979. Improving environmental regulations: final report implementing executive order 12044. Washington, D.C.

U.S. Environmental Protection Agency. 1982. Environmental monitoring catalogue. Washington, D.C.

U.S. Environmental Protection Agency, Office of Policy Analysis. 1983. Guidelines for performing regulatory analysis. EPA-230-01-84-003. Washington, D.C.

U.S. Environmental Protection Agency, Office of Pesticides and Toxic Substances. 1984. User's Guide to AT123D. Washington, D.C.

U.S. Environmental Protection Agency, Office of Research and Development. 1987a. Integrated risk information system. EPA/600/8-86/032b, vols. I, II. Washington, D.C.

U.S. Environmental Protection Agency, Office of Policy Analysis. 1987b. EPA's use of cost-benefit analysis: 1981-1986. EPA-230-05-87-028. Washington, D.C.

Yosie, Terry F. 1987a. EPA's risk assessment culture. _Environmental Science and Technology_ 21:526-531.

Yosie, Terry F. 1987b. Science and sociology: the transition to a post-conservative risk assessment era. Plenary address, Society for Risk Analysis, Houston, Texas.

Zeise, Lauren, Richard Wilson, and Edmund A. C. Crouch. 1986. _The Dose Response Relationships for Carcinogens: A Review_. Cambridge, Massachusetts: Harvard University Press.

Table 1: Summary of EPA's Major Legislative Authorities

Legislation	Health Considerations	Costs and Other Considerations
Clean Air Act		
. Primary National Air Quality Standards (NAAQS)	Set standards for criteria pollutants with an "adequate margin of safety...requisite to protect the public health"	Act specifies only that EPA shall consider public health
. Secondary NAAQS	"Protect the public welfare from any known or anticipated effects"	Act specifies effects on soils, water, crops, vegetation, man-made materials, wildlife, weather, climate, property, transportation, economic values, personal comfort and well-being
. Hazardous Air Pollutants (NESHAPS)	Set standards for hazardous air pollutants that provide "...an ample margin of safety to protect the public health"	Costs and economic feasibility are considered to a limited degree
. New Source Standards	Set standards that reflect "the degree of emission reduction achievable...which has been...demonstrated"	Act specifies consideration of costs, non-air quality health and environmental impacts and energy requirements
. Motor Vehicle Standards	Set standards that reflect "greatest degree of emissions reductions achievable"	Act requires EPA to generally consider costs, e.g...giving appropriate consideration to cost...noise, energy, and safety factors..."
. Aircraft Emissions	Develop standards for any pollutant which "may reasonably be anticipated to endanger the public health or welfare"	EPA must give "appropriate consideration to the cost of compliance"
. Fuel Standards	Analysis depends on grounds for control	
Clean Water Act		
. Private Treatment Works	Act specifies establishment of technology-based effluent limitation guidelines	Technology-based approach with consideration of equipment age, process, non-water quality environmental impacts, and other factors deemed appropriate by the administrator
. Public Treatment Works	Develop guidelines based on the "degree of effluent reduction...through the application of secondary treatment"	
Safe Drinking Water Act		
. Maximum Contaminant Level Goal (MCLG)	Set national primary drinking water goals "...at which...no known or anticipated adverse effects on the health of persons occur and which allows an adequate margin of safety"	MCLG's are strictly health-based criteria for contaminants that may "...have an adverse effect on the health of persons..."
. Maximum Contaminant Level (MCL)	Set national primary drinking water regulations as close to MCLG's as feasible	"Feasible" includes use of the best technology, treatment techniques, taking cost

Table I (continued)

Legislation	Health Considerations	Costs and Other Considerations
. Toxic Substances Control Act	Authorizes EPA to regulate the manufacture, processing, distribution, use, or disposal of any substance that "presents unreasonable risk of injury to the health or the environment"	Directs EPA to consider: i. "Effects...on health and magnitude of the exposure of human beings" ii. "Effects...on the environment and the magnitude of exposure on the environment" iii. "The benefits of such substances...for various uses and the availability of substitutes" iv. "Reasonably ascertainable economic consequences...after consideration...the national economy, small business, technological innovation, the environment, and public health"
. Resource Conservation and Recovery Act (RCRA)	Establish standards for generators and transporters of solid waste and owners and operators of solid waste treatment, storage, and disposal facilities "as may be necessary to protect human health and the environment"	RCRA is generally silent with respect to costs
Comprehensive Environmental Response, Compensation and Liability Act (CERCLA, SARA) . Reportable Quantities	Designate hazardous substances, minimum quantities to report for substances which "may present substantial danger to the public health or welfare or to the environment"	Agency may consider most benefits, but act is silent on costs or economic impact analyses
. National Contingency Plan	Revise NCP for removal of oil and hazardous substances	Revision shall include "means of ensuring that remedial action measures are cost-effective over the period of potential exposure..."
Federal Insecticide, Fungicide, and Rodenticide Act (FIFRA)	Pesticide registration: ensuring that the agent will not have an "unreasonable adverse effect on the environment"	"Taking into account the economic, social and environmental costs and benefits..."
	Data requirements	Must consider effects on "...production and prices of agricultural commodities, retail food prices,..."

Table 2: A Partial List of Environmental Monitoring Programs at EPA

Monitoring Category	Examples of Monitoring Programs
Source Monitoring	
Air	NEDS - National Emissions Data System HATREMS - Hazardous and Trace Emissions System
Water	IFB Organic Data Base Needs Survey
Ambient Monitoring	
Air	SAROAD - Storage and Retreival of Aerometric Data
Water	STORET - Storage and Retreival of Water-Related Data
	NASQAN - National Stream Quality Accounting Network
Groundwater	UIC/HWIS - Underground Injection Control/Hazardous Waste Management Information System
Personal Monitoring	
	TEAM - Total Environmental Assessment Monitoring

Appendix A: Overview of the Linearized Multistage Model

Multistage Model (Zeise et al., 1986). The fundamental assumption inherent in the multistage model is that tumors develop from a single cell which has undergone a number of different transitions or stages. For s different stages, the probability for a cell to undergo transition to a tumorous state is assumed to be

$$I_s(t,d) = \frac{t^{s-1}}{(s-1)!} \prod_{j=1}^{s} \lambda_j(d) \tag{A.1}$$

where λ_j is the probability per unit time for transition to the jth stage. If the probability per unit time for each stage is linear in dose, then

$$I_s(t,d) = \frac{t^{s-1}}{(s-1)!} \prod_{j=1}^{s} (\alpha_j + \beta_j d). \tag{A.2}$$

The probability $p_s(t)$ that any single cell will be tumorous by time t is found by integrating A.2 over time. If dose is constant (i.e., independent of time), then

$$p_s(t,d) = \int_0^t I_s(\tau,d)d\tau \tag{A.3}$$

$$p_s(t,d) = \int_0^t \frac{\tau^{s-1}}{(s-1)!} \prod_{j=1}^{s} (\alpha_j + \beta_j d) \ d\tau = \frac{t^s}{s!} \prod_{j=1}^{s} (\alpha_j + \beta_j d) \tag{A.4}$$

If it is further assumed that tumorous cells act independently, then the number of cells that have undergone transformation may be represented by a binomial probability function,

$$p_k(t) = \binom{N}{k} [Np_s(t)]^k [1 - Np_s(t)]^{N-k} \tag{A.5}$$

which for very large N_s is closely approximated by

$$p_k(t) \sim \frac{1}{k!} [Np_s(t)]^k \exp[-Np_s(t)]. \tag{A.5}$$

The probability of one or more cells being tumorous is then

$$p(t,d) = 1 - \exp[-Np_s(t)] = 1 - \exp\{-N\frac{t^s}{s!} \prod_{j=1}^{s} (\alpha_j + \beta_j d)\} \tag{A.6}$$

For lifetime exposures the time t is taken as a constant,

$$p(d) = 1 - \exp[-(q_0 + q_1 d + q_2 d^2 + \ldots)] \qquad (A.7)$$

in which t and the other terms are collapsed into the constants q_i

Parameter Estimation. In an animal bioassay, N groups of animals are exposed to a chemical, each group at different dose rates. If D_i is the ith dose rate, n_i the number of animals responding at the ith dose rate, and T_i the total number of animals at the ith dose rate, then the probability of observing exactly n_i responses is

$$\text{Prob}(n_i) = \binom{T_i}{n_i} R(D_i;q)^{n_i} \left[1 - R(D_i;q)\right]^{T_i - n_i} \qquad (A.8)$$

where $R(D_i;q)$ is the cancer risk at the ith dose, and q is a vector of parameters to be fit from the data. The risk at a given dose (lifetime risk of cancer at the dose D_i) is represented by the multistage model

$$R(D_i;q) = 1 - \exp\left[-\sum_{k=0}^{N-1} q_k D_i^k \right] \qquad (A.9)$$

where $q_i \geq 0$.

The coefficient vector, q, is estimated by maximizing the log-likelihood function, $L(q) = \prod_{i=1}^{N} \text{Prob}(n_i)$. This involves solving a system of equations $\partial L/\partial q_k = 0$ for $k = 0, 1, 2, \ldots, N-1$. EPA recommends using the 95% upper confidence limit on the linear term, commonly designated q_1^*. q_1^* will produce "plausible upper limits of risk that are consistent with some proposed mechanisms of carcinogenesis." The true risk is unknown and may be as low as zero (although it may also be higher if the assumptions of the multistage model are not met).

In an animal bioassay, the control animals may also develop tumors. If the spontaneous cancers are due to the same basic mechanism as those produced by the chemical being delivered to the animals, then

$$R(D) = R(d + d_b) \approx R(d_b) + d \cdot \left[\frac{\partial R}{\partial d}\right] \cdot d_b = \alpha + \beta \cdot d \qquad (A.10)$$

$$R(D) - R(d_b) \approx \beta \cdot d = \text{constant} \times d \qquad (A.11)$$

where D is the total dose, d_b is the background dose, and d is the experimentally delivered dose. Thus, if there is no threshold and at sufficiently low doses, excess risk will be proportional to the delivered dose.

Appendix B: Pollutant Fate and Transport Modeling: Conservation of Mass

Conservation of mass requires that in an arbitrary volume, the change in mass with time must be equal to the mass flux plus the effects of any mass sources or sinks; that is,

$$\int_V \left[\frac{\partial C}{\partial t} + \nabla \cdot J - Q \right] dV = 0. \tag{B.1}$$

where C is the concentration, J is the net mass per unit time passing through a unit area normal to the direction of flow, and Q is the source/sink density in mass per unit volume per unit time.

Pollutant transport has two components: advective transport and transport by diffusion. Advective transport is mass transport due simply to fluid movement. The advective component of the flux density is given by the product of the concentration and the fluid velocity, U. Transport by diffusion is assumed to be $-K \cdot \nabla C$ where K is the diffusion tensor. The negative sign is to ensure that the mass diffuses from areas of high concentration to regions of lower concentration.

The source/sink term Q is usually given by the sum of a mass source, Q_0 (mass per unit volume per time), which is independent of concentration, and a sink term given by λC where λ is a decay constant in units of time.

The result of these formulations is the advective-diffusion equation,

$$\frac{\partial C}{\partial t} + \nabla \cdot (UC) - \nabla \cdot (K \cdot \nabla C) = Q_0 - \lambda C. \tag{B.2}$$

With appropriate initial and boundary conditions, the advective-diffusion equation serves as the starting point for much of the agency's dispersion modeling work. In various forms, the advective-diffusion equation is used to model pollutant transport in air, surface water, and groundwater. In many of its applications, the advective-diffusion equation leads to various Gaussian plume solutions.

RISK ANALYSIS AND MANAGEMENT OF NATURAL AND TECHNOLOGICAL HAZARDS: A SOCIAL/BEHAVIORAL SCIENCE PERSPECTIVE

Marvin Waterstone

ABSTRACT

Practitioners and theorists in the area of "risk analysis" often approach their subject matter from very divergent perspectives, depending on background, training, inclination, and goals. Some differences are the result of definitional ambiguities; others flow from more fundamental conceptual differences held by physical scientists, engineers, social and behavioral scientists, and public policy makers. This paper attempts to highlight some of these differences, explore their origins, and indicate some policy-relevant implications.

The paper begins with several fundamental definitional matters and then proceeds to an examination of ambiguities in roles for analysts and policy makers. The discussion then turns to the interrelated topics of risk communication, informed consent and rationality, and decisionmaking. Finally, the paper examines the relationship between risk analysis (and/or assessment) and public policy making.

1.   INTRODUCTION

The title of the conference implies both opportunities and challenges for scientists and engineers who are interested in the evaluation and management of natural and technological hazards. The evolving field of risk analysis presents many intriguing avenues of entré to the public policy process, although there are a variety of potential pitfalls as well for the unwary analyst. The conference participants were drawn predominantly from engineering and related disciplines. Therefore, the invitation to prepare this compendium of thoughts and reactions allows me to offer some observations from an under-represented perspective, but one which I believe has some relevance, particularly for the management concerns expressed in the conference title.

It is clear that engineers have a particular view of risk analysis, one that is especially germane to their professional concerns. For those charged with the task of translating nebulous, formative concepts into actual designs and concrete (pun intended) constructions, the notions of reliability, effectiveness, and safety in failure become particularly salient. It is not surprising that these are the most common characterizations of risk developed by engineers. In fact, such conceptions are perfectly congruent with the primary responsibilities of the discipline.

However, it is also not too surprising that these conceptions of risk analysis are not held universally. Furthermore, it is not too surprising that engineers (and many others as well) are frequently puzzled by the lack of fit between such analyses and the public policy

Marvin Waterstone is Associate Director of the Water Resources Research Center, University of Arizona, Tucson, Arizona.

which follows upon their development. Part of this difficulty flows from semantic (or definitional) differences, part stems from more fundamental conceptual differences between physical scientists/engineers on the one hand, social and behavioral scientists/public policy makers on the other. The aims of this brief paper are somewhat limited: to examine some of these points of divergence, and to suggest some means of rapprochement.

The paper begins with definitional matters. Understanding decisionmaking in risky situations is complicated. One factor which adds to confusion is ambiguity in the definition of key terms. What is "risk"? How is it different from "hazard"? What is the difference between "risk" and "uncertainty"? Disagreement or lack of clarity on these concepts, and others, has often led to difficulties in structuring decision problems.

Lack of definitional precision has also resulted in ambiguities in professional and policy-making roles. The implied analytic and decision-making responsibilities for these roles are discussed in the third part of this paper.

In the fourth part of this paper the related topics of risk communication and informed consent are discussed. These are two vital emerging topics for risk analysts and public policy makers alike, and topics that merit some exploration in this discussion.

The fifth section of this paper raises the issue of rationality in decisionmaking. This section specifically examines the differences between expert and lay judgments regarding risk and uncertainty, and explores the implications of these differences.

Finally, the relationship between risk analysis/assessment and public policy making is examined. It is clear that risk analysis/assessment is not equivalent to a decision, but the precise nature of the relationship between the two requires clarification.

2. DEFINITIONS

As the first speaker at the conference admonished, a clear definition of key concepts is essential if discussions about risk are to be productive. Therefore, this section will clarify several critical terms to prevent ambiguity later on.

As indicated in the introduction, the term "risk" has many potential meanings. It is quite common in the literature to use the terms "risk" and "uncertainty" interchangeably. However, the notion of risk involves both uncertainty and the possibility of some adverse effect, damage, or loss. This relationship between risk and uncertainty can be expressed symbolically, as:

$$RISK = UNCERTAINTY + DAMAGE$$

This expresses risk as a measure of uncertainty and the severity of some adverse effect.

It should be noted, however, that the distinction drawn here between these terms differs significantly from the way in which economists differentiate risk and uncertainty. In the economics literature, the term "risk" is used to connote a probability of occurrence less than one, but for which a numerical estimate is possible. "Uncertainty" likewise connotes a probability less than one for which the probability cannot be estimated.

A further clarification is necessary to distinguish "risk" from "hazard," particularly given the conference's concern with hazard management. As the term is used here, a "hazard" (whether natural or technological in its origin) is defined as a source of danger. Therefore, a hazard simply exists as a source of danger or adverse consequences, while risk includes the likelihood of a hazard developing into an actual adverse effect causing loss, injury, or some other form of damage.

Kaplan and Garrick (1981) use the following to illustrate this difference: The ocean is said to be a hazard. If one attempts an ocean crossing in a small rowboat, a great risk is incurred. If the crossing is made in the Queen Elizabeth, the risk is reduced, all else equal. The ocean-going liner is a device used as a safeguard against the hazard. Again the concept can be expressed symbolically:

$$RISK = \frac{HAZARD}{SAFEGUARDS}$$

In addition to illustrating the relationship between risk and hazard, this expression demonstrates the fact that a risk might be diminished by increasing the safeguards, but may never, as a matter of principle, be reduced to zero unless the hazard itself is removed.

Two other terms require clarification: risk analysis and risk assessment. To make the distinction clear, "risk analysis" may be viewed as the process that provides answers to the three following questions: 1) What can happen, or what can go wrong? (the process of hazard identification); and 2) How likely is it that the hazard will occur? (probability estimation); and 3) If the hazard occurs, what will the consequences be? (consequence estimation). "Risk assessment" goes on to answer a fourth question: How important are the consequences if they occur? (consequence evaluation).

In most situations, answering the first three questions is a matter of technical expertise, while answering the fourth is clearly a matter of value judgment. However, it must be recognized that even the first three analytical questions involve assumptions and judgments. For example, in order to estimate the probability of a hazard's occurrence, it is necessary to adopt a specific probability distribution model (e.g., a normal distribution), and this requires an expert judgment.

Similarly, estimating the consequences of a hazardous event may require adopting procedures to extrapolate from past or analogous situations. For example, in an assessment of the carcinogenicity of a particular chemical, the dose-response relationship derived from "high dose, short exposure" experiments with laboratory animals would have to be adapted to more typical "low-dose, protracted exposure" situations for human beings. Again, this requires making assumptions and the application of expert judgment.

A third important term in this regard, and again, one expressed in the conference title, is "risk management." This stage in the process encompasses all of the actions taken to affect, mitigate, and control risk (The Conservation Foundation, 1984). This is closely related to the notion of safeguards, and thus to the controllability of the system in case of failure.

With these definitions in mind, let me now proceed to a discussion of several implications for distinguishing roles in the risk process.

3.   RISK ANALYSIS, RISK ASSESSMENT, AND THE ROLE OF SCIENCE

One useful purpose of clarifying the various stages of the risk analysis/assessment/management process is to identify those portions in which scientists and analysts have an appropriate role, and to distinguish these from areas in which analysts clearly are no better equipped than anyone else to make determinations. To use language common in the risk and decisionmaking literature, this is the issue of fact/value separation.

To elaborate this point, let me use an example which involves all three phases of the risk decisionmaking process: standard setting for environmental contaminants in drinking water. (An analogous case could be made utilizing the 100-year flood, measures of product performance, or many other safety standards.) The first, or analytic, phase in this process entails: (1) the derivation of an appropriate dose-response relationship (which provides information necessary for understanding the nature of the hazard and an estimation of likely consequences if exposure occurs); and (2) an evaluation of likely population exposures. In this phase, the most active participants are scientists and other analysts.

The second phase, risk evaluation, attempts to discern "acceptable risk." For this standard-setting example, this phase would result in a determination of the point on the dose-response curve (i.e., that level of adverse affects) that society would deem acceptable. Are scientists or analysts any more qualified to make this determination (a value judgment) than anyone else? Is there any such thing as an expert in values? The answer is no. Scientists may have better information regarding what is likely to occur and what possible outcomes might result from particular hazard occurrences, but they do not have a more legitimate claim than other members of society on the issue of whether an outcome is desirable or undesirable. This latter determination requires an understanding of, and a decision on, all of the necessary trade-offs.

Such trade-offs can be evaluated in several ways, and even the choice of a method involves a value judgment. One evaluation method would entail a risk/benefit analysis. In our example, this means that the risk incurred by allowing a particular population exposure level (i.e., the standard) must be weighed against the benefits obtained from the particular product or process that produces the regulated substance. A second procedure might employ a risk/risk evaluation. In this process, the risk of exposure is compared with other risks commonly encountered in the society, or with the risks of doing without the regulated substance, or with the risk entailed in the use of a substitute product or process. Finally, the trade-off might be calculated through a risk/cost analysis. In this case, the cost of achieving a more stringent standard would be compared with the resultant reduction in risk (i.e., adverse responses).

This evaluation phase of the process requires the interaction of scientific analysts with public policy makers. The scientific role is primarily one of providing information and placing the risk analysis in context. If trade-offs are to be made, policy makers must understand the nature of the risk and the costs and/or benefits of alternative courses of action. However, the final choice must be made by individuals (in appropriate cases) or by elected or appointed policy makers who are accountable to their constituents. This ultimate social choice cannot be made legitimately by scientists as scientists.

The final phase of the process, risk management, again requires extensive involvement by scientists. In this activity the decisions made regarding acceptable risk levels must be implemented. Risk-reducing measures must be carried out effectively and efficiently. To return to the standard-setting example, in the risk management phase, measures must be designed and carried out to achieve the mandated levels of discharge of the regulated substance. This might require process changes, product substitutions, or other mechanisms, but in any case, scientific and/or engineering involvement will be significant.

In summary, the various activities which comprise the risk analysis/assessment/management process require both technical (or factual) and value judgments. To carry these functions out in an appropriate manner requires that the roles of scientists and officials be clearly distinguished. Too often scientists are asked (in fact, forced) to make value judgments regarding "how safe is safe." Likewise, policy makers are often coerced (either by circumstance or by inclination) into making technical judgments for which they are neither equipped nor trained. This confusion of roles has often led to inappropriate decisions, public suspicion of scientists, and mistrust of politicians. Keeping the roles separated to the fullest extent possible might begin to alleviate some of the more common difficulties.

4.   RISK COMMUNICATION AND INFORMED CONSENT

If public policy regarding risky situations (e.g., regarding hazard management) is to be effective, policy makers (and their constituents) must comprehend the nature of risk, as well as the nature of the risk analysis process. This must include an understanding of the assumptions which govern the analysis and an understanding of the uncertainty in the output of the analysis.

While this argument seems reasonable, the process of communicating risk information is anything but straightforward. Some have argued that the problem of risk communication should itself be structured as a decision problem (Keeney and von Winterfeldt, 1986). However, before risk communication may commence in particular situations, one fundamental issue must be addressed: what is the role of risk communication?

This question underlies the entire issue of the role of government in managing risk and hazards. Is the role simply to provide individuals with enough information to allow them to make their own decisions? Or should governmental policy makers take a more active, protectionist position and actually intervene to reduce hazardous situations and their concomitant risks? Or should government's role lie somewhere between these positions? These policy orientations lead to fundamentally different views of risk management, and to the requirements placed on the risk communication process.

Under the first position, which has sometimes been termed "informed consent," the role of government would be seen as primarily that of providing information and allowing people to make up their own minds regarding the risks they would be willing to incur. This position entails several assumptions about the reliability of information and about human decisionmaking that may prove to be invalid.

What does the term "informed consent" really mean? How informed does someone have to be to be "informed"? Can risks be communicated accurately? Can the risk context be communicated? Are individuals able to trade off the risks and benefits reliably? Are alternatives analyzed and presented so that various courses of action can be evaluated? The concept of informed consent is being challenged on many fronts, and many analysts now believe it to be highly problematic (MacLean, 1982).

In many situations, however, individual decisions are not adequate for preventing or reducing societal risks. In these instances, then, governmental agencies see their role as necessarily more active in the management of hazards and risks. What does the notion of consent mean in these cases? MacLean (1982) argues that in instances of centralized decisionmaking (e.g., in governmental management of risks and hazards), we necessarily move away from the explicit consent that is possible in individual choice (at least theoretically), and toward more implicit, or even hypothetical, consent. His argument indicates that centralized decisions that entail placing members of society at risk require some kind of justification. Consent (either implied or hypothetical) provides that legitimation.

However, arriving at an understanding of society's implied consent is extremely difficult methodologically. Reliance on market mechanisms (revealed preference techniques) or upon responses obtained in structured situations (expressed preferences) frequently forms the basis for an understanding of implied consent. Behaviors observed in the market or in laboratory settings are believed to be transferrable to other similar situations. In other words, the assumption is made that these behaviors reveal a form of consent (or a level of accepted risk) in those circumstances, and that this level of risk acceptance can be applied in new situations as well. Much to our chagrin as analysts, however, these assumptions often prove erroneous.

5.   RATIONALITY AND RISK IN CONTEXT

To understand this situation, we must re-examine the notions of rationality, or risk in context, and of consent. Many studies have documented the disparity between expert and lay judgments of risk. All too frequently scientists and others are heard lamenting the lack of rationality in public assessments of risks and in the resultant public policy choices. However, attribution of apparently aberrant decisions to irrationality more often represents a failure of analysis to capture all of the relevant factors included in a particular decisionmaker's calculus.

We scientists often proclaim that if only the public could be educated and informed, their perceptions of risk would more closely match our own. Very-high-probability events are overlooked, while the significance of very-low-probability events is exaggerated. Risk levels accepted in one context are not tolerated in other situations. For example, it is difficult to understand why an individual will conscientiously fasten his or her seatbelt on the way to go hang-gliding. Or why, at a more aggregated level, U.S. society tolerates nearly 50,000 automobile-related deaths per year and abhors nuclear power.

Why do these disjunctures occur? One question that must be asked is: are these really inconsistencies? Or is it more likely, as MacLean suggests (1982), that we have begun to substitute unrealistic rules of rationality in those risky situations for which we do not have explicit consent on the part of the public? In areas where centralized decisions are made on the basis of implied or hypothetical consent, there is a tendency to rely much more heavily on rationality (defined narrowly) to justify those decisions. The argument proceeds from an analysis of the level of risk accepted by the public under one set of circumstances, and posits that rational people would accept the same level of risk under other circumstances. To behave otherwise is to fall into irrationality.

However, it is more often the case that risk assessments vary depending upon the context (i.e., the mix of perceived risks, benefits, and alternatives) in which the risk is presented, upon the perceived (personal) controllability of the risk, upon the magnitude of the likely adverse effects (mundane vs. catastrophic), and upon a whole host of other considerations. It is quite unlikely that analysis will be able to capture all of these components. What is clear is that it is highly inappropriate to attribute these differential assessments to irrationality because they do not appear (at least on their face) to adhere to such axioms of rationality as consistency and transitivity.

6.   RISK ASSESSMENT AND PUBLIC POLICY

In this final section I wish to explore briefly the relationship between risk analysis/assessment and public policy making. Frequently, risk analysts undertake a careful investigation of a risky situation and follow all the rules (such as they are) of risk communication to transmit their findings to the appropriate policy makers, only to have their conclusions apparently ignored or discounted as the policy wanders off in another direction. Not only is this a frustrating experience, but it is often incomprehensible as well.

Some of this mystery can be resolved if we return to a few of the points made in the previous section. First, it must be accepted that a risk analysis is not equivalent to a decision, much as we analysts might desire this. If a risk analysis is not a decision, then what is it? It is one piece of information taken into account by the policy maker. It must be integrated with a variety of other considerations that usually fall outside the boundaries of the formal analysis. Such intangible factors as political feasibility and weighting of interest groups' desires are quite difficult to include in formal analyses, but must be included in the political decisions that make up public policy in areas like hazard management.

Understanding this point does not make it less frustrating to have formal analyses downplayed in the policy process, but it may provide some insight into how the role of these analyses in formulating public policy may be enhanced. To a greater degree than ever before, the public is demanding a direct voice in decisions that affect their health and safety. Demands for community and worker right-to-know ordinances, the Freedom of Information Act, and the increasing use of public referenda to decide complex technical issues all demonstrate a lack of trust in public officials (and in scientists as well) to decide these issues without public scrutiny.

One of the ways in which the public's confidence might begin to be restored is through a perception that the processes which produce decisions are themselves technically sound, fair, and equitable. In order for this change in perception to take place, it is possible that policy makers will have to demonstrate, to a greater degree than they do at present, that their decisions are based on the most accurate and careful analysis of the risk involved and upon an assessment of the implications of those risks for public consequences. If this turns out to be the case, the link between risk analysts and public policy makers will necessarily require strengthening. In that event it will be more critical than ever that analysts understand the constraints on policy makers and vice versa.

7.   SUMMARY

The characterization of risk analysis by engineers and other physical scientists as failure analysis is much narrower than the construction of this concept by either social/behavioral scientists or public policy makers. This difference in concepts leads to different expectations of the risk analysis/assessment/management process, and often leads to confusion and frustration on both sides. Formal risk analysts must come to understand the role of their analyses in the policy-making process in order to increase the likelihood that these analyses will play a significant (rather than a symbolic) part in public decisionmaking for hazard management. Likewise, policy makers must base their decisions on careful, comprehensive analyses to a greater degree than in the past if public confidence in decision processes is to be restored. It is not sufficient for risk analysts to perfect their procedures in a vacuum. Neither will the public tolerate risk decisions which appear to be capricious or arbitrary.

REFERENCES

Kaplan, S. and B. J. Garrick. 1981. On the quantitative definition of
risk. Risk Analysis 1 (1).

Keeney, R. and D. von Winterfeldt. 1986. Improving risk communication.
Risk Analysis 6 (4): 417-424.

MacLean, D. 1982. Risk and consent: philosophical issues for centralized decisions. Risk Analysis 2(2): 59-67.

The   Conservation   Foundation.   1984.   State of the Environment: An
Assessment at Mid-Decade. Washington, D.C.: The Conservation Foundation.

RISK ANALYSIS OF DAM FAILURE AND EXTREME FLOODS

Raja Petrakian, Yacov Y. Haimes, Eugene Stakhiv, and David A. Moser

ABSTRACT

In this work, the partitioned multiobjective risk method (PMRM) is applied to a real but somewhat idealized dam safety case study. During the course of the analysis, useful relationships are derived that greatly facilitate the application of the method, not only to dam safety but also to a variety of other risk-related problems. Apart from theoretical investigations, the practical usefulness of the PMRM is examined in detail through its use in the evaluation of various dam safety remedial actions. It is shown that the PMRM allows the decision-makers to enhance their understanding of a problem's characteristics, especially those of an extreme and catastrophic nature.

Examination is also made of the sensitivity of results generated by the PMRM to variations in the value of the return period of the probable maximum flood (PMF) and to changes in the probability distribution used to describe the frequency of extreme floods. Four different distributions (log-normal, Gumbel, log-Gumbel, and Weibull) are tested, reflecting a range of density functions which are commonly used in describing the probabilistic behavior of extreme flood flows. Of these distributions, the ones with the thinnest tails are found to produce the largest conditional expected damages in the low-probability/high-consequence risk domain and consequently to yield the most conservative choices. Also, the low-probability/high-consequence conditional expectation exhibited significant sensitivity to variations in the return period of the PMF.

1. INTRODUCTION

This study is aimed primarily at illustrating how the partitioned multiobjective risk method (PMRM)--a new risk analysis approach that is designed to aid the evaluation of low-probability/high-consequence events--can be applied to the problem of dam safety and extreme floods.

Raja Petrakian is a Ph.D. student with the Industrial Engineering and Operations Research Department, University of California, Berkeley, CA.

Yacov Y. Haimes is Lawrence R. Quarles Professor of Engineering and Applied Science, and Director, Center for Risk Management of Engineering Systems at the University of Virginia, Charlottesville, VA.

Eugene Stakhiv is Program Manager for Risk Analysis Research at the Institute for Water Resources, U.S. Army Corps of Engineers, Fort Belvoir, VA.

David A. Moser is an economist with the Institute for Water Resources, U.S. Army Corps of Engineers, Fort Belvoir, VA.

First, it is important to define the word "safe." By saying that a structure is safe, it is meant that risks associated with this structure are "acceptable" to society. Risk is defined as a measure of the probability of occurrence of a potentially hazardous event and of the event's consequence to society. It is important to understand that risks can never be reduced to zero, and therefore it becomes necessary to determine a risk level that can be considered acceptable.

Risk analysis can be a useful tool to assist the decisionmaker in evaluating the impact of various policies and remedial actions, in this case on dam safety. Risk analysis can also help the decisionmaker determine the amount of protection that should be added to a dam given the construction and maintenance costs needed to modify its characteristics to the desired level. Any decision involves the consideration of a trade-off between somewhat more certain expenditures and relatively uncertain benefits and economic losses.

Dams are designed, in part, to control the extreme variability in natural hazards (floods and droughts), but they simultaneously impose an even larger, though much less frequent, technological hazard: potential dam failure. Therefore, a low-probability/high-consequence (LP/HC) risk analysis of dams is the most appropriate approach to tackle the issue of dam safety. But risk analysis of LP/HC events is still an evolving field, and it must cope with complex, incommensurable issues. In general, information related to LP/HC events is scarce. Different statistical tools are required to make full use of the sparse information. Such events as nuclear power plant accidents, dam failures, and toxic chemical spills constitute major LP/HC events, and many risk assessment studies of these events have been done. These studies have, in general, relied on the application of traditional risk analysis tools, which do not appropriately characterize the hazards associated with LP/HC events (as will be shown in the next section).

In a joint study on dam failures and accidents [1975], the American Society of Civil Engineers (ASCE) and the United States Committee on Large Dams (USCOLD) showed that overtopping accounts for 26% of all reported dam failures and that the principal reason for overtopping is inadequate spillway capacity. Thus, the evaluation of adequate spillway capacity is a vital issue in dam safety analysis, and it comprises the focus of this analysis. Two types of corrective or remedial actions will be considered here: widening the spillway and increasing the dam's height.

The main function of a spillway is to protect the dam itself during extreme floods. Spillways help to avoid dam failure by passing excess water--that is, water beyond the design flood volume--which might otherwise cause the dam to be overtopped or breached. The hazards posed by inadequate spillways might approach or even exceed damages that would have occurred under natural flood conditions without the existence of the dam.

A large number of studies have been done to determine the statistical distribution that makes best use of the existing records of annual floods and extrapolates beyond the records with maximum confidence. But since most flood records in the U.S. do not exceed 50 to 100 years of data, it becomes very difficult to determine which distribution is most appropriate for extreme floods larger than the 100-year flood.

A National Research Council report on dam safety [1985] disapproved of the extrapolation of frequency distributions of rare floods from flood-frequency distributions derived from data less than 100 years old. The report discusses whether the return period of the probable maximum flood (PMF) can be estimated with reasonable accuracy. It was observed that if antecedent conditions had little impact on the PMF values, then the required calculations could be simplified and could provide credible solutions, but otherwise estimates of the return period of the PMF could be quite unreliable.

Appendix E of the National Research Council report presented a procedure for estimating extreme flood frequency based on interpolating the assumed flood-frequency distribution between the 100-year flood and the PMF. This approach does not in general yield credible results, however, unless the return period of the PMF can be estimated. Since the return period of the PMF is often difficult to estimate, the report affirms that the use of either 10^6 or 10^4 years can give reasonable results. The report also recommends that research be done on the impact and advisability of using different flood-frequency distributions.

Stedinger and Grygier [1985] have studied the sensitivity of the results of a dam safety risk analysis to the value of the assigned return period of the PMF and to the choice of the flood-frequency distribution used to extend the frequency curve to the PMF. They found that the results could be easily influenced by either a change in the return period of the PMF or by the choice of the distribution. Therefore, it is concluded that any risk analysis on dam safety should include a sensitivity study of these factors, by which the range of uncertainty could be somewhat bounded.

Consequently, in this study one of the key issues is to perform a sensitivity analysis of decision situations by varying both the frequency distribution of the inflow events and the return period of the PMF (also see Petrakian [1986] and Haimes et al. [1988]). This analysis includes an adequate representation of the wide variety of flood-frequency distributions used in different studies: the log-normal, Gumbel, log-Gumbel, and Weibull distributions. Of particular interest are the log-Pearson type III distribution (which with zero skew becomes the log-normal distribution), recommended by Bulletin 17B of the Interagency Advisory Committee of Water Data [1982], and the Gumbel distribution (type-I extreme value distribution), which is characterized by its thin tail. The values of 10^4 and 10^6 years have been assigned to the return period of the PMF, thus following the recommendation of Appendix D of the 1985 National Research Council report on dam safety.

2. THE PARTITIONED MULTIOBJECTIVE RISK METHOD AND DAM SAFETY

Kaplan and Garrick [1981] stated that expressing risk as probability times consequence implicitly reduces risk to expected damage, in turn leading to equating low-probability/high-damage alternatives with high-probability/low-damage ones, which are clearly not equivalent events unless the decisionmaker is risk-neutral. Kaplan and Garrick insisted that risk is not the mean of the risk curve but rather the curve itself. They stated that "a single number is not a big enough concept to communicate the idea of risk. It takes a whole curve." Moreover, if the uncertainties due to our incomplete knowledge must be considered, then a whole family of curves might be needed to express the idea of risk.

Vohra [1984] even presented a quantitative definition of risk that accounts for the higher impact of extreme events on society. In many other reports and papers, scientists have expressed reluctance or discomfort when using the traditional expected value method to evaluate risks associated with extreme events. For example, the 1985 NRC report on dam safety borrows this quotation from Raiffa [1968]:

The issue is how much members of society are willing to pay to avoid such unlikely events. It is highly plausible that they are ready to pay more than the expected cost.

There is obviously a strong need for a risk analysis method that would allow us to consider explicitly the low-frequency/high-damage domain. The partitioned multiobjective risk method, or PMRM (Asbeck [1982], Asbeck and Haimes [1984], Leach [1984], Leach and Haimes [1987]), provides the capability to quantify risks for extreme events.

The PMRM attempts to avoid the problems associated with the concept of traditional expected value by collapsing the risk curve into a set of points that represent the conditional expected values for the different damage domains. These points are obtained by partitioning the exceedance probability axis into different ranges, and then taking for each range the expected value for damages that have their exceedance frequencies lying within that range. This method allows us to represent a distribution by a number of points instead of just one point, as in the traditional expected value method, and therefore preserves more information about the risk curve. Ideally, we would like to keep the whole risk curve, but the PMRM is still an improvement on the method of traditional expected value. Through an appropriate partitioning of the probability axis, we can calculate the conditional expected value for damages that correspond to the low-probability/high-consequence events, thus quantifying the risk of extreme events.

The PMRM has another advantage: it avoids the explicit use of utility functions to represent the decisionmaker's preferences. Utility theory has often been criticized because it is based on assumptions about individual behavior that are sometimes inconsistent with reality. For example, Slovic and Tversky [1974] showed that Savage's independence principle, which is at the heart of expected-utility theory axioms, is not always satisfied. MacCrimmon and Larsson [1975] and Shoemaker [1980] argued along similar lines.

Moreover, the PMRM does not replace the judgment of the decision-maker; it is merely a tool to help the decisionmaker express individual preferences through the consideration of trade-off information among the different objectives. The surrogate worth trade-off (SWT) method and its extensions (Haimes and Hall [1974], Chankong [1977], Haimes and Chankong [1979], and Haimes [1980]) are used to develop trade-offs and, through interaction with the decisionmaker, to obtain a preferred solution. This means that the SWT allows implicit expression of the decisionmaker's utility function through the use of information about trade-offs among the several objectives. In the case of dam safety, these objectives could be, for example, the desire to reduce risks associated with moderate or extreme floods, and the simultaneous desires to minimize the cost of remedial actions or to minimize loss of life.

It is a distinctive characteristic of the PMRM that it allows analysis of a safety problem in a multiobjective framework. Haimes [1984] has illustrated the advantages of performing risk assessment using a multiobjective approach. First, more than one objective can be taken into consideration, and therefore a better approximation of the real decisionmaking process is obtained. Also, the analyst can limit the scope of the work to such areas as system modeling, the quantification of risks and objectives, and the calculation of trade-offs. The actual decisionmaking process is left to the decisionmaker, who uses subjective preferences and judgment, interprets the results, and determines appropriate policies.

3. AN OVERVIEW OF THE PARTITIONED MULTIOBJECTIVE RISK METHOD

The PMRM is based on the use of conditional expectation, which is defined as follows. Given $p_X(x)$, the marginal probability distribution of the random variable X, and assuming that

$$p_X(x) \begin{cases} \geq 0 & \text{for } 0 \leq x \leq \infty \\ = 0 & \text{for } -\infty \leq x < 0 \end{cases}$$

the conditional expectation of an event $D = \{x / x \varepsilon [a,b]\}$ is given by

$$E[X \mid D] = \frac{\int_b^a x \, p_X(x) \, dx}{\int_b^a p_X(x) \, dx}$$

Finding Marginal Probability Density Functions

The use of the PMRM requires knowledge of the probability density function of losses, and this function is often dependent on the policy option. Let this function be denoted by $h_X(x, s_j)$, where x is the magnitude of losses and s_j (j=1,2,...,n) is the policy option or scenario.

Given $h_X(x,s_j)$, it is possible to calculate the exceedance probability function, defined as

$$1 - H_x(x,s_j) = 1 - \int_0^x h_x(y,s_j)\, dy \quad j=1,2,\ldots,n$$

Partitioning the Probability Axis

The next step is to partition the exceedance probability axis into a set of ranges. These ranges should be compatible with the nature of the problem of interest. The partitioning should be done in such a way that it enhances the decisionmaker's understanding of the problem. In other words, the analyst should try to capture the subtleties of the problem by adequately determining the number of ranges, m, into which he will partition the probability axis and the positions of the partitioning points, α_i (i=1,2,...,m+1). The partitioning points on the risk curves should be exactly the same for all the various policy options. Note that partitioning the cumulative probability axis would produce the same final results as partitioning the exceedance axis.

Mapping Partitions to the Damage Axis

Before the conditional expectations are used, the values of the partitioning points of the probability axis should be mapped onto the damage axis. Therefore, for each partitioning point α_i (i=1,2,...,m+1) and each policy option s_j (j=1,2,...,n), it is necessary to find an $a_{ij} \geq 0$ such that

$$1 - H(a_{ij}, s_j) = \alpha_i$$

Notice that a_{ij} is actually the projection of the partitioning point α_i on the damage axis. Furthermore, when the exceedance probability is plotted against the damage axis, α_1 and α_{m+1} correspond to 0 and 1.

These values of a_{ij} (i=1,2,...,m+1 ; j=1,2,...,n) will be used to calculate the conditional expectations for the m domains of the damage axis that correspond to the m ranges of the cumulative probability axis. These domains have been defined in the following way:

$$D_{1j} = [a_{1j}, a_{2j}] \qquad j=1,2,\ldots,n$$

$$D_{ij} = (a_{ij}, a_{i+1,j}] \qquad j=1,2,\ldots,n$$

$$i=2,3,\ldots,m$$

(Here $c \ \varepsilon \ (a,b]$ means that $a < c \leq b$)

Finding Conditional Expectations

The conditional expectations can be computed using

$$E[X \mid D_{ij}; s_j] = \frac{\int_{a_{ij}}^{a_{i+1,j}} x \, h_X(x, s_j) \, dx}{\int_{a_{ij}}^{a_{i+1,j}} h_X(x, s_j) \, dx} \quad \begin{array}{l} i=1,2,\ldots,m \\ j=1,2,\ldots,n \end{array}$$

The denominator can be reduced to

$$\int_{a_{ij}}^{a_{i+1,j}} h_X(x, s_j) \, dx = \alpha_i - \alpha_{i+1} \quad i=1,2,\ldots,m$$

Since the partitioning points are invariant for all policy options, the value of the denominator, which can be considered a weighting factor, will be unaffected by the policy option. On the other hand, the damage regions $[a_{ij}, a_{i+1,j}]$ ($i=1,2,\ldots,m$; $j=1,2,\ldots,n$) vary with the policy options.

Generating Functional Relationships

A set of risk functions $f_i(s_j)$ can be generated from the conditional expectations by setting

$$f_{i+1}(s_j) = E[X \setminus D_{ij}; s_j] \quad \begin{array}{l} i=1,2,\ldots,m \\ j=1,2,\ldots,n \end{array}$$

For a given policy s_j, each $f_i(s_j)$ represents the particular risk associated with the corresponding partitioning range $[\alpha_{i-1}, \alpha_i]$.

In addition to the m risk functions defined above, the unconditional expectation (or expected value of damages), which still has some use, should be computed for all policy options. It will be utilized to form the m + 1st risk function, denoted by $f_{m+2}(s_j)$ ($j=1,2,\ldots,n$).

Moreover, the cost function is a vital part of the analysis in many problems. It represents the costs associated with the different policies. The notation $f_1(s_j)$ ($j=1,2,\ldots,n$) will be used to represent the cost function.

If it can be assumed that the decision policy is continuous between s_j and s_{j+1} ($j=1,2,\ldots,n-1$) and if, in addition, it can be assumed that the m + 2 risk functions defined above are continuous in a simple way, then by the use of regression, it is possible to fit a smooth curve $f_i(s)$ to the points $(s_j, f_i(s_j))$ ($j=1,2,\ldots,n$) for $i=1,2,\ldots,m+2$.

Employing the Surrogate Worth Trade-Off Method

At this stage, the proposed decisionmaking problem involves m + 1 risk objective functions and one cost objective function. Since as little information as possible should be lost in the analysis, we need to try to make use of all the objective functions. Only a multiple-objective decisionmaking methodology would be appropriate in this case. The surrogate worth trade-off (SWT) method is one such method. Its advantage over other methodologies is that it allows the decisionmaker to express his preferences during the decisionmaking process. Basically, the SWT method provides the decisionmaker with the Pareto optimal policies and the associated trade-offs among the various objectives.

Of particular interest are the trade-offs between f_1, the cost function, and the m + 1 risk objectives.

Haimes et al. [1975] define the noninferior solution as a solution in which no decrease can be obtained in any of the objectives without causing a simultaneous increase in at least one of the other objectives. The trade-off rate function between $f_1(s)$ and $f_i(s)$ can be defined as

$$\lambda_{i1}(s) = - \frac{\delta f_i(s)}{\delta f_1(s)} = \frac{1}{\lambda_{1i}(s)} \qquad i=2,3,\ldots,m+2$$

Notice that the trade-off rate functions $\lambda_{i1}(s)$ are defined as the negative of the partial derivatives of f_i relative to f_1. Thus, if the decision variable is discrete, partial derivatives do not exist and consequently the trade-off rate functions cannot be used. In this case, the concept of total trade-off is introduced, and it will be used in this work for discrete problems. When it is stated that the total trade-off between f_1 and f_i from s_j to s_h is $\lambda_{i1}(s_j,s_h)$, this means that by using policy s_h instead of policy s_j, a change of magnitude $\lambda_{i1}(s_j,s_h)$ in f_i will correspond to a change of one unit in f_1 (see Figure 1). The total trade-off is defined by

$$\lambda_{i1}(s_j,s_h)= - \frac{f_i(s_j)-f_j(s_h)}{f_1(s_j)-f_1(s_h)} \qquad \text{for } i=2,\ldots,m+2$$

For convenience, before calculating the total trade-offs, the alternatives should be reordered to obtain an increasing function f_1.

In this work, a heuristic procedure based on computer simulations was used to reorder the alternatives and to determine all noninferior solutions.

Other concepts central to this work must now be defined. The "indifference band" is defined as a subset within the set of noninferior solutions in which positive improvement of one objective equals (in the mind of the decisionmaker) a corresponding negative change in another objective. The "preferred solution" is any noninferior solution that belongs to the indifference band.

The aim of any multiobjective optimization procedure is to determine the preferred solution. To achieve this purpose, if the continuity

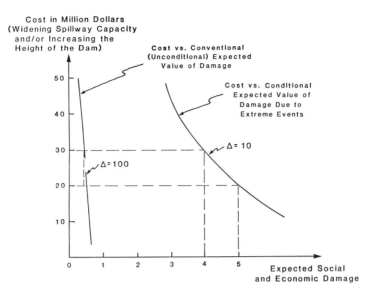

Figure 1 Trade-offs between cost and conditional and unconditional expected value of risk

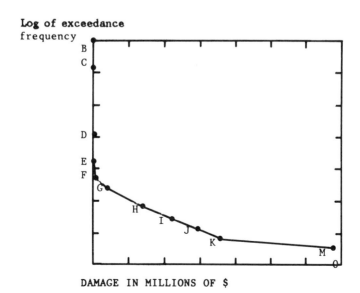

Figure 2 Risk Curve for scenarios s_1

assumptions hold, the SWT uses the surrogate worth function (SWF). To determine the surrogate worth function, the analyst should ask the decisionmaker to assess the value of the surrogate worth function at a certain number of points (corresponding to noninferior solutions), given the trade-offs involved and the achievement levels among the objectives. The surrogate worth function is then constructed, possibly by regression.

If the continuity assumptions do not hold, the above discussion must be modified. Essentially, the total trade-off concept must be used rather than the partial trade-off concept. It is still possible to employ a function similar to the surrogate worth function.

4.   SIMULATION OF DAM OPERATIONS

A computer model developed at the U.S. Army Institute of Water Resources (IWR) was used to simulate the routing of various inflow events through dams. Moser [1987] provided a detailed description of the IWR dam safety risk/cost analysis model, which is used extensively in the rest of this chapter. He followed the assumption in the Corps' guidelines that overtopping in excess of the assumed safe amount would cause the dam to fail with certainty. Other circumstances that might cause the dam to fail were ignored for the sake of simplification.

Two preventive remedial actions are of interest: widening the spillway and raising the dam's height. Inherent to each of these actions is a trade-off between two situations. For example, widening the spillway reduces the chances of a failure caused by rare floods with high magnitudes that overtop the dam; but on the other hand, greater damage is incurred downstream by medium-sized floods that pass through the spillway. Similarly, augmenting the dam's height reduces the likelihood of a dam failure but increases the severity of downstream damages in the event of failure. This reflects an incommensurable trade-off in risk reduction. Each alternative can meet a stated design objective, but the damages occur in different parts of the frequency spectrum. The expected-value approach cannot capture this risk-reduction. Sixteen remedial actions which variously combine changes in the spillway's width and the dam's height will be considered here.

The peak rate of inflow as a percentage of the probable maximum flood (PMF) must be specified. The PMF was assumed here to be 432,000 cfs for the hypothetical Tomahawk dam (data corresponding to a real dam was used here).

When the computer model is executed, the hydrologic model routes the specified peak inflow event through the dam and calculates the corresponding peak outflows for cases of both dam failure and non-failure.

For each of the 16 alternatives combining changes in the spillway's width and the dam's height, 15 inflow events in sizes varying from 0 cfs to 432,000 cfs (the assumed PMF) were routed through the reservoir.

In the simulations, the following assumptions have been made concerning flood conditions antecedent to the routed flood inflow event:

- No antecedent floods could either cause the dam to be overtopped or produce any downstream damages. This assumption is contradicted by the results of some simulations that have been performed, and it tends to reduce risk estimates artificially.

- If at the arrival of the routed inflow event, antecedent conditions have caused the pool to be filled to the spillway's crest, then it is assumed that the outlet will be opened to 75% of its capacity. If antecedent floods have not filled the reservoir, it is assumed that the routed inflow event will not result in any downstream damages. This assumption will have the same negative effect on risk estimates as the previous one.

Given these assumptions, it can be said that only inflow flood events following antecedent flood conditions that have filled the pool to the spillway's crest will produce downstream damages. Therefore it is possible to write:

$$h_X(x) = \text{PFP} \times \Pr(x\,|\,\text{full pool})$$
$$+ (1\text{-PFP}) \times \Pr(x\,|\,\text{no full pool})$$

$$\text{and } \Pr(x\,|\,\text{no full pool}) = \begin{cases} 0 & \text{for } x \neq 0 \\ 1 & \text{for } x = 0 \end{cases}$$

then

$$h_X(x) = \begin{cases} \text{PFP} \times \Pr(x\,|\,\text{full pool}) & \text{for } x \neq 0 \\ \text{PFP} \times \Pr(x\,|\,\text{full pool}) + (1\text{-PFP}) & \text{for } x = 0 \end{cases}$$

where

x : downstream damages in millions of US$

$h_X(x)$: probability function of damage

PFP   : probability of having antecedent flood conditions filling the pool to the spillway's crest

Because a full pool has been assumed when inflow events have been routed in the simulations, it will be possible to estimate $\Pr(x\,|\,\text{full pool})$.

It was stated earlier that some of the assumptions made above were at odds with simulations that were performed, resulting in risk estimates that are smaller than their actual values. These errors can be compensated for in an _ad hoc_ manner by specifying an artificially large value to the probability of having a full pool (PFP).

5.   FLOOD-FREQUENCY DISTRIBUTION

In this section the flood-frequency distribution of the inflows to the hypothetical Tomahawk Reservoir will be derived. Different procedures will be used, depending on whether the floods of interest are smaller or larger than the 100-year flood. To simplify the analysis, no regional or historical information was taken into consideration in our calculations.

Flood-Frequency Distribution for Ordinary Floods

For floods smaller than the 100-year flood, systematic data records on maximum yearly inflows into the Tomahawk Reservoir were used. It has been assumed that peak flood inflows smaller than the 100-year flood follow a log-normal distribution. Note that the log-normal distribution is essentially the log-Pearson type III with zero skew coefficient, which was recommended in Bulletin 17B (Interagency Advisory Committee on Water Data [1982]). The method of moments was used to fit the log-normal distribution to the available data; the statistics for the available 24 years of record were computed. Let us denote the estimates of the mean and the standard deviation of $\ln Q_i$ (Q_i is the maximum inflow for data year i) by m_1 and s_1, respectively, where

$$m_1 = \sum_{i=1}^{24} (\ln Q_i)/24$$

$$s_1 = \sqrt{\sum_{i=1}^{24} \frac{\left[\ln Q_i\right]^2}{24} - \left[\frac{\sum_{i=1}^{24} \ln Q_i}{(24)}\right]^2}$$

It was found that $m_1 = 9.5954$ and $s_1 = 0.3511$.

The log-normal distribution function has the form

$$f_Q(q) = \frac{1}{q\, a\, (2\pi)^{1/2}} \exp\left\{-\frac{1}{2}\left[\frac{\ln q - m}{a}\right]^2\right\}, \quad q \geq 0$$

In this case m_1 and s_1 can be used respectively as the best estimators of m and a.

$$\Longrightarrow \begin{cases} m = 9.5954 \\ a = 0.3511 \end{cases}$$

Using the standard normal tables, the 100-year flood q_{100} can be found

$$F_Q(q_{100}) = 0.99 \Longrightarrow \frac{\ln q_{100} - m}{a} = 2.326$$

$$\Longrightarrow q_{100} = 33{,}258 \text{ cfs.}$$

Flood-Frequency Distribution For Rare Floods

For floods larger than the 100-year flood, the recommendations of the National Research Council [1985] on dam safety will be followed concerning the estimation of the flood-frequency distribution. Since there is much uncertainty about the form of the real flood-frequency distribution, the analysis here has considered four probability distributions that have often been used in the literature: log-normal, Pareto, Weibull, and Gumbel. The impact of the various assumed distributions on the decisionmaking process will be studied.

The log-normal distribution has been widely used as a flood-frequency distribution, in particular for floods with moderate return periods. The Pareto distribution (Pearson type IV), which has a tail similar to that of the log-Gumbel, is often used by the Bureau of Reclamation as a flood-frequency distribution. The Weibull distribution is widely employed in reliability models; it takes on shapes similar to the gamma distribution. The Weibull distribution is also known as the extreme value type III distribution of the smallest value. The Gumbel (or extreme value type I) distribution is still very popular among European scientists and particularly among British scientists, who use it extensively as a flood-frequency distribution. The Gumbel distribution is also the limiting form to which the probability distributions of extreme values (largest values) from initial distributions with exponential tails converge. It seems, therefore, that the Gumbel distribution might be proper for representing maximum yearly floods, which can be considered the extreme values of daily floods. Moreover, the Gumbel distribution has a thinner tail than the other distributions considered in this analysis.

The cumulative distribution derived from the assumed flood-frequency distribution between the probable maximum flood (PMF) and the 100-year flood will be interpolated, but first it will be necessary to estimate T, the return period of the PMF. This task involves many uncertainties and, in general, yields inaccurate estimates. The return period of the PMF is sometimes estimated to be as low as 10^4, but the American Nuclear Society [1981], for example, has estimated it to be larger than 10^7. Therefore, it was decided to perform a sensitivity analysis on the value of the return period of the PMF; the values 10^4, 10^5, 10^6, and 10^7 were examined. The following notation will be used: $T_4 = 10^4$, $T_5 = 10^5$, $T_6 = 10^6$, $T_7 = 10^7$.

The distribution parameters for the four assumed flood-frequency distributions and the four assumed return periods of the PMF can be easily derived. For example, in the case of the Gumbel distribution, the following was done:

Since the Gumbel probability functions are

$$f_Q(q) = \left[\frac{1}{d}\right] \exp\left[-\frac{q-c}{d}\right] \exp\left\{-\exp\left[-\frac{q-c}{d}\right]\right\}$$

$$\text{and } F_Q(q) = \exp\left\{-\exp\left[-\frac{q-c}{d}\right]\right\}$$

it is possible then to write

$$\begin{cases} F_Q(q_{100}) = 0.99 \\ F_Q(PMF) = 1 - (1/T) \end{cases} \implies \begin{cases} F_Q(33,258) = 0.99 \\ F_Q(432,000) = 1 - (1/T) \end{cases}$$

$$\implies \begin{cases} c - 33,258 = d \ln(- \ln 0.99) \\ c - 432,000 = d \ln \{- \ln [1-(1/T)]\} \end{cases}$$

When these equations are solved, the following is obtained:

	T=T 4	T=T 5	T=T 6	T=T 7
c=	-364620	-232087	-165787	-125995
d=	86492.4	57681.9	43269.3	34619.2

6. THE PROBABILITY FUNCTION OF DAMAGES

For each inflow event routed through the Tomahawk Dam, the IWR model is used to compute downstream damages. If the dam is overtopped, the model will then determine damages for both cases of nonfailure and failure.

For each alternative s_j (j=1,2,...,n), the 15 damage values y_{kj} that correspond to the 15 routed inflows q_k (k=1,2,...,K; K=15) have been calculated assuming that the reservoir is filled to the spillway's crest prior to design flood. There are therefore 15 data points (q_k, y_{kj}) for each of the n alternatives. The following notation will be used in the rest of this work:

s_j : alternative j   j=1,2,...,m; here m=16

q_k : inflow event k   k=1,2,...,ℓ; here ℓ=15

$y(q_k, s_j) = y_{kj}$:downstream damages in millions of US$
        for inflow q_k under alternative s_j
        (assuming full pool and outlet
        open at 75%)

$x(q_k, s_j) = x_{kj}$:downstream damages in millions of US$
        for inflow q_k under alternative
        s_j

f(q) : probability density function of inflow q

F(q) : cumulative probability function of inflow q

$\phi(y)$: conditional probability density function of damages given that antecedent floods have filled the reservoir

$\bar{\phi}(y)$: cumulative probability function of damages given that antecedent floods have filled the reservoir

$h(x)$: probability density function of damages

$H(x)$: cumulative probability function of damages

Given that

$$h(x) = \begin{cases} \text{PFP } \phi(x)+0 \times (1\text{-PFP})= \text{PFP } \phi(x) & \text{for } x \neq 0 \\ \text{PFP } \phi(0)+1 \times (1\text{-PFP}) & \text{for } x=0 \end{cases}$$

it will be shown how, for a given alternative s_j, $h(x)$ can be derived from $F(q)$, given the K points (q_k, y_{kj}). It is assumed that q_k and y_{kj} are related by the function g_j in the following way:

$$y_{kj} = g_j(q_k) \text{ or } q_k = g_j^{-1}(y_{kj})$$

Note that $q_k < q_{k+1} \Longrightarrow y_{kj} < y_{k+1,j}$ for all values of k; therefore, it will be assumed that g_j is a continuous strictly monotone increasing function of the inflow variable q. Such properties guarantee that the mapping of $g_j^{-1}(y)$ on the set of images of $g(q)$ to the domain of $g(q)$ is a one-to-one function.

For $y_{kj} \leq y \leq y_{k+1,j}$, g_j can also be approximated by G_j, using piecewise linearization:

if $y_{kj} = g_j(q_k)$ and $y_{k+1,j} = g_j(q_{k+1})$

then $y = y_{kj} + \left[\dfrac{y_{k+1,j}-y_{kj}}{q_{k+1}-q_k} \right] (q - q_k) = G_j(q)$

$$\text{for } q_k \leq q \leq q_{k+1}$$

Let a constant K be defined as

$$\overset{\curlyvee}{K} = \left[\dfrac{q_{k+1}-q_k}{y_{k+1,j}-y_{kj}} \right]$$

Note that $q = G_j^{-1}(y) = q_k + \overset{\curlyvee}{K}(y-y_{kj})$ for $y_{kj} \leq y \leq y_{k+1,j}$ given that

$$\Phi(y) = Pr(Y \leq y) = Pr(G_j(Q) \leq y)$$
$$= Pr(Q \leq G_j^{-1}(y)) = F(G_j^{-1}(y))$$

Since $G_j^{-1}(y)$ does not have one closed form relationship, it will be easier to use the following approximation for the derivations:

$$\phi(y) \simeq \begin{cases} \dfrac{\Phi(y_{k+1,j}) - \Phi(y_{kj})}{y_{k+1,j} - y_{kj}} & \text{for } y_{kj} \leq y \leq y_{k+1,j} \\ & \text{where } k=1,\ldots,K-1 \\ 0 & \text{otherwise} \end{cases}$$

But since

$$\Phi(y_{kj}) = F(G_j^{-1}(y_{kj})) = F(q_k) \qquad \text{for } k=1,\ldots,K$$

then the above equality becomes

$$\phi(y) \simeq \begin{cases} \dfrac{F(q_{k+1}) - F(q_k)}{y_{k+1,j} - y_{kj}} & \text{for } y_{kj} \leq y \leq y_{k+1,j} \\ & \text{where } k=1,\ldots,K-1 \\ 0 & \text{otherwise} \end{cases}$$

7. ALTERNATIVES AND COST ESTIMATES

The 16 alternatives we are studying combine the remedial actions of raising the dam's height and increasing the spillway's width. They are described in detail in Table 1.

Table 1. Description of the alternatives s_j ($j=1,2,\ldots,16$)

INCREASE IN DAM HEIGHT	SPILLWAY WIDTH (1 UNIT = 620 FT.)			
	1	1.5	2	2.4
0 FT.	s_1	s_5	s_9	s_{13}
3 FT.	s_2	s_6	s_{10}	s_{14}
6 FT.	s_3	s_7	s_{11}	s_{15}
10 FT.	s_4	s_8	s_{12}	s_{16}

If the dam's height is raised by 10 feet to an elevation of 920 ft above sea level and if the present spillway width is maintained, the dam will safely pass the PMF. Similarly, if the present dam's height is kept and if the spillway is widened to 2.4 times its current size, the dam will also safely pass the PMF. Alternatives such as increasing the spillway's width by more than 2.4 times or raising the dam's height by more than 10 feet were disregarded, since corresponding added construction costs only ensure that the dam would pass floods larger than the PMF. Floods of such large magnitude are considered to be very unlikely, however, and have generally been ignored by analysts in the field of dam safety.

Cost estimates of remedial actions for the Tomahawk Dam were derived by the U.S. Army Corps of Engineers [1983 and 1985]. These values have been used to obtain the cost estimates for the 16 alternative actions (see Table 2). It has been assumed that if the dam's height is raised by less than three feet, then a concrete parapet wall will be used. On the other hand, if the dam's height is to be increased by more than three feet, then in addition to a three-foot concrete parapet wall, earthfill will be used to consolidate the dam. If the

Table 2.   Construction costs for the remedial actions
(in millions of US$)

INCREASE IN DAM HEIGHT	SPILLWAY WIDTH (1 UNIT = 620 FT.)			
	1	1.5	2.0	2.4
0 FT.	0	19.32	25.88	31.12
3 FT.	0.8	20.12	26.68	31.92
6 FT.	5.15	22.02	27.93	32.64
10 FT.	20.83	36.14	42.04	46.76

spillway is also to be widened, then the material from the spillway's excavations would be utilized as the stabilizer earthfill, to reduce costs. In fact, while construction of the concrete parapet wall is relatively cheap, the use of earthfill can be very costly, particularly if it is not available at a location close to the dam. Furthermore, widening the spillway requires extremely costly modifications.

8.   APPLICATION OF THE PARTITIONED MULTIOBJECTIVE RISK METHOD

In section 3 of this paper, the partitioned multiobjective risk method (PMRM) was described. In subsequent sections, all the information needed to apply the PMRM to the case study was assembled. In this section, the PMRM is applied to our problem, following the procedure for discrete decision variables, also outlined in section 3.

Finding the Probability Distribution of Damages

In section 5, the probability distribution function of the inflows, $f(q)$, was derived along with the corresponding cumulative probability function, $F(q)$, for the four assumed distributions. Section 6 contained the derivation of the relationships between the probability distribution of damages $h(x)$ and $F(q)$. It was found that the probability distribution of damage is

$$h(x) \simeq \begin{cases} PFP \left[\dfrac{F(q_{k+1})-F(q_k)}{y_{k+1,j}-y_{kj}} \right] & \text{for } y_{kj} \leq x \leq y_{k+1,j} \\[2mm] & \text{where } k = 1,\ldots,K-1 \\[2mm] 1-PFP & \text{for } x = 0 \\[2mm] 0 & \text{otherwise} \end{cases}$$

where y_{kj} represent the damages resulting from inflow q_k under alternative s_j assuming that antecedent floods have filled the pool to the spillway's crest. These values of y_{kj} were computed through computer simulations based on the IWR model.

Partitioning the Probability Axis

Next, it is necessary to partition the exceedance probability axis (or, alternatively, the cumulative probability axis) into various ranges that enhance the understanding of the different risk-related aspects of the problem. The analyst should perform the partitioning only after he carefully studies the risk curves (exceedance probability curves) for the various alternatives. For example, Figure 2 represents the risk curve that corresponds to alternative s_1 (the status quo). Note that only the relevant part of the risk curve is shown in these figures.

After examining the available information, we decided to partition the probability axis into four ranges, representing: (1) no hazards, (2) high-probability/low-consequence (HP/LC) risk, (3) intermediate risk, (4) and low-probability/high-consequence (LP/HC) risk. We will be using the following notation for the partitioning points:

$[\alpha_1,\alpha_2]$: no damages range

$[\alpha_2,\alpha_3]$: HP/LC risk range

$[\alpha_3,\alpha_4]$: intermediate risk range

$[\alpha_4,\alpha_5]$: LP/HC risk range

Since the full exceedance probability axis is being partitioned, α_1 will be set equal to 1 and α_5 will be set to zero. Moreover, the range $[\alpha_1,\alpha_2]$ corresponds to the no-damage domain, which in turn corresponds to the case where antecedent floods do not fill all of the empty reservoir; therefore, α_2 will be set equal to the PFP (where the PFP is the probability of having antecedent floods filling the pool to the spillway's crest). In the following analysis it will be assumed that PFP = 1, thus yielding $\alpha_2=1$. Therefore, the range $[\alpha_1,\alpha_2]$ becomes the range $[1,1]$ and the number of ranges is then reduced to three. For the sake of simplicity, the values of α_3 and α_4 will be chosen from among the values of $[1-\Phi(y_{kj})]$ (k=1,2,...,K) or, equivalently, from among the values of $[1-F(q_k)]$ (k=1,2,...,K).

A sensitivity analysis will also be performed on each of α_3 and α_4, but for convenience, they will be allowed to take values only among $[1-F(q_k)]$ (k=1,2,...,K).

Mapping Partitions to the Damage Axis

Because of the above simplification, mapping the partitions on the probability axis onto the damage axis is simplified as well.

If α_i (for an i, i=2,3,4) is set equal to $[1-F(q_{k'})]$ (for certain k' such that k'=1,2,...,K), then the following relations hold:

Since

$$F(q_{k'}) = \Phi(y_{k',j}) = [H(x_{k',j})-(1-PFP)]/PFP$$

$$= H(x_{k',j}) \text{ (because PFP = 1)},$$

it follows that

$$1-\alpha_i = H(x_{k',j})$$
$$\Longrightarrow H^{-1}(1-\alpha_i) = x_{k',j}$$
$$\Longrightarrow a_{ij} = x_{k',j}$$
$$\Longrightarrow a_{ij} = y_{k',j}$$

This value of k' will be denoted by γ_i.

Since $q_1=0$, then $F(q_1) = 0$, and since $\alpha_2 = 1$

$$\Longrightarrow \alpha_2 = 1 -F(q_1) \text{ and } \gamma_2 = 1$$

From the above results, it follows that

$$a_{2j} = y_{1j}$$

For $\alpha_5 = 0$, a_{5j} can be approximated by $y_{15,j}$ and $\gamma_5=15$.

The three domains corresponding to the three ranges of the probability axis are

domain I   : $[y_{1j}, a_{3j}]$ HP/LC risk

domain II   : $[a_{3j}, a_{4j}]$ intermediate risk

domain III : $[a_{4j}, y_{15j}]$ LP/HC risk

Finding Conditional Expectations

For alternative s_j and for the damage domain D_{ij} defined by the interval $[a_{ij}, a_{i+1,j}]$, the conditional expectation is

$$E[X \setminus D_{ij}; s_j] = \frac{\int_{a_{ij}}^{a_{i+1,j}} x\, h_X(x)\, dx}{\int_{a_{ij}}^{a_{i+1,j}} h_X(x)\, dx}$$

$$= \frac{\sum_{z=\gamma_i}^{(\gamma_{i+1}-1)} \int_{y_{zj}}^{y_{z+1,j}} x\, h_X(x)\, dx}{\sum_{z=\gamma_i}^{(\gamma_{i+1}-1)} \int_{y_{zj}}^{y_{z+1,j}} h_X(x)\, dx}$$

$$= \frac{\sum_{z=\gamma_i}^{(\gamma_{i+1}-1)} \int_{y_{zj}}^{y_{z+1,j}} x\, \text{PFP}\left[\frac{F(q_{z+1})-F(q_z)}{y_{z+1,j}-y_{zj}}\right] dx}{\sum_{z=\gamma_i}^{(\gamma_{i+1}-1)} \int_{y_{zj}}^{y_{z+1,j}} \text{PFP}\left[\frac{F(q_{z+1})-F(q_z)}{y_{z+1,j}-y_{zj}}\right] dx}$$

$$= \frac{\text{PFP} \sum_{z=\gamma_i}^{(\gamma_{i+1}-1)} \left[\frac{F(q_{z+1})-F(q_z)}{y_{z+1,j}-y_{zj}}\right] \int_{y_{zj}}^{y_{z+1,j}} x\, dx}{\text{PFP} \sum_{z=\gamma_i}^{(\gamma_{i+1}-1)} \left[\frac{F(q_{z+1})-F(q_z)}{y_{z+1,j}-y_{zj}}\right] \int_{y_{zj}}^{y_{z+1,j}} dx}$$

$$= \frac{\displaystyle\sum_{z=\gamma_i}^{(\gamma_{i+1}-1)} [F(q_{z+1}-F(q_z)] \left[\frac{y_{z+1,j} + y_{zj}}{2}\right]}{\displaystyle\sum_{z=\gamma_i}^{(\gamma_{i+1}-1)} [F(q_{z+1})-F(q_z)]}$$

$$= \left[\frac{1}{\alpha_i - \alpha_{i+1}}\right] \sum_{z=\gamma_i}^{(\gamma_{i+1}-1)} [F(q_{z+1})-F(q_z)] \left[\frac{y_{z+1,j} + y_{zj}}{2}\right]$$

This derivation is valid for the three domains, I, II, and III, which correspond in the above equations to $i = 2$, 3, and 4, respectively. We computed the conditional expectations of domains I, II, and III for all alternatives.

The probability that damage falls in the interval $[a_{ij}, a_{i+1,j}]$ is

$$Pr(x\varepsilon D_{ij}) = \alpha_i - \alpha_{i+1}$$

This probability is the weight coefficient for the conditional expectation, and it represents the relative importance of this conditional expected value.

Generating Functional Relationships

The expected conditional values $E[x\backslash D_{2j};s_j]$, $E[x\backslash D_{3j};s_j]$, and $E[x\backslash D_{4j};s_j]$ (where $j=1,2,\ldots,16$) will be used to define the set of risk objective functions f_2, f_3, and f_4, where

$$f_i(s_j) = E[x\backslash D_{ij};s_j] \qquad (i=2,3,4)$$

Thus $f_2(s_j)$ will correspond to domain I or HP/LC risk, $f_3(s_j)$ will correspond to domain II or intermediate risk, and $f_4(s_j)$ will correspond to domain III or LP/HC risk.

Another risk objective function is the unconditional expected value of damages, which will be denoted by $f_5(s_j)$; it can be approximated by

$$f_5(s_j) \approx \sum_{i=2}^{4} [\alpha_i - \alpha_{i+1}] f_i(s_j)$$

Figure 3 contains an example of the graph of the risk objective functions f_3, f_4, and f_5.

The cost function $f_1(s_j)$ was also constructed from Tables 1 and 2. The graph of the cost function is shown in Figure 4.

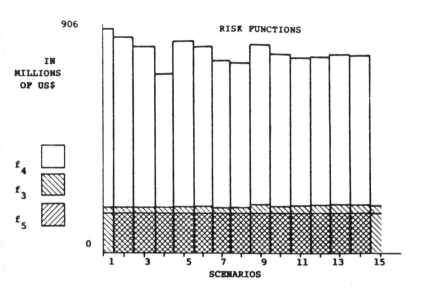

<u>Figure 3</u>   Risk objective functions $f_3(s)$, $f_4(s)$, and $f_5(s)$

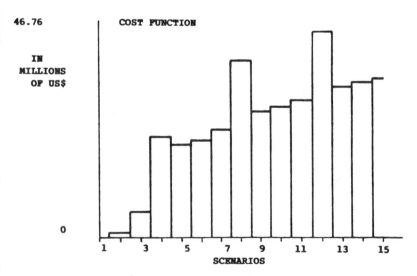

<u>Figure 4</u>   Cost objective function

Employing the Surrogate Worth Trade-off Method

So far, five objective functions have been determined. To find the decisionmaker's preferred solution, a multiobjective optimization problem will have to be solved. Since the decision variable is discrete, the modified version of the SWT method that makes use of total trade-offs will be used, as discussed in Section 3.

Although the main interest here is in the trade-offs between the cost function f_1 and the LP/HC risk objective function f_4, the trade-offs between f_1 and the risk objective function f_5, which represents the unconditional expected value, will also be analyzed.

All noninferior solutions were determined through an exhaustive search. The modified version of the surrogate worth function (see section 3) can then be used to determine the preferred solution. A decision support system based on this function was developed to assist the decisionmaker in this crucial task.

9. ANALYSIS OF RESULTS

This section contains a discussion of the results obtained by applying the partitioned multiobjective risk method (PMRM) to the dam safety problem described in section 4. In particular, a sensitivity analysis will be performed on the distribution used to extrapolate the frequency curve to the PMF, the return period of the PMF, and on the partitioning points.

Explanation of the Section's Figures

Figure 5 contains two tableaus. In the first one, the objective functions $f_i(s_j)$ (i=1,...,5) are shown for all the alternatives, while in the second one, the total trade-off functions $\lambda_{i1}(s_j,s_h)$ (i=2,...,5) are listed for the noninferior solutions. In the figure, the value of -1 was assigned to the total trade-off functions whenever the value of these functions was negative. When the total trade-off functions are assigned the symbol "**" for some alternative action, it means that the alternative corresponds to a noninferior solution but that the total trade-off function cannot be computed because there are no noninferior solutions with higher costs to compare with it.

Notice that the alternatives are ordered according to a continuously increasing cost function. The total trade-off functions were calculated using this order. Also, all the values in the first tableau are in units of 10^6 US\$. Moreover, the figure contains a display of the values that were assigned to the various parameters, and it contains the yearly probabilities of an event belonging to various risk domains.

Trade-offs in the Dam Safety Problem

The main advantage that the PMRM has over other risk analysis methodologies is that it does not collapse the risk curve into one point, the yearly expected value. Instead, it represents this curve by

S SCENARIO	$f_1(S)$ COST FUNC.	$f_2(S)$ HP/LC RISK	$f_3(S)$ INTER.RISK	$f_4(S)$ LP/HC RISK	$f_5(S)$ EXPE. VALUE
1	0	159.9796	209.5453	1260.525	161.7427
2	0.8	159.9796	209.0715	1038.298	161.5697
3	5.15	159.9796	208.652	899.4269	161.4593
5	19.32	159.9796	217.3968	1028.719	161.7225
6	20.12	159.9796	216.9271	834.9428	161.5707
4	20.83	159.9796	208.2824	678.3266	161.2892
7	22.02	159.9796	216.5687	721.5613	161.4802
9	25.88	159.9796	223.0323	908.4608	161.7421
10	26.68	159.9796	222.6253	746.3894	161.6148
11	27.93	159.9796	222.3008	744.2904	161.6071
13	31.12	159.9796	226.3684	758.2729	161.6955
14	31.92	159.9796	225.946	756.9135	161.6863
15	32.64	159.9796	225.6023	755.37	161.6786
8	36.14	159.9796	216.1759	718.7776	161.4706
12	42.04	159.9796	221.9517	741.6782	161.5984
16	46.76	159.9796	225.2417	753.775	161.6705

RETURN PERIOD OF PMF = 1E 4   THERE IS 0.00000% CHANCE THAT F0 OCCW
GUMBEL DISTRIB. FOR RARE FLOODS   THERE IS 98.00605% CHANCE THAT F2 OCC
THE PARTITIONING POINTS ARE D AND K THERE IS 1.92024% CHANCE THAT F3 OCCW
CONDITION. PROBABI. OF FAILURE=100% THERE IS 0.07371% CHANCE THAT F4 OCCW
DM'S ESTIMATE OF PR(FULL POOL)=100%

S SCENARIO	$f_1(S)$ COST FUNC.	LAM(2,1) HP/LC RISK	LAM(3,1) INTER RISK	LAM(4.1) LP/HC RISK	LAM(5,1) EXPE. VALUE
1	0	**	0.5921555	277.784900	0.2161406
2	0.8	-1.0000000	0.0964531	31.9242800	0.0253857
3	5.15	-1.0000000	0.0235674	4.3075570	0.0108466
5	19.32	-1.0000000	-1.0000000	-1.0000000	-1.0000000
6	20.12	-1.0000000	-1.0000000	220.586500	-1.0000000
4	20.83	-1.0000000	**	**	**
7	22.02	-1.0000000	-1.0000000	-1.0000000	-1.0000000
9	25.88	-1.0000000	-1.0000000	-1.0000000	-1.0000000
10	26.68	-1.0000000	-1.0000000	-1.0000000	-1.0000000
11	27.93	-1.0000000	-1.0000000	-1.0000000	-1.0000000
13	31.12	-1.0000000	-1.0000000	-1.0000000	-1.0000000
14	31.92	-1.0000000	-1.0000000	-1.0000000	-1.0000000
15	32.64	-1.0000000	-1.0000000	-1.0000000	-1.0000000
8	36.14	-1.0000000	-1.0000000	-1.0000000	-1.0000000
12	42.04	-1.0000000	-1.0000000	-1.0000000	-1.0000000
16	46.76	-1.0000000	-1.0000000	-1.0000000	-1.0000000

Figure 5   Risk objective functions (x10^6) and total trade-off function

a number of points that correspond to the yearly conditional expected values. It will be demonstrated that this advantage improves the decisionmaking process for our dam safety problem. For this demonstration, the numbers in Figure 5 will be examined and Figure 6 will be used to illustrate the trade-offs involved.

Traditionally, risk analysis has relied heavily on the concept of the yearly expected value, which corresponds in Figure 5 to $f_5(s_j)$. First, note that f_5, the yearly expected damage, takes unusually high values (on the order of 161×10^6 US\$ to 162×10^6 US\$). This is due to the assumptions concerning antecedent floods; in particular, the assumptions that the reservoir is filled to the spillway's crest and that the outlet is open to 75% of its capacity. Therefore, any small inflow into the reservoir will cause large damages, on the order of 160×10^6 US\$. These two assumptions were made to comply with the guidelines and recommendations established by the U.S. Army Corps of Engineers.

It is also apparent that when the dam's height is increased ($s_1 \rightarrow s_2 \rightarrow s_3 \rightarrow s_4$, $s_5 \rightarrow s_6 \rightarrow s_7 \rightarrow s_8$, ...), f_5 decreases, but by less than 0.3%. Next, when the spillway's width is increased ($s_1 \rightarrow s_5 \rightarrow s_9 \rightarrow s_{13}$, $s_2 \rightarrow s_6 \rightarrow s_{10} \rightarrow s_{14}$, ...), f_5 increases in general, and when it decreases, it does so by less than 0.02%. These observations could lead the decisionmaker to conclude that increasing the spillway's width is not an attractive solution because any investment in such an action will mainly increase the risks. By looking at λ_{51}, the decisionmaker could also find incentives not to invest money to raise the dam, since under alternative s_2, an investment of 10^6 US\$ will not reduce the expected yearly damages by more than \$25,386.

But if the decisionmaker takes into consideration the rest of the risk objective functions, in particular $f_4(s_j)$ and $f_3(s_j)$, then his picture of the problem might radically change. First, he will notice that f_4 decreases greatly when the spillway's width is increased, but that f_3 increases. In other words, the decisionmaker will be able to see that by increasing the spillway's width, he is reducing risks in the low-probability/high-consequence (LP/HC) domain, because spillway widening reduces both the probability of dam failure and the damages in case of failure. On the other hand, the decisionmaker will also see that the risks associated with less extreme events are increasing, because floods which are relatively frequent will cause more downstream damages. Moreover, even when compared to increasing the dam's height, spillway widening could still be an attractive solution. For example, s_6, which would have been disregarded if traditional risk analysis methods were used, becomes a noninferior solution if the risk objective f_4 is considered. Thus, by using the PMRM, the decisionmaker can better understand the trade-offs among risks that correspond to the various risk domains.

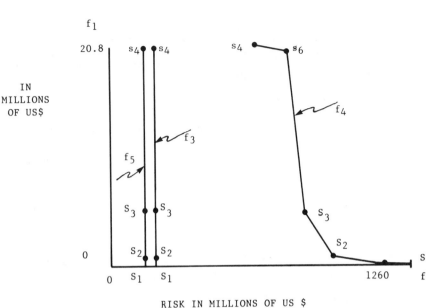

Figure 6   Pareto Optimal frontiers

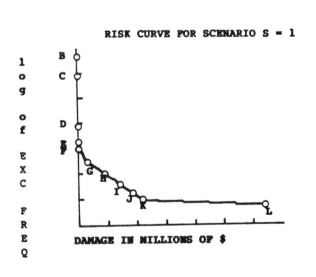

Figure 7   Risk curve for scenario s_1

Moreover, regarding the alternative of increasing the dam's height, the use of f_4 allows explicit quantification of risks in the LP/HC risk domain, and this might induce the decisionmaker to invest money in some situations where he might have been reluctant to do so if he had just used f_5. Using the same example as above, investing 10^6 US$ under alternative s_2 only reduces the expected yearly damages by $25,386. It is apparent that if f_4 is included, then, in the case of an extreme event, up to $31,924,280 in yearly damages might be saved with a probability of 7.371×10^{-4}.

Notice that for this problem, because smaller inflows caused the same amount of damages for all alternatives, $f_2(s_j)$ is constant for all alternatives and therefore is of no interest to the decisionmaker. This can be interpreted to mean that the high-probability/low-consequence (HP/LC) risk domain provides no information for this problem in the decisionmaking process, and for this reason will be disregarded in the rest of this section.

It is obvious that by using the PMRM, the decisionmaker is able to grasp certain aspects of the problem that would have been completely ignored had he simply used the yearly expected value of damages. These aspects were mainly associated with LP/HC risks in this case, but this is not a general restriction.

Sensitivity to the Flood Frequency Function

Sections 1 and 5 called attention to the fact that there is very little knowledge concerning the type of probability distribution function that should be used to extrapolate the flood-frequency curve beyond the 100-year flood to the PMF. Moreover, Stedinger and Grygier [1985] showed that the results of their risk analysis could be influenced by the choice of this probability distribution function. Thus, a sensitivity analysis was performed to try to determine the impact of this choice on the decisionmaking process in this case study.

The approach that has been used here to facilitate the application of the PMRM (see section 8) does not allow partitioning of the probability axis at the same points for all the distributions. Therefore, to be able to compare the results for the various distributions, linear interpolation has been used to approximate the risk objective functions. Partial results for a return period of 10,000 years are partially listed in Tables 3, 4, 5, and 6 for the noninferior scenarios.

By studying the results of this sensitivity analysis, one notices that f_5 increases by less than 1% when the flood-frequency distribution function of rare floods is changed using, alternatively, the Pareto, log-normal, Weibull, and Gumbel distributions. But it is also to be

Table 3 Objective functions for the

Pareto distribution

SCE. s	f1(s) COST FUN	f3(s) INTE RIS	f4(s) LP/HC RI	f5(s) EXP VALU
s1	0	166.676	1147.482	160.162
s2	0.8	166.619	998.343	160.154
s3	5.15	166.572	905.176	160.148
s5	19.32	167.851	991.702	160.178
s6	20.12	167.793	859.103	160.170
s4	20.83	166.527	665.780	160.136

Table 4 Objective functions for the

log-normal distribution

SCE. s	f1(s) COST FUN	f3(s) INTE RIS	f4(s) LP/HC RI	f5(s) EXP VALU
s1	0	170.103	1264.684	160.238
s2	0.8	170.015	1055.099	160.226
s3	5.15	169.941	932.060	160.218
s5	19.32	171.877	1042.946	160.262
s6	20.12	171.786	871.349	160.251
s4	20.83	169.871	684.473	160.204

observed that the total trade-off function λ_{51} increases dramatically when the Gumbel distribution is used instead of the Pareto distribution. Therefore, if only those traditional risk analysis methods that focus on f_5 are used, then the use of the Gumbel distribution is likely to give results different from those ones obtained using other distributions.

From our numerical results, it is apparent that when the distributions are changed in the same order as above, f_4 increases up to 37%, but λ_{41} does not vary much until the Gumbel distribution is used. Because of its thin tail, the Gumbel distribution puts more weight on the extreme range, therefore making some alternatives (which might have seemed attractive to the DM had f_5 only been used) less attractive in the LP/HC domain.

Table 5 Objective functions for the
Weibull distribution

SCE. s	f (s) 1 COST FUN	f (s) 3 INTE RIS	f (s) 4 LP/HC RI	f (s) 5 EXP VALU
s 1	0	174.112	1356.605	160.321
s 2	0.8	173.670	1193.864	160.303
s 3	5.15	173.566	1011.948	160.292
s 5	19.32	176.175	1164.029	160.352
s 6	20.12	176.050	913.897	160.337
s 4	20.83	173.468	707.526	160.276

Table 6 Objective functions for the
Gumbel distribution

SCE. s	f (s) 1 COST FUN	f (s) 3 INTE RIS	f (s) 4 LP/HC RI	f (s) 5 EXP VALU
s 1	0	202.479	1438.588	160.891
s 2	0.8	199.313	1451.709	160.827
s 3	5.15	198.179	1259.544	160.794
s 5	19.32	205.791	1389.484	160.953
s 6	20.12	204.729	991.343	160.912
s 4	20.83	197.888	751.035	160.764

It can also be shown that f_3 and λ_{31} increase dramatically if the distributions are changed in the same order as described above. Here, too, the use of the Gumbel distribution has great impact on the results.

It is clear, therefore, that the decisionmaking process in the PMRM is also sensitive to a change in distributions. In particular, the use of the Gumbel distribution tends to give high estimates of risk which might induce the decisionmaker to choose conservative and expensive remedial actions. Therefore, it is recommended that all studies on dam safety include a sensitivity analysis that examines the effects of changes in the selection of a flood-frequency distribution function on the results. In the sensitivity analysis, the Gumbel distribution should be used in addition to any other distribution that does not have a thin tail, such as the Pareto distribution, the log-normal distribution, or the log-Gumbel distribution.

Notice that by using the PMRM, it was possible to see how the choice of the probability distribution for rare floods affects the risk estimates in the various risk domains, and therefore a better understanding of the problem was achieved.

Sensitivity of Risk/Cost Analysis to the Return Period of the PMF

The problems associated with estimating the return period of the PMF and the uncertainties that characterize this parameter have already been discussed. In fact, Stedinger and Grygier [1985] also found that their results were sensitive to changes in the return period of the PMF. Thus, in this section, there will be an attempt to determine how changes in T, the estimate of the return period of the PMF, influence the choices of the decisionmaker. The PMRM has been used, assuming, alternatively, 10^4, 10^5, 10^6, or 10^7 to be the value of the return period of the PMF.

Here, too, the risk objective functions were approximated by linear interpolation because the structure of the problem does not allow partitioning of the probability axis at the same points for all values of the return period of the PMF. Partial results obtained for the Gumbel distribution are shown in Tables 7, 8, 9, and 10 for the non-inferior scenarios.

When T is increased from $T_4 = 10^4$ to $T_7 = 10^7$, the following occurs: (1) f_5 decreases by less than 0.4%, but λ_{51} decreases by more than 190%--thus, a DM using traditional risk analysis will tend to take more conservative actions if he assumes a lower return period of the PMF; (2) f_4 decreases by more than 100% and λ_{41} decreases by more than 150%--thus it is obvious that changes in T impact most on LP/HC risks; (3) f_3 also decreases but not as drastically as f_4, and λ_{31} decreases as well.

It can be seen that here, too, that although the use of the PMRM did not improve the robustness of the results, it added more insight to the problem.

Table 7   Objective functions for $T = 10^4$

SCE. s	$f_1(s)$ COST FUN	$f_3(s)$ INTE RIS	$f_4(s)$ LP/HC RI	$f_5(s)$ EXP VALU
s 1	0	178.709	1320.635	160.682
s 2	0.8	178.537	1177.850	160.638
s 3	5.15	178.389	1076.94	160.606
s 5	19.32	181.853	1148.869	160.695
s 6	20.12	181.680	998.460	160.648
s 4	20.83	178.254	714.677	160.499

Table 8   Objective functions for $T = 10^5$

SCE. s	$f_1(s)$ COST FUN	$f_3(s)$ INTE RIS	$f_4(s)$ LP/HC RI	$f_5(s)$ EXP VALU
s 1	0	166.968	616.034	160.238
s 2	0.8	166.911	569.755	160.226
s 3	5.15	166.867	541.796	160.218
s 5	19.32	168.304	600.075	160.262
s 6	20.12	168.243	561.865	160.251
s 4	20.83	166.820	488.039	160.203

Table 9   Objective functions for $T = 10^6$

SCE. s	f (s) 1 COST FUN	f (s) 3 INTE RIS	f (s) 4 LP/HC RI	f (s) 5 EXP VALU
s 1	0	163.359	413.191	160.105
s 2	0.8	163.337	397.940	160.102
s 3	5.15	163.320	389.547	160.100
s 5	19.32	163.993	431.163	160.123
s 6	20.12	163.968	420.515	160.121
s 4	20.83	163.301	379.498	160.097

Table 10   Objective functions for $T = 10^7$

SCE. s	f (s) 1 COST FUN	f (s) 3 INTE RIS	f (s) 4 LP/HC RI	f (s) 5 EXP VALU
s 1	0	161.615	314.332	160.051
s 2	0.8	161.608	310.009	160.050
s 3	5.15	161.603	307.570	160.049
s 5	19.32	161.885	337.099	160.063
s 6	20.12	161.877	334.029	160.062
s 4	20.83	161.596	305.175	160.049

This short exposition concludes with the recommendation that sensitivity analysis be performed on the return period of the PMF for all risk analyses on dam safety.

Where and How to Partition

In this section, the emphasis is on determining the partitioning points on the probability axis. It was shown that the partitioning points α_3 and α_4 could be chosen from a specified set of points $[1-F(q_k)]$ (where $k=1,2,\ldots,K$). In Figure 7, which represents the risk curve for s_1, this set of points corresponds to the set of points A, B, C,..., O.

Ideally, the objective is to partition the probability axis in a way that would allow isolation of extreme risks from ordinary risks. More specifically, an attempt is being made to construct an LP/HC risk domain that corresponds to dam failure and an HP/LC risk domain that corresponds to damages caused by floods smaller than the 100-year flood.

Notice that the probability of dam failure is largest for the status quo alternative (alternative s_1). If, through a certain partitioning, we include all failure damages in the LP/HC risk domain, this domain will then contain all failure damages for all of the remaining alternatives. Therefore, if the analyst partitions the probability axis, the risk curve for alternative s_1 would be most useful.

A study of the graph in Figure 7 reveals that by adopting the following partitioning of the probability axis, $\alpha_3 = [1-F(q_4)] = .02$ (point D in the graph) and $\alpha_4 = [1-F(q_{11})]$ (point K in the graph), which for $T = 10^4$ gives $\alpha_4 = 0.29 \times 10^{-3}$, most of the risks associated with floods smaller than the 100-year flood can be isolated in domain I, and all the risks associated with dam failure plus some less extreme risks can be isolated in domain III. This partitioning seems to be the best approach to achieve the objective stated above, albeit partially.

Sensitivity to the Partitioning

Since the choice of the partitioning points on the probability axis is a somewhat arbitrary process, it is necessary to examine the sensitivity of the results to changes in the partitioning points. The PMRM has therefore been applied using various partitioning points in the neighborhood of D and K.

A study in which the partitioning points varied from (D,I) to (D,J), (D,K), (D,L), and (D,M) revealed that the magnitudes of the risk objective functions f_3 and f_4 increase--especially f_4, whose magnitude increases for some alternatives by as much as 58%. But it is notable that these increases are smaller when the partitioning point is moved in the neighborhood of k (less than 32% increase for f_4). Moreover, since the DM uses the probabilities of the risk domains to implicitly weigh the importance of the corresponding risk objective functions, these increases in f_4 are partially compensated for by decreases in the probability that LP/HC events will occur. A study of the total trade-off functions makes it clear that the set of noninferior solutions does not change when the partitioning point is varied in the neighborhood of K. But if the partitioning point chosen is further away from K, then this set can vary greatly.

When the partitioning points varied between (C,K), (D,K), and (E,K), it was seen that f_2 increases by less than 3% while f_3 increases by as much as 34% for some alternatives. But here, too, the increases in f_3 are matched by a very important decrease in the probability of intermediate risks (domain II).

Therefore, it can be said that for this problem, the results of the decisionmaking process would be relatively stable in the neighborhood of the partitioning points. But additional sensitivity analyses showed that the results of the decisionmaking process become sensitive to the partitioning if the magnitude of failure damages is not much larger than the magnitude of nonfailure damages.

Since these partitioning points are arbitrary points, it might be more appropriate to obtain more robust results. The partitioning of the damage axis might be an adequate solution to this problem. In particular, it would allow very stable results to be obtained for the LP/HC risk domain. But, on the other hand, the risk objectives would be somewhat invariant with the different alternatives. In fact, more theoretical research needs to be done to investigate the partitioning of the damage axis approach.

Why Are LP/HC Risk Estimates So Sensitive?

It has been apparent throughout this section that f_4 is quite sensitive to changes in the parameters and in the partitioning points. This issue needs elaboration to understand the mechanism behind the behavior of f_4. Figure 8 represents an approximate sketch of the exceedence probability function of damages for the three alternatives s_1, s_2 (increase the dam's height by three feet), and s_6 (widen the spillway to 1.5 times its present size and increase the dam's height be three feet).

It can be shown how, by a gradual decrease in α_4, s_6 becomes noninferior, after which s_2 becomes inferior in the LP/HC risk domain. Notice that this problem arises only if there is no first-degree stochastic dominance among the alternatives (for a discussion on stochastic dominance, see Zeleny [1982]). Imagine that someone is moving downward a horizontal line from the actual position of α_4. For each alternative, the value of f_4 can be visualized as the product of $(1/\alpha_4)$ and the area bounded by the X-axis, the Y-axis, the horizontal line passing through α_4, and the risk curve. First, it can be seen that since α_4 is invariant for all alternatives, only the magnitude of the area defined above will determine which decision situations (or alternatives) are inferior

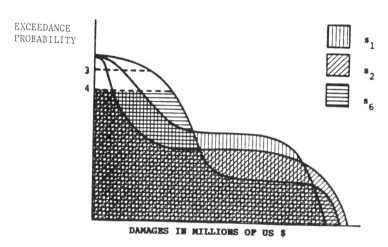

Figure 8   Risk curves for s_1, s_2
s_6. Partitioning point is H
(not on scale)

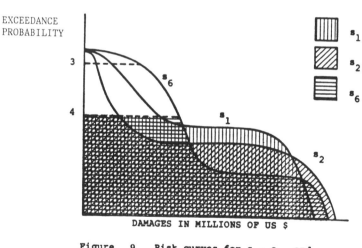

Figure 9   Risk curves for s_1, s_2, and
s_6. Partitioning point is k
(not on scale)

in the LP/HC risk domain. These areas will be called A_1, A_2, and A_6 for alternatives s_1, s_2, and s_6, respectively.

Figures 9 and 10 help make clear that when α_4 gradually decreases, first $f_4(s_6)$ is larger than $f_4(s_1)$, but then it becomes smaller, and therefore s_6 becomes a noninferior solution in the LP/HC risk domain. But on the other hand, $f_4(s_2)$, which at first is smaller than $f_4(s_1)$, will become larger eventually, and thus s_2 will become an inferior solution in the LP/HC risk domain.

10. CONCLUSION

The major objective of this work was to investigate the usefulness of the partitioned multiobjective risk method (PMRM) in the context of dam safety. Of particular interest was the analysis of the sensitivity of results generated by the PMRM (1) to variations in the value of the estimate of the return period of the PMF, and (2) to changes in the type of probability distribution function used to describe the frequency of unusually high floods. Another major goal was identifying problems and suggesting corrections to any deficiencies that the method might exhibit when applied to dam safety problems. In this section, in addition to summarizing and evaluating the results of these various inquiries, recommendations will be made regarding possible directions for future applications of the PMRM to dam safety problems.

Evaluation

In this work, the partitioned multiobjective risk method (PMRM) was applied to a case study, based on realistic simulations of dam failure using the U.S. Army Institute of Water Resources (IWR) computer model. The parameters of the model were calibrated to allow a good representation of a real dam.

In sections 5, 6, 7, and 8, the PMRM was used in the context of dam safety, and it was shown that it could be done with relative ease and few complications. In fact, the approach that was developed here to apply the PMRM to dam safety problems can be used for a large category of problems having the following general structure: (1) damages should be a function of a random event, unaffected by the scenarios, and denoted by the variable Q, (2) the probability distribution function of Q should be available, (3) a set of couples (Q_k, X_{kj}), where X_{kj} is the value of damages caused by the random event Q_k under scenario s_j, should be obtained either by using simulation techniques or from historical records. Nuclear power plant safety, mitigation of contamination from the spill of toxic chemicals, and prevention of damages caused by earthquakes are examples of the kind of problems to which the PMRM can be applied through the approach outlined in this paper.

Moreover, it was shown that the PMRM is not a substitute for the decisionmaker. In fact, it is merely a tool that provides useful information, such as estimates of the risks associated with the various risk domains and of the trade-offs between the cost objective and the risk objectives. A modified version of the surrogate worth trade-off

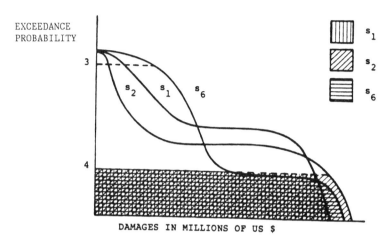

Figure 10 Risk curves for s_1, s_2, and s_6. Partitioning is $\hat{N}$ (not on scale)

(SWT) method was used to help the decisionmaker express his preferences in a multiobjective framework. In addition, an interactive decision support system based on the PMRM was developed, designed to be user-friendly. This DSS was used to perform all the sensitivity analyses of section 9, which will be discussed next.

It was noticed that the PMRM gives results that are sensitive to changes in the value of the return period of the PMF and in the type of distribution function used to extrapolate the flood frequency distribution beyond the 100-year flood. In particular, the use of the Gumbel distribution gives larger estimates of LP/HC risks. These high estimates have the effect of inducing conservative decisions in the decisionmaking process. If the expected value of damages only had been used, similar conclusions to the ones obtained here would have been derived. Therefore, the sensitivity of the PMRM to changes in the estimate of the return period of the PMF and in the type of flood-frequency distribution used shows that the PMRM does not lose information. It is important to notice here that it is a desirable feature for a method to be sensitive to changes in parameters that are external to the method and that are inherent to the problem being studied. Also, the PMRM gives the decisionmaker more information on the impact of the changes discussed above on the various risk domains. Thus, by using the PMRM the DM can perceive some aspects of the problem that would not have been evident otherwise.

This statement is supported by observations that when the PMRM is used, some solutions which would have been considered inferior under the expected values approach prove to be noninferior when the LP/HC risk domain is considered too. Increasing the spillway's width is one of these solutions which would have been absolutely disregarded had we had not used the PMRM.

There were some problems associated with the use of the PMRM: the results were not as insensitive to the choice of the partitioning points as one would have desired. Here, sensitivity of the method to the partitioning points is undesirable because this kind of problem will arise only if the PMRM is used. The problem becomes more acute for dam problems where the magnitudes of failure damages are neither much larger nor slightly larger than the magnitudes of nonfailure damages.

Recommendations for Future Studies

To improve the robustness of the results obtained by the PMRM against changes in the partitioning points, partitioning the damage axis in addition to the probability axis is recommended. But before this damage partitioning approach can be used, more theoretical work needs to be done.

Since the use of the Gumbel distribution to represent the frequency of extreme floods definitely induces the decisionmaker to adopt conservative solutions, it is recommended, to be on the safe side, that risk analysts always include this distribution in their work in, addition to any other distributions. Similarly, it was shown that changes in the return period of the PMF also have a large impact on the decisionmaking process. Therefore, unless more precise estimates are available, any risk analysis of dam failure should be performed for a return period of the PMF set at 10^4 and 10^6 (or 10^7).

Karlsson [1986] has developed analytical relationships that allow the combination of the PMRM with the statistics of extremes. In fact, he has derived an expression that relates the risk objective function f_4 (corresponding to the LP/HC risk domain) to the partitioning points on the probability axis and to the characteristic largest value. This would allow the analyst to accurately compute f_4 for continuous damage functions without going through time-consuming numerical integrations.

Karlsson also showed that if the flood-frequency is a normal, log-normal, or Weibull distribution, it is possible to derive a closed form expression for f_4 where both the partitioning point and the decision policy are variables. Notice that he applies his results to a simplified case of dam safety where his decision variable was the magnitude of the critical flood and where the damage function was continuous.

Mitsiopoulos and Haimes [1987] expanded on these results and even derived distribution-free results concerning the magnitude of f_4 and distribution-free estimates of the sensitivity of f_4 to partitioning choices on the exceedance probability axis (see Mitsiopoulos and Haimes [1988]).

There is no doubt that if the above works are combined to result in the practical approach taken here and applied to dam safety problems, more accurate results and a better understanding of the risks involved in dam safety decisionmaking can be gained.

Acknowledgments

The research documented in this paper culminates the results of a collaborative study on dam safety among several institutions and individuals. Some of these results have been previously published in:

(a) Petrakian, R., "Risk Analysis of Dam Failure and Extreme Floods: Application of PMRM," M.S. Thesis, Department of Systems Engineering, Case Western Reserve University, Cleveland, Ohio, 1986 and

(b) Haimes, Y. Y., R. Petrakian, G., P.-O. Karlsson and J. Mitsiopoulos, 1988. Multiobjective Risk-Partitioning: An Application to Dam Safety Risk Analysis. Prepared for U.S. Army Institute for Water Resources by Environmental Systems Management, Inc. IWR Report 88-R-4.

We thank Per-Ola Karlsson and Jim Mitsiopoulos for contributing to this effort; Virginia Benade and Susan Hitchcock for their valuable editorial work; and Melanie Farrish for her typing and secretarial assistance.

REFERENCES

American Nuclear Society. 1981. American National Standard for Determining Design Basis Flooding at Power Reactor Sites. Report ANSI/ANS-2.8-1981. La Grange Park, Illinois.

American Society of Civil Engineers (ASCE), United States Committee on Large Dams. 1975. Lessons from Dam Incidents, USA, American Society of Civil Engineers. New York.

Asbeck, E. 1982. The Partitioned Multiobjective Risk Method. M.S. Thesis, Department of Systems Engineering, Case Western Reserve University. Cleveland, Ohio.

Asbeck, E., and Y. Haimes. 1984. The Partitioned Multiobjective Risk Method. Large Scale Systems 6.

Chankong, V. 1977. Multiobjective Decision Making Analysis: the Interactive Surrogate Worth Trade-off Method. Ph.D. Dissertation. Department of Systems Engineering, Case Western Reserve University. Cleveland, Ohio.

Chankong, V., and Y. Y. Haimes. 1983. Multiobjective Decision Making: Theory and Methodology. New York: North Holland Publishing.

Haimes, Y. Y. 1980. The Surrogate Worth Trade-off (SWT) Method and its Extensions. In G. Fandel and T. Gal (eds.), Multiple Criteria Decision Making: Theory and Applications. New York: Springer-Verlag.

Haimes, Y. Y. 1984. Integrated Risk and Uncertainty Assessment in Water Resources Within a Multiobjective Framework. Journal of Hydrology 68, pp. 405-417.

Haimes, Y., and V. Chankong. 1979. Kuhn-Tucker Multipliers as Trade-offs in Multiobjective Decision-Making Analysis. Automatica 15:1, 59-72.

Haimes, Y., and W. Hall. 1974. Multiobjectives in Water Resources Systems Analysis: The Surrogate Worth Trade-off Method. Water Resources Research 10:4, 615-624.

Haimes, Y. Y., W. A. Hall and H. T. Freedman. 1975. Multiobjective Optimization in Water Resources Systems: The Surrogate Worth Trade-Off Method. New York: Elsevier Publishing.

Haimes, Y. Y., R. G. Petrakian, P.-O Karlsson and J. Mitsiopoulos. 1988. Multiobjective Risk Partitioning: An Application to Dam Safety Risk Analysis. Prepared for U.S. Army Institute for Water Resources by Environmental Systems Management, Inc. IWR Report 88-R-4.

Interagency Advisory Committee on Water Data. 1982. Guidelines for Determining Flood Flow Frequency. Bulletin 17B of the Hydrology Subcommittee, U.S. Department of the Interior. Geological Survey. Reston, Virginia.

Kaplan, S., and B. J. Garrick. 1981. On the Quantitative Definition of Risk. Risk Analysis 1:1.

Karlsson, P.-O. 1986. Theoretical Foundations for Risk Assessment of Extreme Events: Extensions of the PMRM. M.S. Thesis. Department of Systems Engineering, Case Western Reserve University. Cleveland, Ohio.

Karlsson, P.-O., and Y. Y. Haimes. 1988a. Risk-Based Analysis of Extreme Events. Water Resources Research 24:1.

Karlsson, P. O., and Y. Y. Haimes. 1988b. Probability Distributions and their Partitioning. Water Resources Research 24:1.

Leach, M. R. 1984. Risk and Impact Analysis in a Multiobjective Framework. M.S. Thesis. Department of Systems Engineering, Case Western Reserve University. Cleveland, Ohio.

Leach, M. R., and Y. Y. Haimes. 1987. Multiobjective Risk-Impact Analysis Method. Risk Analysis 7:225-241.

McCann, M. W., J. B. Franzini, E. Kavazanjian, and H. Shah. 1984. Preliminary Safety Evaluation of Existing Dams. Department of Civil Engineering, Stanford University, Stanford, California.

MacCrimmon, K. R., and S. Larsson. 1975. Utility Theory: Axioms Versus "Paradoxes." In M. Allais and O. Hagen (eds.), Rational Decisions Under Uncertainty.

Mitsiopoulos, J., and Y. Y. Haimes. 1987. Generalized Quantification of Risk Associated with Extreme Events. Presented at the annual meeting of the Society for Risk Analysis.

Mitsiopoulos, J., and Y. Y. Haimes. 1988. Approximating Catastrophic Risk Through Statistics of Extremes. Center for Risk Management of Engineering Systems, University of Virginia. Technical report 01-88. Charlottesville, Virginia.

Moser, D. 1987. Dam Safety Risk Cost Analysis Model. Draft Report. U.S. Army Engineer Institute for Water Resources.

National Research Council (NRC), Committee on Safety Criteria for Dams. 1985. Safety of Dams--Flood and Earthquake Criteria. National Academy Press. Washington, D.C.

Newton, D. W. 1983. Realistic Assessment of Maximum Flood Potential. Journal of Hydraulic Engineering 109:6, 905-918.

Petrakian, R. G. 1986. Risk Analysis of Dam Failure and Extreme Floods: Application of the PMRM. M.S. Thesis. Department of Systems Engineering, Case Western Reserve University. Cleveland, Ohio.

Raiffa, H. 1968. Decision Analysis. Reading, Massachusetts: Addison-Wesley.

Shoemaker, P. 1980. Behavioral Issues in Multiattribute Utility Modeling and Decision Analysis. In J. N. Morse (ed.), Organizations: Multiple Agents with Multiple Criteria. Springer-Verlag.

Slovic, P., and A. Tversky. 1974. Who Accepts Savage's Axiom? Behavioral Science. 19:378-373.

Stedinger, J. and J. Grygier. 1985. Risk-Cost Analysis and Spillway Design. In H. Torno (ed.), Computer Applications in Water Resources, ASCE. New York.

U.S. Army Corps of Engineers. 1983. Design Memorandum for Correction of Spillway Deficiency for Mohawk Dam. Prepared by the Huntington District, Corps of Engineers. Huntington, West Virginia.

U.S. Army Corps of Engineers. 1985. Justification for Correction of Spillway Deficiency for Mohawk Dam. Prepared by the Huntington District, Corps of Engineers. Huntington, West Virginia.

Vohra, K. G. 1984. Statistical Methods of Risk Assessment for Energy Technology. In R. A. Waller and V. T. Covello (eds.), Low-Probability/High-Consequence Risk Analysis: Issues, Methods, and Case Studies. New York: Plenum Press.

Zeliny, M. 1982. Multiple Criteria Decision Making. New York: McGraw-Hill Inc.

MARINE POLLUTION: PERCEPTIONS OF RISKS

Howard Levenson

ABSTRACT

Public attitudes and perceptions about the risks associated with marine pollution frequently do not correspond with quantitative or qualitative analyses of such risks. This article examines several situations illustrating this theme and concludes with a plea to place the results of risk assessment studies in a broader social context.

1. INTRODUCTION

The Office of Technology Assessment (OTA) recently concluded, in a comprehensive report on marine pollution, that estuaries and coastal waters around the country will continue to degrade or begin to do so during the next few decades unless additional protective measures are taken (U.S. Congress, 1987). Critical policy decisions about the scope and magntiude of such measures will hinge on society's perceptions about the relative risks of waste disposal and other activities to human health and marine environmental quality.

Risk assessments provide one means of evaluating these risks. The quantitative results of risk assessments often are used at the local, site-specific level to identify the most important problems or the effectiveness of different control strategies. At the state and national levels, they are important inputs into the procedures leading to guidelines, standards, and other control requirements for individual pollutants or sources of pollutants.

Many of today's choices at the national level, however, are about which broad problems to address. At this level of generality, the risks from cumulative pollution and the trade-offs of different control policies are not easily quantified. As a result, decisions at the national level are often made in a more nebulous manner, mixing multiple objectives and decision criteria--mostly without the benefit of quantitative risk assessments.

Public perceptions of risks are always a major driving factor in the ultimate decisionmaking process, at any governmental level. Furthermore, these perceptions are important whether or not they involve a quantitative element. Not surprisingly, public perceptions frequently dominate decisionmaking at the national level, especially in debates about broad directional questions.

Howard Levenson is a Senior Analyst in the Oceans and Environment Program, Office of Technology Assessment (OTA), a non-partisan analytical support arm of the United States Congress. The views expressed in this article are entirely those of the author and are not necessarily those of OTA.

While OTA's analyses did not involve the use of risk assessment methodologies, a common observation about such assessments did emerge during the study. That is, general attitudes and perceptions displayed by the public about marine pollution often do not correspond with quantitative or qualitative analyses of the risks associated with different polluting activities.

This observation--while not news to practitioners of risk assessment--has important ramifications. When public perceptions of risks differ from the results of quantitative risk assessments, those results can be rendered almost useless. For example, risk assessments often are conducted at specific sites to estimate the risks associated with particular polluting activities, and these estimates can be used to generate guidelines or public health advisories. In some cases, however, the risk assessment results simply may not be used by an agency or elected officials. Or, even if they are used to generate guidelines or advisories, these non-enforceable provisions frequently can be ignored.

This article examines several such situations--involving fish consumption by urban fishermen in New Jersey, ocean incineration of hazardous waste, general estuarine and coastal water quality, and a workshop attempt to rank marine problems. It concludes with a plea to place the results of risk assessment studies within a broader social context.

2.   URBAN FISH CONSUMPTION

Risk assessments can be conducted to estimate the probability of developing cancer in response to exposure to specific pollutants. The assumptions used in such assessments often are challenged by interested parties, leading to considerable controversy in many cases; witness the ongoing controversy regarding risks from consuming white croaker caught in Santa Monica Bay, California.

People who actually catch and consume fish perceive risk assessment results in varying ways. One noteworthy case study concerns residues of polycholorinated biphenyls (PCBs) in striped bass and bluefish caught in the waters of New York and Newark Bays (Belton et al., 1986). Sampling studies indicated that over one-half of the finfish and shellfish taken from these waters had detectable levels (greater than 0.1 ppm) of PCBs in their tissue. Furthermore, about 11 percent of the finfish had levels exceeding the Food and Drug Administration tolerance limit of 2 ppm. For people consuming 37 grams of striped bass daily over a lifetime (about 2 meals a week, not an unreasonable amount), it was estimated that 38 out of 100,000 might develop some form of liver cancer; this is a relatively high cancer risk.

As a result, New Jersey issued public health advisories and/or closed fisheries for selected species and areas. Urban fishermen then were surveyed to see how the fishing public perceived these risks/advisories. About one-half knew about the advisories. Of these, 40 percent misinterpreted the advisories as referring to size restrictions, catch limits, or other fishery management techniques. Of all fishermen (whether they were aware or not of the advisories), two-thirds considered the fish totally safe to eat and one-fifth viewed them as slightly polluted but not harmful.

Those respondents who comprised the "no risk" group attributed the advisories and closures to misinformation among regulators or excessive caution by local officials. The authors did not find this surprising: most contaminated fish look normal, other fishermen were eating their catch, and no health problems had appeared yet.

This anecdote illustrates one situation in which people do not pay attention to quantitative estimates of risks. In this and other cases, people perceive environmental quality in terms of how they expect to use a particular resource. The authors considered this a clear example of the difficulty in convincing people to alter behavior when a hazard takes a long time to appear, is invisible, and has a low probability of occurrence.

3. OCEAN INCINERATION OF HAZARDOUS WASTES

Few waste treatment concepts have generated as much controversy and polarization as has ocean incineration, which involves the burning of liquid hazardous wastes in incinerators mounted on ocean-going vessels. More than 6,000 people attended a 1983 public hearing on ocean incineration in Texas, and currently the activity is halted pending the issuance of regulations by the Environmental Protection Agency (EPA). Public concerns focus primarily on the risks to human health and environmental quality from catastrophic spills and from stack emissions.

The technical, environmental, and social issues associated with ocean incineration have been examined by OTA (U.S. Congress, 1986). The following discussion summarizes some issues that illustrate the relationship between public perceptions and quantitative estimates of risk; the discussion is drawn directly from the report.

OTA examined ocean incineration within the context of a general waste management hierarchy that accords preferred status to methods that reduce risk. Its highest tiers include reducing the generation of waste and recycling waste, while its lowest tier includes disposal practices that contain or disperse wastes in the environment. The technology of ocean incineration occupies a middle tier--superior to disposal practices such as landfilling but inferior to most reduction, recycling, and advanced treatment technologies.

Spill Rates and Sizes

The handling and transport of incinerable hazardous waste typically involves many steps. For ocean incineraton, in comparison with land-based incineration, an extra step is required to bring wastes to the dock, load them onto the vessel, and transport them to the incineration site. This step increases the risk of accidental release of wastes and has received a great deal of public attention.

EPA has estimated the probability of a spill from an ocean incineration vessel. While subject to several limitations, the estimated rates for an incineration vessel were seven- to ten-fold lower than the historical spill rates for all tank ships of comparable size operating worldwide between 1969 and 1982. This is not surprising, because the estimate accounts for special safety and operational features of incineration vessels (e.g., double hull, shallower draft, smaller tanks, greater maneuverability). Estimating spill size is more difficult, but the average spill size from an incineration vessel is expected to be smaller than that resulting from a typical tanker accident, where both tank and total cargo size tend to be larger.

Transportation of Hazardous Waste and Nonwaste Materials

Ocean incineration critics also contend that transporting hazardous waste is riskier than transporting hazardous nonwaste materials. Most wastes suitable for ocean incineration result from industrial processes that use chemicals in pure form. All major categories of ocean-incinerable wastes are represented among materials routinely transported by tank ships in liquid form -- crude oil, petroleum products, petrochemicals, liquefied gases, and nonpetroleum-based chemicals. However, mixing and contamination during industrial processing can render the resulting wastes more complex than their precursor nonwaste materials, and the concentrations of particular toxic constituents can differ from those of the raw materials.

Since environmental toxicity is a function of both concentration and composition, generalizing about the relative toxicities of wastes and pure or raw materials is impossible; a case-by-case analysis is required, since some ocean-incinerable wastes will be more toxic and others less toxic. For example, much attention has focused on the transport of polychlorinated biphenyls (PCBs), which are very persistent and have a high potential to accumulate in exposed organisms. For this reason, special regulations have been developed for PCBs.

The quantities of hazardous wastes that might be transported during ocean incineration are easier to place in context. Table 1 summarizes some data on the annual tonnages of hazardous materials and petroleum passing through various U.S. ports and compares these amounts to the quantities that would be carried by one or 30 incineration vessels. Even with a fleet of 30 vessels, marine transport of hazardous materials would be expected to increase by less than one percent.

Incinerator Emissions

The potential emissions from ocean incineration include unburned waste, organic chemicals from incomplete combustion, metals, and acid gases. Emissions typically are released to the atmosphere and then settle over the ocean surface.

Some data are available on the potential emissions of unburned wastes, particularly for PCBs in the Gulf of Mexico. Averaged over the entire Gulf, the data indicate that an incineration vessel operating at a destruction efficiency of 99.9999 percent (the current standard) would

Table 1.--Annual Tonnages of Hazardous Materials and Crude Petroleum
Passing Through Various U.S. Ports in 1984

Location	Quantity Transported in 1984 (millions of metric tons)			Quantity normalized to one Vulcanus vessel
	Hazardous materials[a]	Crude petroleum	Total	
Total for all U.S. ports (1983)[b]	--	--	1,384	21,290
Port of New York	104	8	112	1,723
Delaware River/Bay	26	46	72	1,108
Port of Mobile, AL	3	3	6	92
Port of Lake Charles, LA	20	7	27	415
Houston Ship Channel, TX	46	11	57	877
San Francisco Bay, CA	25	26	51	785
			Annual quantity[c]	
One incineration vessel				
Vulcanus			0.065	1
Apollo			0.100	1.5
30 incineration vessels				
Vulcanus			2.0	30
Apollo			3.0	46

[a] Includes the following commodities:

Sodium hydroxide	Basic chemicals	Jet fuel
Crude tar, oils, gas	Paints	Kerosene
Dyes, pigments	Gum, wood chemicals	Distillate fuel oil
Alcohols	Insecticides, disinfectants	Residual fuel oil
Benzene and toluene	Miscellaneous chemicals	Lubricating oil and grease
Sulfric acid	Gasoline	Naptha, petroleum solvents

[b] This 1983 quantity is cited in the preamble to EPA's proposed Ocean Incineration Regulation, 50 FR 8226, Feb. 28, 1985. The data are originally derived from the Waterborne Commerce Statistics of the U.S. Army Corps of Engineers. A national total for 1984 was not available at the time of publication of this report.

[c] Estimates based on information obtained from vessel owners.

SOURCE: Office of Technology Assessment (1986)

cause a 0.02 to 0.2 percent increase in the quantity of PCBs entering the water from the atmosphere. At the upper end of this range, an increase in the number of vessels operating in the Gulf or a decrease in the achieved destruction efficiency could result in a significant increase above background levels.

PCBs also enter marine waters from waste discharges, dumping, and rivers. Table 2 lists several estimates of direct PCB inputs to marine waters from various sources. The data indicate that ocean incineration used on a modest scale would cause an incremental increase in the total input of PCBs to marine waters. The relative magnitude and significance of such an increase would vary with respect to location.

Expected emissions of metals from ocean incineration may be compared with inputs of metals into marine waters from other sources. According to one estimate, land-based sources annually deposit about 5600 metric tons of seven metals in the Gulf of Mexico. If incinerated wastes contained the (proposed) maximum allowable amounts of these metals, each incineration vessel operating in the Gulf would increase the input of these seven metals by about 2.6 percent.

EPA examined how metals from emissions might mix in the open ocean, estimated the concentrations of four metals under three different scenarios (Table 3), and compared these concentrations with natural background levels. For mixing to 60 meters, three of the four metals would be well below background; only cadmium would exceed its low background concentration. Even if metals from emissions were confined to the upper one meter of water, only cadmium would exceed its background level, but in this case by 100-fold.

The third scenario, in contrast, indicates potential problems with metals in the microlayer. The microlayer supports high densities of plants and animals that may play an important role in marine food chains, and appears to serve as essential habitat for the embryonic life stages of many fish and crustaceans. It also exhibits enriched concentrations of organic matter, metals, and organic chemicals. If mixing were entirely confined to the microlayer, all four metals would far exceed background levels; the ecological significance of this substantial increase, however, is unclear.

In addition to metals, large amounts of hydrogen chloride gas are produced during incineration. In a worst-case scenario, EPA estimated that the gas emissions would decrease the alkalinity of seawater in the mixing zone (about 20 meters) by about 1.3 percent, well below the ten percent change allowed by the proposed standards. This scenario has been criticized, however, because it ignores potential impacts of the higher concentrations of acid that would exist before initial mising was achieved, particularly in regions of the surface microlayer that came in direct contact with the incinerator emission plume.

Conclusion

This brief review of some semi-quantitative and qualitative information regarding the risks of ocean incineration is purposely cursory and does not address the overall benefits and costs of ocean incineration.

Table 2.--Estimated Inputs of PCBs
to Various Marine Waters

Affected waters: source	Annual PCB loading (kg/yr)
New York Bight:[a]	
Sewage sludge dumping	600-2,000
Dredge materials dumping	3,500
POTW discharges	200-1,000
Upstream sources	3,100
Southern California Bight:[b]	
Sewage	2,000
One incineration vessel at 99.9999% DE[c]	18
One incineration vessel at 99.99% DE[c]	1,800

[a]J. O'Connor, J. Klotz, and T. Knelp, "Sources, Sinks and Distribution of Organic Contaminants in the New York Bight Ecosystem," Ecological Stress and the New York Bight, G. Mayer (ed.) (Charleston, SC: Estuarine Research Federation, 1982), pp. 631-653.
[b]M. Connor, "Statement on Incineration of Hazardous Waste At Sea," In Hearing Before the Subcommittee on Fisheries and Wildlife Conservation and the Environment and the Subcommittee on Oceanography of the House Committee on Merchant Marine and Fisheries, 98th Cong., 1st sess., Dec. 7, 1983, Serial No. 98-31 (Washington, DC: U.S. Government Printing Office, 1984).
[c]Assumes a throughput of 50,000 metric tons per year of 35% PCB-laden waste. EPA has proposed the higher DE of 99.9999% for ocean incineration of PCBs.

SOURCE: Office of Technology Assessment (1986)

Table 3.--Metal Concentrations Resulting From
Ocean Incineration, Under Three Different Scenarios
for Mixing of Emissions in Seawater

	Resulting concentration[a]	
Scenario 1: All metals are deposited within the surface microlayer, represented by the upper 0.1 millimeter of the ocean surface in the affected area	320,000	ppt[b]
Scenario 2: All metals are evenly mixed in the upper 1 meter of the affected area	32	ppt
Scenario 3: All metals are evenly mixed in the upper 60 meters[c] of the affected area	0.53	ppt

[a]Assumes that four metals (arsenic, cadmium, chromium, and nickel) are present in the incinerated waste at 100 ppm each.
[b]ppt = parts per trillion.
[c]EPA's proposed Ocean Incineration Regulation (50 FR 8245, Feb. 28, 1985) would define the release zone for incinerator emissions as comprising the upper 20 meters of surface water, this represents an estimate of the depth of the surface thermocline, above which the initial mixing would be expected to occur.

SOURCE: U.S. Environmental Protection Agency, Office of Policy, Planning and Evaluation, "Background Report IV: Comparison of Risks From LandBased and Ocean-Based Incineration: Appendix I," Assessment of Incineration as a Treatment Method for Liquid Organic Hazardous Wastes (Washington, DC: 1985).

The point is not whether ocean incineration is a bad or good technology. In comparison with other practices, it clearly poses some lesser risks, some roughly equivalent risks, and a few greater risks. However, the general public usually does not place these varying levels of risk within a broad context; for example, the daily transport of hazardous nonwaste materials occurs with little fanfare, while potential ocean incineration activities draw enormous attention.

4.   THE QUALITY OF ESTUARIES AND COASTAL WATERS

OTA's recent report (U.S. Congress, 1987) focused on the overall quality of our marine environments. Estuaries and coastal waters in particular are extremely valuable from ecological, commerical, recreational, and esthetic perspectives. Despite this, OTA found that these waters bear the brunt of marine waste disposal activities and that many are in trouble around the country. Among disposal activities, and as a generalization, industrial and municipal discharges are at least as important as dumping in causing damages; urban and agricultural run-off, although not disposal activities, also are of equivalent importance.

The report documented a wide range of impacts. For example, high levels of organic chemicals, metals, and pathogens are present in many estuaries and coastal waters, along with excess nutrients and low oxygen levels. Reported human illness from eating contaminated shellfish or swimming in contaminated waters is rising. Commercial harvests from about one-third of productive shellfish areas are not prohibited or partially restricted because of contamination.

The regulatory system designed to protect estuaries and coastal waters was found to be lacking. The programs and procedures developed under the Clean Water Act and the Marine Protection, Research, and Sanctuaries Act have helped reduce the amounts of some pollutants (particularly suspended solids, fecal coliform bacteria, and other "conventional" pollutants). As implemented, however, they have not protected some estuaries and coastal waters from degradation. Moreover, even with total compliance, which is unlikely, existing regulations were not considered sufficient to maintain or improve the health of all estuaries and coastal waters.

Public debates about marine pollution during the last 20 years have not focused on overall marine environmental quality, however, but rather on discrete issues such as the dumping of sewage sludge, ocean incineration of hazardous wastes (see above), disposal or loss of plastics from ocean-going vessels, and the use on boats of anti-fouling paints containing tributyltin.

While these discrete issues are important and deserve attention, this author considers them to be a subset of the larger problem of general estuarine and coastal water quality. Of course, concern about these discrete issues is related to an overall public concern about marine environments, but the public has not focused on the issues within a broader context.

Fortunately, the broader issue of estuarine and coastal water quality is now receiving more attention from the media, from local, state, and federal legislators and agencies, and from the public at large. This welcomed increase in attention partly reflects a growing recognition of widespread and chronic problems and of the economic and recreational opportunities that are lost as a result of such problems. It is also in response to an unfortunate series of highly publicized incidents in 1987.

- the infamous garbage barge "Mobro" that wandered around six states and three countries with a load of solid waste before returning to New York;
- deaths of hundreds of dolphins along the east coast; and
- the washing ashore of hospital wastes on New Jersey beaches.

As a result, the public is now increasingly worried about the safety of consuming fish and shellfish caught in marine waters, about the property values of coastal real estate, and about the economic viability of towns that depend on seasonal influxes of tourists to their beaches. We are beginning to realize that the problems in estuaries and coastal waters generally do not arise from single sources of pollution but rather from a combination of many activities, not all of which are related to disposal: industrial and municipal pipeline discharges, urban and agricultural run-off, dumping, coastal development, diversion of freshwater, habitat destruction, and overfishing.

Maintaining or improving the quality of estuaries and coastal waters will require additional long-term management efforts. The federal and state governments have realized this need and established what OTA calls "waterbody management" programs; examples include the Chesapeake Bay Program and the Puget Sound Water Quality Authority. These programs basically attempt to bring together the appropriate governmental and citizen parties (including ones with authority to deal with non-point source pollution and land-use issues), identify the most important problems, and devise management plans that specify goals and that discuss the most effective means of achieving those goals.

While this site-specific approach is promising, such programs do not exist for most estuaries and coastal waters. The critical missing link is a systematic framework for deciding when and how to provide additional, site-specific management for estuaries and coastal waters.

Such a framework might not be necessary if public and governmental attention to estuarine and coastal water quality was sufficient. However, sufficient attention probably will not be forthcoming until our primary focus is placed on the overall quality problem, rather than on discrete problems. Even then, the discrete problems would still be addressed. In any case, solving the overall problems in estuaries and coastal waters will take years of effort and the maintenance of long-term political support, which in turn will depend on public perceptions about the most important risks and problems.

Risk assessment has not been mentioned yet in this section. Risk assessments certainly can play a role in developing solutions to marine pollution problems, particularly at the site-specific level, where they can help developers of waterbody management programs to decide which problems are most critical. In contrast, decisions at the state or federal level regarding the development of a more systematic approach to developing management programs will not depend on risk assessment studies. Such decisions, however, could have a significant influence on the extent to which risk assessment studies are used by individual management programs; for example, if national guidelines for management plans are developed, they could explicitly require or suggest the use of risk assessment studies where appropriate.

5. NOAA'S MARINE POLLUTION PLAN

A final example illustrates how the activities of a different sector of the public--federal agencies--often do not correspond to "expert" opinions regarding risks and problems. The National Marine Pollution Program was established by the National Oceanic and Atmospheric Administration (NOAA) to prepare a comprehensive five-year plan for coordinating federal research and monitoring efforts in the sphere of marine pollution (U.S. Department of Commerce, 1985). An interagency committee established by NOAA allows the other federal agencies to participate in the planning process. The primary federal agencies involved in the program are the EPA, NOAA, Department of Agriculture, Coast Guard, Food and Drug Administration, Corps of Engineers, Minerals Management Service, Fish and Wildlife Service, and Department of Energy. In 1985, these agencies administered about 650 projects related to marine pollution, with a total budget of about $121 million.

As part of the five-year plan, NOAA is charged with identifying the major marine pollution issues facing the country. In 1984, NOAA conducted a workshop to identify national needs and problems related to marine pollution and determine their relative priorites. Using an approach known as "Decision Analysis by Paired Comparisons" and a set of weighted evaluation critieria, 65 expert participants evaluated each of 50 issues. With these evaluations and additional information, the workshop coordinators then ranked the 50 issues and grouped them into a set of 13 national problems and five national needs. Each national problem then was assigned to one of three categories based on relative threat to the marine environment (Table 4).

The problems identified as being of highest priority involved waste disposal (both municipal and industrial), nonpoint source pollution (e.g., agricultural and urban runoff), and habitat alterations in wetlands and estuaries and coastal waters. These threats are in general agreement with the conclusions reached by OTA in its assessment of marine pollution (see above).

Interestingly, some issues with a history of controversy and general public concern received only a medium or low priority ranking-- dredged material disposal, oil spills, low-level radioactive waste disposal, oil and gas development on the outer continental shelf, and ocean incineration. The public's perceptions regarding important marine pollution problems thus did not correspond well with qualitative rankings deriving from a group of experts.

Table 4. National Marine Problems and Needs
Identified in NOAA's 1984 Workshop

High Priority Problems

    Nonpoint Source Pollution
    Habitat Loss and Modification
    Sewage Effluent and Sludge Disposal
    Industrial Waste Disposal

Medium Priority Problems

    Oil and Gas Extraction
    Accidental Discharges of Oil, Haz. Materials
    Radioactive Waste Disposal
    Dredged Materials Disposal

Low Priority Problems

    Marine Transportation
    Marine Energy Development
    Deep Seabed Mining
    Sand, Gravel, and Shell Mining
    Strategic Petroleum Reserve Brine Disposal

Priority National Needs

    Monitoring Environmental Status
    Coordination with Fundamental Marine Science Research
    Development of Measurement Methods
    Information Dissemination
    Quality Assurance

Source: U.S. Department of Commerce (1985)

Furthermore, and perhaps of more significance, the workshop's ranking of priorities did not correlate with the funding allocated by the federal agencies for research and monitoring (Table 5). Three pollution problems that ranked in the medium priority category at the workshop received the largest amounts of funding for research and monitoring: oil and gas mining ($30 million), dredged material disposal ($10 million), and radioactive waste disposal ($9 million). In contrast, the problems that ranked in the high priority category each received $6 million or less: nonpoint source pollution ($3 million), habitat loss and modification ($3 million), municipal sewage waste disposal ($6 million), and industrial waste disposal ($5 million).

Funding allocations, of course, may reflect considerations other than mere rankings (even if we assume universal agreement with the rankings, which is unlikely). An agency may lack the statutory authority to address a particular problem, which could skew the funding comparison. For some problems, the costs of research and monitoring simply may be higher. Besides these considerations, though, a better test of the situation may be to examine whether subsequent funding (i.e., post-1985) begins to correspond with the identified priorities.

6.   CONCLUDING REMARKS

These examples were chosen to illustrate situations in which public perceptions and behavior were not greatly influenced by risk assessment results or in which problems were too general or diffuse to be amenable to risk assessment methodologies. In these types of situations, decisions are often made on the basis of public perceptions of what is most critical, whether or not the perceptions correlate with technical judgements about risks and priorities.

This may be particularly true of decisionmaking at the national level. In a pluralistic society, decisions about national policy tend to be made when there is a convergence of interests and ideas about an issue and when it is politically expedient to formulate a solution. A general public perception that a problem is critical can create an opportunity for convergence and expediency.

Public perceptions, though, are not easily categorized. In some situations, the public pays more attention to the benefits than to the costs of an activity (e.g., urban fish consumption). In others, the public may focus more attention on associated risks and be willing to pay the costs of using alternative strategies (e.g., with ocean incineration). Perceptions about risks also are influenced by other factors not explicitly discussed here: the involuntary or voluntary nature of exposure to the risks, the probability and potential magnitude of the risk, and the relative imminence of occurrence.

Another important factor is the role of the media, especially television. The media can dramatize any situation, whether it be of great significance in any context or of significance only when viewed in isolation. Media attention to an issue can galvanize and shape public opinions and perceptions and, in some cases, lead to great pressure for governmental action.

Table 5. National Marine Pollution Program Expenditures By
          Polluting Activity: FY 1984

POLLUTING ACTIVITY	THOUSANDS OF DOLLARS
Marine Waste Disposal	
Dredged Material	10,435
Industrial Waste	4,915
Municipal Waste	5,888
Radioactive Waste	9,084
Brine (Strategic Petroleum Reserve)	2,580
Other	157
Marine Mining	
Deep Seabed	1,106
Oil and Gas	30,274
Sand, Gravel, and Shell	148
Other	65
Marine Energy/Ocean Thermal Energy/Conversion	797
Marine Transportation	1,861
Accidental Discharges	2,173
Nonpoint Sources	
Agriculture/Forestry	1,901
Urban/Coastal Development	788
Terrestrial Mining	10
Other	104
Habitat Modification	
Water Diversion	79
Dredge or Fill	1,313
Other	1,588
Miscellaneous	710
Not Specific to a Polluting Activity	45,123
Total	121,099

SOURCE: U.S. Department of Commerce (1985)

These considerations, however, cannot downplay the value of risk assessments. Even in a highly charged atmosphere, risk assessments can be an important tool or catalyst in the development of environmentally sound and cost-effective management solutions to pollution problems. This is particularly true in situations involving local problems or the development of uniform national regulations for specific pollutants, where the use of quantitative risk assessments to specify levels of human and environmental risks is essential.

To be more useful, however, risk assessments must be placed in a broader social context when possible, one that considers the legitimate public role in decisionmaking and that considers how the public perceives different problems. In addition, the public needs to be educated about the benefits and limitations of risk assessments and about how technical analyses can best be used in decisionmaking. Given the role of the media in communicating information and shaping public perceptions, it may be time for the risk assessment and media communities to jointly discuss how information about risks, public perceptions, and social priorities can best be communicated and explained.

REFERENCES

Balton, T., R. Roundy, and N. Weinstein. 1986. Urban fishermen: managing the risks of toxic exposure. Environment 28(9):19-20, 30-37.

U.S. Congress, Office of Technology Assessment. 1986. Ocean Incineration: Its Role in Managing Hazardous Waste. Report OTA-O-313. Washington, D.C.: United States Government Printing Office.

U.S. Congress, Office of Technology Assessment. 1987. Wastes in Marine Environments. OTA-O-334. Washington, D.C.: United States Government Printing Office.

U.S. Department of Commerce. 1985. National Marine Pollution Program, Federal Plan For Ocean Pollution Research, Development, and Monitoring, Fiscal Years 1985-1989. Washington, D.C.: National Oceanic and Atmospheric Administration.

INTERACTIVE MULTIOBJECTIVE PROGRAMMING AND THE EVALUATION OF ALTERNATIVE OCEAN DISPOSAL SITES

Thomas M. Leschine, William A. Verdini, and Hannele Wallenius

ABSTRACT

Problems of environmental policy frequently involve complex trade-offs among risks to the environment, risks to human health, and the costs of pollution control. The decision aids most likely to help policy makers resolve these questions appear to be those which emphasize search and learning and which integrate well with norms of individual behavior. In this paper the ability of a free-search, interactive, multiobjective programming method known as Pareto Race to help decisionmakers resolve ocean dumpsite selection problems is explored. Trials with role-playing decisionmakers suggest that methods like Pareto Race can indeed assist decisionmakers in their attempts to resolve ill-structured problems in environmental decisionmaking.

1. INTRODUCTION

The necessity of ocean-dumping waste materials has been a subject of increasing debate worldwide over the past two decades. Despite rising public and scientific concern over the possible deleterious effects of pollutants introduced by ocean dumping, however, the availability of cost-effective and environmentally safe land-based alternatives for waste disposal has not been demonstrated. Land disposal has also proven to be a politically divisive issue. As a result, waste disposal in the sea has gained a degree of political acceptability over the past decade. This idea has been supported by arguments in the marine scientific community that the sea has an assimilative capacity to absorb a certain amount of pollutants safely (See, e.g., Hirvonen and Cote, 1986).

Assimilative capacity-based waste management strategies attempt to select a disposal method in a way which balances detrimental impacts in the human and natural environments against the cost and need for the disposal. In effect, the assimilative capacity strategy treats water quality management problems as allocation problems, and thus finds correspondences in the resource economics literature (Baumol and Oates, 1975). If the logic of the assimilative capacity concept is followed, then ocean dumping decisions can be reduced to a problem of determining

Thomas M. Leschine is a research associate professor at the Institute for Marine Studies, University of Washington, Seattle.

William A. Verdini is on the faculty of the Department of Decision and Information Systems, Arizona State University, Tempe.

Hannele Wallenius is on the faculty of the Department of Economics, Jyvaskyla University, Jyvaskyla, Finland.

This work was supported by the National Science Foundation, Program on Decision, Risk and Management Science, Grant No. SES-8510463.

suitable volumes, input rates, and locations of disposal operations to maintain pollutant concentrations below threshold levels, with thresholds selected to reflect social and economic goals for the use or quality of the receiving waters.

Such rationalistic, assimilative capacity-based approaches to ocean waste disposal management have by and large failed to achieve successful implementation, however (Leschine, in press). The difficulties may lie with the nature of ocean waste management problems. They are typically of the kind Thomas and Samson (1986) call "ill-structured." That is, they are characterized by high stakes, a structure complicated enough to exceed the ability of any one expert to describe, a need for multiple viewpoints for their resolution, and the requirement that decisions be justified to regulatory authorities, interest groups, and the public at large (Thomas and Samson, 1986). As with other ill-structured problems, optimization-oriented decision aids have had but limited success in influencing the development of policy.

If one were to follow the line developed by Thomas and Samson, then multiobjective analytic approaches intended to aid in the resolution of waste disposal problems would be more in tune with the philosophy of "decision support" than with the traditional ideas behind optimization. The primary emphasis would be on helping decisionmakers understand better the structure of the problems they face, and on helping them understand their own preferences in relation to them. Thomas and Samson would go a step further, suggesting that analytic approaches need to be judged "[on their] contribution to organizational process rather than on specifically recommending an action and getting it adopted" (Thomas and Samson, 1986, p. 253).

An analytic framework for decision support studies would ideally blend elements from behavioral decision theory, from the organizational theories associated with James March, Herbert Simon, and others (see, e.g., March and Simon, 1958), and from the theory of optimization. While interactive multiobjective programming methods seem to offer more potential than other optimization techniques for providing the type of decision support considered here, they too have been criticized on grounds which suggest that their utility for addressing ill-structured problems might be limited (French, 1984). While Wallenius (1975), Klein, Moskowicz, and Ravindran (1986), and others have found generally positive results in experimental trials with interactive methods, these experiments do not really address many of the issues raised here. As French notes, interactive optimization methods do not permit decisionmakers to behave in ways behavioral studies suggest they frequently do behave.

In this paper we discuss the results of an experiment conducted to evaluate the utility of a much different interactive MOP procedure-- Pareto Race (Korhonen and Wallenius, in press). Pareto Race is a dynamic, visual, interactive procedure for solving decisonmaking problems with multiple, conflicting objectives. Unlike other interactive optimization procedures, Pareto Race permits direct, free search of the efficient frontier. Because it emphasizes search and learning rather than convergence toward optimal solutions, Pareto Race appears to be more in tune with the precepts of behavioral decision theory than other interactive methods. And, because Pareto Race is a freeform tool, it appears to offer the experimenter more opportunities to probe the intricacies of decisionmaker behavior than other MOP procedures.

The problem of study is that of applying an assimilative capacity-based approach to resolving ocean disposal problems where both costs to municipalities and pollution of the marine environment are to be minimized. More specifically, we investigate the interactions of role-playing decisionmakers with an assimilative capacity-based model of the sewage sludge disposal problem in the New York Bight region. Subjects are able to access the model through Pareto Race.

A mathematical model developed by Verdini and Leschine (1984) facilitates the evaluation of pollutant discharges from both environmental and economic standpoints, and enables one to consider also the contributions uncontrolled sources make to pollution at different sites. This model was modified for use in this study. Multisite disposal, involving one or more of three different disposal sites, is considered. Problem objectives include the costs of disposal operations to municipalities and pollutant accumulation in both the near-shore and the more distant marine environment. The Multi-Attribute Trade-off System (MATS) (Brown and Valenti, 1983) is also used to generate information for evaluating solutions to the decision problem. Fifteen marine policy students were the experimental subjects.

2. PROBLEM DESCRIPTION

The waters of the New York Bight are among the most intensively utilized in the world. Concern for water quality in the Bight region is long-standing, particularly for waters in the inner portion of the Bight. Even so, sewage sludge has been dumped at the so-called 12-mile site since 1924 (Figure 1). In recent years, some 6.4 million metric tons of sludge have been dumped there annually. Now the communities using the site are in the process of shifting to the more distant 106-mile site, under orders issued in 1985 by the U.S. Environmental Protection Agency (EPA).

Opinion in the region has long been split over the fate of the 12-mile site. Highly publicized pollution-related episodes which have occurred over the past decade have had a lasting impact on public opinion. These include fish kills due to low dissolved oxygen, closure of shell beds due to bacterial contamination, and occasional wash-ups of sewage-derived materials onto Long Island beaches. Scientific studies have generally concluded, however, that no clear relationship exists between such incidents and dumping at the 12-mile site. Marine scientists generally agree that for most pollutants, sludge dumping activities account for only one to ten percent of the total contamination of the inner region of the Bight.

The U.S. EPA has for the last several years been considering a number of alternative ocean disposal sites. The sites pictured in

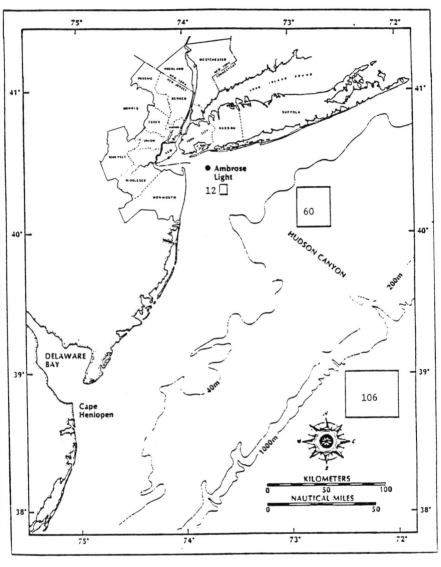

Figure 1.   The New York Bight and Sites which have been Proposed for
             Sewage Sludge Disposal.   The Bight Apex encompasses
             roughly the area off Monmouth Co., N. J. and Nassau Co.,
             N. Y.

Figure 1 have all received consideration. Generally speaking, inshore pollution is reduced as disposal is pushed further to sea, but at the cost of possibly increasing pollution-related damages to such offshore coastal resources as fisheries. Use of the most distant site causes concern for as yet unknown impacts associated with broadcasting pollutants into the deep ocean at very low concentrations. In addition, costs to the New York Bight region increase rapidly as dumping is pushed further offshore.

In this study we reexamine the 1985 EPA decision in a way that permits simultaneous multi-site dumping. This is in contrast to requiring exclusive use of a single site, as the Agency has done. Following instead a line of reasoning based on an assimilative capacity argument, use of less expensive near-shore sites would be permitted up to the point where pollutant accumulations reach levels associated with deleterious impacts. This alternative decision strategy remains highly relevant, particularly if the EPA decision to permit continuation of ocean disposal leads to increased demand for ocean dumping services. The problem formulation is also relevant to the regional dredge spoil disposal policies of the U. S. Army Corps of Engineers, which frequently involve multisite operations subject to capacity, environmental, and cost constraints (Ford, 1984).

Formulation of the Dumpsite Selection Model

Verdini and Leschine (1984) have formulated a linear programming model which provides a framework for ocean waste disposal management. The model optimizes the total variable costs of disposal subject to the constraints of disposal requirements, capacity limitations of the barges, and the limitations of the assimilative capacity of the ocean. The cost of transportation of sewage sludge from source to dumpsite is a function of the disposal method, energy and labor cost at the source, the distance from source to dumpsite, and the volume of sludge transported. The assimilative capacity restriction is estimated from the existing levels of pollution and the estimated near- and far-field effects of the disposal activities. A Markov analysis is used to model the oceanic transport and decay processes which are fundamental to the concept of assimilative capacity. (The model does not consider other alternative disposal methods, such as land dumping, and does not take into account the additional capital expenditures for increased capacity to the present disposal methods, or for the implementation of new technology.)

The constraint set of the model contains three parts: 1) the transportation component which specifies the barging of sludge from sources to disposal sites; 2) the two-phase Markov component, which first models the distribution of pollutants soon after the sludge is discharged and then models the subsequent movement of residual pollutants throughout the ocean; and 3) the environmental constraints which limit the amount of residual pollutants following disposal. This third component links the first two components of the model.

There are two types of constraints on disposal operations. All waste material generated at a source must be disposed of. Disposal operations are limited by the number and type of vessels available to a particular source and the sites chosen for disposal. The farther a site is from a source, the fewer the trips that can be made.

The short- and long-run effects of ocean dumping, as well as on-site and off-site impacts on the environment, are also considered. Initially, much of the material sinks quite rapidly to the bottom, then eventually some of it is dissipated by spreading, erosion, and biochemical action. Some of the residual pollutants may remain in the water column and then, depending on the conditions at the time of the dumping, may dissipate or sink to the bottom.

The environmental constraints of the model can be used to restrict both the initial impact of the waste, as well as the long-term impact of residual pollutants, at the dumpsite or other locations. Near-surface and near-bottom cells at particular locations in the ocean can be used to monitor the environmental impacts of sludge disposal. For the experiments described below, two hypothetical monitoring stations were selected to account for "inshore" pollution, such as might affect bathing beaches and recreational shellfish harvest, and "offshore" pollution, such as might affect commercial fisheries of the mid-continental shelf.

The parameters related to disposal operations used in the model are calculated from data collected in 1981 (Leschine and Broadus, 1985). Three alternative disposal sites are considered in the model: the 12-mile site, the 60-mile site, and the 106-mile site (see Figure 1). We assume that a combination of the above sites is a possibility, such that all three sites could be used at the same time in different proportions. All vessels are assumed to travel to dumpsites fully loaded.

If the original version of the optimization model is run to minimize the total cost of disposal (regardless of the pollution effects), then all sludge is dumped at the 12-mile site at a total annual cost of $10.07 million. The total and proportionate distribution of costs among dumpers are in accordance with actual costs of disposal operations in the New York Bight region in 1981. When the environmental constraints are tightened and become binding, total costs increase at an increasing rate as dumping is forced first to the 60-mile site and then to the 106-mile site. In this way the cost of "not" polluting the ocean can be assessed in terms of the increased disposal costs. Verdini and Leschine (1984) can be consulted for details.

Pareto Race as a Decision Aid for Ill-Structured Problems

Pareto Race (Korhonen and Wallenius, in press) is a dynamic, visual, interactive procedure for solving decisionmaking problems with multiple, conflicting objectives. With it the decisionmaker can freely search the set of efficient solutions by controlling his or her "velocity" and "direction" of motion along the efficient frontier. The values of the objectives are represented both numerically and as bar graphs on a computer display, and the length of the bars changes dynamically in real time as the decisionmaker explores the set of efficient outcomes.

Using Pareto Race to explore the set of possible outcomes for a decisonmaking problem has been compared to driving a car over the region where the outcomes lie in view on the terrain. The decisionmaker can also pause to examine the values of the decision variables at any time during the "race." Mathematically, Pareto Race is based on projecting certain reference directions on the efficient frontier (Korhonen and Wallenius, in press).

Much of the emphasis in previous studies has been on the relative merits of prior versus progressive methods of articulating preference information for solving multiobjective decision problems (see, e.g., Wallenius, 1975; Sarin, Dyer, and Nair, 1979; Klein, Moskowitz, and Ravindran, 1986). But Pareto Race dispenses with explicit formulations of preference information altogether, relying instead on whatever guidance the decisionmaker's internal (and perhaps poorly formed) preference structure provides in helping to differentiate the worth of efficient outcomes. Pareto Race thus represents a significant departure from these other optimization-oriented multiple objective decisionmaking methods.

Pareto Race shares with its cousins an emphasis on efficiency. Otherwise, however, it is for all intents and purposes the "archetypal" interactive method which French (1984) refers to in his criticism of interactive multiobjective programming: The decisionmaker is presented with a trial solution, determines through interactive dialogue (with the computer and/or analyst) a new "improved" solution, and continues this process until "converging" to a solution which, in comparison to other solutions he or she has found, is sufficiently satisfactory to warrant its acceptance.

French's criticism of the traditional multiobjective programming approaches is based on what he sees as the failure of their underlying assumptions to be supported by the findings of behavioral decision theory. He allies himself with the notions of Thomas and Samson (1986) in arguing that "a good decision aid should help the decisionmaker explore not just the problem, but also himself, ... bring[ing] to his attention possible conflicts and inconsistencies in his preferences so that he can think about their resolution" (French, 1984, p. 833). RACE, properly used, would appear to offer opportunities for that kind of introspection.

The progressive articulation (interactive) optimization approach, RACE, can be contrasted with the value function measurement approach, MATS, which was also used in the experiment described below. The fundamental difference between the two approaches is that MATS establishes value trade-off weights among objectives representing cost and pollution, while RACE permits direct examination of the outcome set (efficient set).

The underlying behavioral assumptions of the two procedures are very different. RACE is an unstructured, free-search procedure which makes no assumptions about the properties of a decisionmaker's value function. MATS assumes transitivity of preferences and preferential independence among the criteria. Like other trade-off methods, it also relies upon hypothetical comparisons.

The findings of French (1984) discussed above suggest that subjects would favor a procedure that does not force them to exhibit consistent, transitive behavior. In addition, they would favor a procedure that allows them to terminate at a satisfactory solution, rather than at an "optimal" solution forced upon them. Finally, they would favor techniques that do not force convergence the way the mid-value splitting techniques of MATS or the interactive procedures of Wallenius (1975) and Klein et al. (1986) do. (See also Kok, 1986.) These observations lead to an expectation that subjects would favor RACE to MATS, a finding noted in Wallenius et al. (1987).

Because RACE allows its users the freedom to deviate substantially from the consistent, convergent, and utility-enchancing behavior that the other decision aids we have discussed assume or demand, the arguments French and others make can be extended to argue that it has value as a decision aid in the more general sense outlined above. It is relevant to ask, therefore, what benefits decisionmakers actually derive from its use.

Effective decision aids should still influence decisionmakers to move toward problem solutions or problem-solving strategies that are in some sense "better" than those they would otherwise employ. Returning to RACE's car driving analogy, we can ask whether the trip provides significant enough vantage points to really reveal the character of the surrounding terrain. Is the map provided so poor or the car so balky to drive that, having stumbled into unfamiliar terrain (either by inadvertence or by design), motorists are unable to return to more familiar ground?

One approach to answering such questions is to ask how RACE-generated solutions compare to the initial holistic preferences of the experiment's subjects, to ask what impact the use of RACE has on influencing the nature of their final preferences, and to compare the pattern of initial and final preferences over the group as a whole to see if the experiment fosters a convergence of views. Results of our inquiry along these lines are discussed below.

3. DESIGN OF THE EXPERIMENT

The dumpsite selection linear programming model described above was modified to incorporate three objective functions. The first objective of the model is to minimize the total variable costs of sewage sludge disposal to New York City and communities in New Jersey and on Long Island. Minimizing the inshore and offshore pollution levels are the two other objective functions. The pollution objectives were expressed as percentages of worst possible pollution levels (as predicted by the model) accumulating at hypothetical inshore and offshore monitoring stations. (In later trials with research scientists and pollution control managers in Finland, actual levels of pollutants were used rather than percentage accumulations. Both percentages and volumes of pollutants serve as proxy attributes for the real impacts of pollution, however. Subjects had some difficulty in relating to either proxy, a point which is elaborated upon below.)

The decision variables in the model are the number of trips made from source i to site j by disposal method k. In the form of the model used in the experiments, the total variable costs of disposal could vary between ten and 50 million dollars ($10 million if all sludge was dumped at the 12-mile site and $50 million if all was dumped at the 106-mile site); inshore pollution accumulation between 2.7 percent and 100 percent (or 4.3 and 162.9 metric tons of actual pollutants, respectively); and offshore pollution accumulation between 25 percent and 100 percent (or between 12 and 30.4 metric tons of pollutants). No distinctions were made among disposal methods.

Pollution is highest at the inshore monitoring station when all sludge is dumped at the 12-mile site and highest at the offshore station when all dumping occurs at the 60-mile site. Compared to the inshore monitoring station, the highest attainable level of pollution at the offshore station is associated with a much lower volume of actual pollutants. This is due to the increased dispersion of pollutants at sites more distant from shore in the underlying model. Pollution levels at both the inshore and offshore monitoring stations are lowest when all sludge is dumped at the 106-mile site. As might be expected, costs of disposal operations are minimized at the 12-mile site and maximized at the 106-mile site.

The subjects in the experiment were 15 marine policy students at the University of Washington. Each subject was asked to role-play a U.S. EPA decisionmaker who must make a decision on where to locate one or more sludge disposal sites in the New York Bight. They were instructed to act on behalf of the communities in the New York Bight region, taking into account both pollution impacts and the costs communities might have to bear as a result of their decision.

The RACE and MATS methods were implemented on an IBM PC. They were used in a diverse order to avoid learning bias, and subjects were assisted in computer operations by the instructors, who performed all keyboard operations according to their wishes. The flexibility of RACE together with its ability to provide considerable information on demand about the current problem solution, made it the much more preferred method. MATS, on the other hand, was found much easier to understand conceptually (Wallenius et al., 1987).

In line with the framework developed above, here we ask: To what extent does RACE influence subjects to change their initial preferences? Does it stimulate them to search for and find solutions which enhance their welfare in terms of their stated preferences? What do individuals learn about problem structure through the use of RACE? And, does such learning influence individuals with differing preferences in ways which lead toward a common view of how best to resolve the problem at hand?

We focus here on the extent to which the use of RACE helped subjects find a problem solution they found preferable to their initial holistic choice, or which had a higher utility to them (as expressed by their MATS-generated weights). We examine also the distribution of initial and final preferences across subjects as a means of assaying the ability of the methods employed to generate converging preferences.

All subjects were furnished with written instructions for the methods as well as with background information on the sewage sludge disposal problem in the New York Bight. They were then asked to rank-order seven possible dumping patterns, using flash cards prior to working with the decision aids (Table 1). At the end of the session, all subjects were presented with an exit interview in which they were asked what effect the two methods taken together had on their overall preferences.

Methods and Results

The RACE user controls the speed and direction of search on the efficient frontier with the computer's F keys and space bar. It is possible to fix temporarily the value of one or more of the objectives, and the Num key can also be used to change the direction of search as the value of a particular objective is increased. (If one proceeds in the projected direction to the point where the boundary of the efficient frontier is reached, it becomes necessary to turn in order to avoid passing into the domain of infeasibility.)

MATS* (Brown and Valenti, 1983) is a computerized procedure for assessing a decisionmaker's additive value function. It is based on the well-known mid-value splitting technique (Keeney and Raiffa, 1976). In our experiment, it was used to calculate the scaling constants for the objectives. These in turn were used as the objective function coefficients of the linear programming model for determining the value to each subject of the initial preferred solution (obtained via flash card ranking) and the final solution obtained with RACE.

Although the LP model coupled with the MATS weights can in theory be solved to obtain each subject's most preferred dumping pattern, because of the way the LP model distributes pollutants following dumping, this solution in effect double-counts the reduction-in-pollution objectives. For nearly all subjects, this resulted in a model-determined preference for all disposal to occur at the 106-mile site, the dumping pattern which minimizes total pollutants (see Table 1). This solution was never selected as the dumping pattern of first choice in the exit interviews.

Do subjects using RACE find a solution they prefer to their initial holistic choice?

Because subjects using RACE can in theory find and examine infinitely many different solutions to the ocean dumping decision problem, it would seem fairly likely that they will find a solution preferable to the one they selected initially from among the seven preselected solutions of Table 1. Leaving out one subject who used RACE to find his initial solution, we find that 11 of the remaining 14 subjects did indeed succeed in finding a dumping pattern they liked better than their initial selection.

---

*We thank the Bureau of Reclamation, Denver, Colorado, for making this software available to us.

Table 1. Dumping Patterns Displayed on Flash Cards, Together with their Objective Values.*

	DISTRIBUTION OF SEWAGE SLUDGE DUMPING	TOTAL COST (mil. of $)	POLLUTION INSHORE (%)	POLLUTION OFFSHORE (%)
1.	All as 12-mile site	10.1	100.0	63.6
2.	Split between 12- and 60-mile sites	16.7	56.4	83.5
3.	All at 60-mile site	29.1	17.9	100.0
4.	Split between 60- and 106-mile sites	37.4	10.8	65.4
5.	Split between 12- and 106-mile sites	26.6	48.3	43.2
6.	Split among 12-,60-, and 106-mile sites	25.9	39.1	67.1
7.	All at 106-mile site	50.0	2.7	25.1

*Subjects were not shown the objective values associated with the patterns until the end of the experiment.

Do subjects using RACE find a solution of higher utility than their initial holistic preference?

The MATS weights can be used to calculate the utility of both the dumping pattern initially selected from Table 1 and the stopping point reached via RACE. (MATS produces normalized weights based on subjects' willingness to trade to avoid losses among successive pairs of the three cost and pollution objectives. The attribute levels associated with each dumping pattern preference were also normalized to the model-determined minimum and maximum values of each objective. A linear additive value function was then used to calculate each subject's overall value for each dumping pattern.) Subjects whose preferences are indeed consistent with their weights might be expected to use RACE to find a solution which increases their utility relative to their initial preference.* For ten of the 14 subjects, this proved to be the case. For some subjects spectacular gains in utility were made when the utility of the initial selection was low. Other subjects proved very adept at using RACE in ways that enhanced utility even when the initial choice had fairly high overall utility. (Subjects did not know the utilities of either their initial choices or the solutions they explored during the RACE trials. No connections were made between the products of the MATS and RACE portions of the experiment during its course. Subjects were encouraged to serve only their own overall preferences in using RACE.)

Do subjects actually prefer a solution which has higher utility to them?

While nine of 14 subjects preferred the solution of higher utility (regardless of whether they found it using RACE), five subjects emerged from the experiment preferring a solution of lower utility. It is of course possible that the MATS weights failed to reflect these subjects' true preferences. Careful inspection of the results suggests that more than that may be going on, however. Three of these five failed to come up with a higher utility solution using RACE, but preferred their RACE-generated solution anyway. Some of these individuals appeared to get "lost." But one of these three, one of the remaining two, and the subject who used RACE to guide his way back to his initial preference apparently changed their decision rules while using RACE. One decided he would rather minimize total pollutants than trade between inshore and offshore pollution (but failed to recognize that this could also be done using the model), another didn't want to dump in areas where there were no monitoring stations (the case with the 106-mile site the way the problem was constructed), while the third decided she did not want to employ more than two sites simultaneously regardless of other factors.

For some individuals, RACE's "hands-on" character may have made RACE-generated solutions seem more attractive than they should have been. RACE's occasional quirks (crossing edges may cause the solution to "jump"; one may also get "stuck" trying to execute a turn) may have

---

*RACE trials were begun at the same starting point, a dumping pattern in which some sludge was delivered to each of the three available dumpsites. Subjects would thus not necessarily encounter their initial selection during their RACE trial.

caused others to get "lost." (Robin Hogarth has characterized dynamical systems as either "wicked" or "kind" depending on whether they behave in ways that frustrate attempts to understand how the underlying causal system they model operates. RACE's occasional lapses may cast doubt on whether the system exhibits congruence between cause and effect, an important cue to causality. While mostly kind, RACE may occasionally be wicked.) For still others, however, "failures" to find solutions judged better by the traditional criteria of utility may represent the effects of learning through the model new facets of problem structure important to decisionmaking about it.

<u>Did subjects as a group converge toward a common view of the dumpsite selection problem as a result of the experiment?</u>

Figures 2 and 3 show the distribution of initial and final dumping pattern preferences of the 15 subjects. Initial preferences represent their selections from the flash cards of Table 1. Final preferences represent their choices at the end of the experiment, usually the RACE solution. Some seemingly substantial shifts in preferences do occur. Where the initial preferences are nearly uniformly spread over the seven alternatives, the final preferences show gravitation toward dumping patterns which involve simultaneous use of two of the three sites. Five individuals selected the exclusive use of some site as their preferred alternative initially, but only one did so in the final choice. Also the seven subjects who preferred to make no use of the 12-mile site initially were reduced to four at the end of the experiment.

Subjects had by and large used RACE effectively to discover how the underlying model works. The way to balance the problem's objectives, regardless of one's relative preferences among them, is to engage in multisite dumping. Upon learning this through the use of RACE, subjects for the most part changed their overall preferences accordingly.

4. CONCLUSIONS

Ocean waste disposal problems, like many other problems of environmental policy, present difficult-to-resolve multiobjective, multi-party choice situations. As Kunreuther (1983) has observed, there is often little in the internal dynamics of such situations to lead parties to convergent views of how best to achieve problem resolution. Computerized decision aids can play an important role in helping the parties find acceptable solutions, but studies of their use in realistic problem settings suggest that for them to do so effectively requires that they better account for the realities of individual and organizational behavior than they have generally done in the past.

Pareto Race, a free-search, interactive, multiobjective opimization tool, suggests a new direction for the development of MOP-based decision aids. Its use in a decisionmaking experiment aimed at a re-examination of the controversial sewage sludge disposal problem of the New York Bight region suggests that it illuminated for subjects the connections between their own preferences among problem objectives and potential problem solutions. They responded in a variety of ways, many seeking and finding alternatives which increased their own utility, others by

Initial Preferences

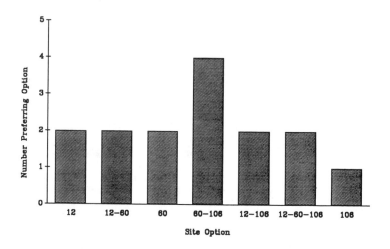

Figure 2.   Distribution of Subjects' Initial Preferences for
Allocation of Sludge Dumping among Disposal Sites.

Final Preferences

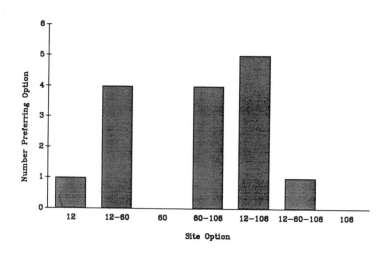

Figure 3.   Distribution of Subjects' Final Preferences for
Allocation of Sludge Dumping among Disposal Sites.

redefining the objectives they would pursue in selecting ocean disposal
sites. As one result, subjects tended to abandon initial preferences
for exclusive use of a single disposal site in favor of multisite
disposal patterns which make better usage of the ocean's assimilative
capacity. We can speculate that, had these subjects been engaged in a
group decisionmaking process, the use of Pareto Race might have assisted
them in finding a mutually acceptable solution.

Much work needs to be done to refine the tools described here.
Among difficulties identified in this and later experiments is the use
of levels of pollutants in the environment as a proxy for the actual
levels of environmental or human health effects likely to be associated
with them. As Fischer et al. (1987) note, the effect may have been to
cause subjects to overweight the proxy attributes, in this case the
levels of pollution inshore and offshore associated with particular
dumping patterns. As a result they may have directed more disposal than
they otherwise would have to the 106-mile site, where pollution is
minimized but costs are highest.

REFERENCES

Baumol, W.J., and E. Oates. 1975. The Theory of Environmental Policy. Englewood Cliffs, New Jersey: Prentice-Hall.

Brown, C. A., and T. Valenti. 1983. Multi-attribute trade-off system (MATS). U. S. Department of the Interior.

Fischer, G. W., N. Damodaran, K. Laskey, and D. Lincoln. 1987. Preferences for proxy attributes. Management Science 33(2): 198-214.

Ford, D. T. 1984. Dredged material disposal management model. Water Resources Planning and Management 110(1): 57-74.

French, S. 1984. Interactive multi-objective programming: its aims, applications and demands. Journal of Operational Research Society 35(9): 827-834.

Hirvonen, H., and R. P. Cote. 1986. Control strategies for the protection of the marine environment. Marine Policy 10: 19-28.

Keeney, R., and H. Raiffa. 1976. Decisions with Multiple Objectives: Preferences and Value Tradeoffs. New York: John Wiley & Sons.

Klein, G., H. Moskowitz, and A. Ravindran. 1986. Comparative evaluation of prior versus progressive articulation of preference in bicriterion optimization. Naval Research Logistics Quarterly 33: 309-323.

Kok, M. 1986. The interface with decision makers and some experimental results in interactive multiple objective programming methods. European Journal of Operational Research 26: 96-107.

Korhonen, P., and J. Wallenius. A Pareto Race. Naval Research Logistics, forthcoming.

Kunreuther, H. 1983. A multi-attribute multi-party model of choice: descriptive and prescriptive considerations. In P. Humphrey et al. (eds.), Analyzing and Aiding Decision Processes. Amsterdam: North-Holland.

Leschine, T. M. 1988. Ocean waste disposal management as a problem in decision making. Ocean and Shoreline Management, forthcoming.

Leschine, T. M., and J. M. Broadus. 1985. Economic and operational considerations of offshore disposal of sewage sludge. In D. Kester et al. (eds.), Wastes in the Ocean. 5. Deep-Sea Waste Disposal. New York: John Wiley & Sons.

March, J. G., and H. A. Simon. 1958. Organizations. New York: John Wiley & Sons.

Rowe, M. D., and B. L. Pierce. 1982. Some tests of analytical multi-objective decisionmaking methods. Socio-Economic Planning Science 16(3): 133-140.

Sarin, R., J. Dyer, and K. Nair. 1979, A comparative evaluation of three approaches for preference function assessment. UCLA Graduate School of Management, working paper no. 352, Los Angeles, California.

Thomas, H., and D. Samson. 1986. Subjective aspects of the art of decision analysis: exploring the role of decision analysis in decision structuring, decision support, and policy dialogue. Journal of the Operational Research Society 37: 249-265.

Verdini, W. A., and T. M. Leschine. 1984. Modeling the effects of environmental restrictions on operational ocean waste disposal decisions. Arizona State University, working paper DIS 84/85-7, Tempe, Arizona.

Wallenius, H., T. M. Leschine, and W. A. Verdini. 1987. Multiple criteria decision methods in formulating marine pollution policy: a comparative investigation. Proceedings of the University of Vaasa, research paper no. 126, Vaasa, Finland.

Wallenius, J. 1975. Comparative evaluation of some interactive approaches to multicriterion optimization. Management Science 21: 1387-1396.

UNCERTAINTY IN ENVIRONMENTAL RISK ANALYSIS

Istvan Bogardi, Lucien Duckstein, Andras Bardossy

ABSTRACT

The classical formulation of environmental risk has two main types of deficiencies: (1) forcing the use of classical probabilities where there is no frequency basis to estimate probabilities, and (2) using single preferred values for other uncertain elements. Classical formulations of risk are extended according to two principles: (1) use probabilities whenever a frequency-based estimation can be made or, alternatively, when subjective probabilities can be assessed, and (2) use fuzzy sets to consider imprecision in non-frequency probability and other elements. Risk estimates are then calculated as fuzzy numbers and ranking methods are used for risk management. Composite programming as a possible risk management tool is extended to consider imprecise risk and cost estimates. A groundwater contamination example illustrates the proposed risk management technique.

1. INTRODUCTION

The purpose of the paper is to develop a methodology for the joint consideration of stochastic and fuzzy uncertainty in environmental risk analysis. "Fuzzy uncertainty" or "imprecision" refers to vagueness in the boundaries of a set. This definition corresponds to a generalized notion of imprecision, which is characterized by so-called "soft" tolerance interval boundaries in tolerance analysis (Dubois and Prade, 1980). This is in contrast with the usual notion of imprecision, where tolerance interval boundaries are well-defined (so-called "crisp") numbers. Fuzzy uncertainty can occur in different elements of a system:

- imprecision of definitions (for example, failure event definition),

- imprecision in knowledge (ignorance), reflected mostly in empirical formulas and approximate reasoning (for example, dose-response relationships in health hazard analysis),

- imprecision in data (for example, doubtful measurements, linguistic descriptions of events, scenarios, probabilities, or consequences).

Istvan Bogardi is a professor in the Department of Civil Engineering, University of Nebraska-Lincoln.

Lucien Duckstein is a professor in the Department of Systems Engineering, Case Western Reserve University, Cleveland, Ohio.

Andras Bardossy is a professor in the Department of Hydrology and Water Resources, University of Karlsruhe, West Germany.

In the present context, stochastic uncertainty refers to the stochastic character of elements in risk analysis. This uncertainty is usually caused by small sample size and/or lack of prior information about the statistical parameters of the probability distribution functions describing elements of risk analysis. Thus, stochastic uncertainty belongs to the realm of probability theory or statistical decision theory and may be analyzed by a Bayesian approach (DeGroot, 1970; Benjamin and Cornell, 1970; Davis et al., 1972; Bogardi et al., 1982; Berger, 1985; Bernier, 1987).

Risk analysis, in general, has four main elements:

1.   the exposure x to a hazard,

2.   the probability $PF = P(x > x_o)$ where x_o is a threshold value reflecting the capacity or resistance of the system,

3.   the consequence of exposure,

4.   the perception of the consequence.

This paper addresses the first three elements of risk analysis. As a simple example, the common formulation of risk to public health considers the environmental risk cost corresponding to event A as an expected value:

$$\text{Risk } (A) = \int_{d_o}^{+\infty} [\text{Resp } (k(x)) \times g(x)] \, dx \times \text{Mort} \times \text{Value} \tag{1}$$

where:

x: the exposure, say the amount of given carcinogen, taken as a random variable.

$g(x)$: the probability density function (pdf) of x.

Resp (x): response function value based on animal experience; that is, given exposure x, the probability of event A, say cancer development.

d_o: the "safe" dose (safe exposure) pertaining to (a) a low probability response value, say $P(A/d_o) = 10^{-6}$ representing background risk, or (b) the resistance of the system, such as a hazardous waste containment.

$k(x)$: a safety function; usually kx where $k < 1$ is the safety factor for translating animal response to human.

Mort: the mortality rate given event A.

Value: the economic value of human life.

Even this seemingly simple case is fraught with both types of uncertainty; namely, g(x): reflects stochastic uncertainty both in times and space; however, a frequency-base may be available to estimate g(x), and a Bayesian approach can be used to account for small sample size.

Resp (x): in higher dose region, points may correspond to frequencies, but the low dose extrapolation is highly subjective.

The mortality rate can be frequency-based. In Equation 1, further imprecision is present: the safety function, the value of human life, and the number of people affected must often be determined subjectively. As a consequence of these imprecise elements, the "classical" risk formulation has two main types of deficiencies.

1. Forcing the use of classical probabilities where there is no frequency basis to estimate probabilities (say, in the low dose response region).

2. Using a single index to account for other uncertain elements, such as a safety factor.

These deficiencies have resulted in the fact that the calculated risk cost according to Equation 1 is far from being reproducible and may not serve as a basis for realistic risk assessment or management. In the simple example of Expression 1, risk management corresponds to the following problem:

$$\text{Risk } (A,z) = \int_{d_o}^{+\infty} [\text{Resp}(k(x),z) \times g(x,z)] \, dx \times \text{Mort} \times \text{Value} \qquad (2)$$

where z is the symbol for various policies. One may then select an optimum policy by a benefit-cost type of analysis. Expression 2 is used in cases where a single value function such as the value of human life can be postulated (Graham and Vaupel, 1981; Sharefkin et al., 1985). Often there is no such single value function to express the overall consequence of exposure, in which case multicriterion decisionmaking (MCDM) may be necessary to seek trade-offs (Haimes and Stakhiv, 1986).

2. BACKGROUND

The following review will focus on the methodology available to cope with fuzzy uncertainty as defined in this paper. Fuzzy uncertainty can be treated mathematically using the notion of fuzziness and techniques of fuzzy se theory (Hipel, 1982). Fuzziness represents situations where membership in sets cannot be defined on a yes/no basis (Zadeh, 1965); in other words, the boundaries of sets are vague. The membership function is a central concept in fuzzy set theory. The value of the membership function of an element in an ordinary set is either 0 or 1; i.e., it is not or it is a member of the set. However, an element may be a member of a fuzzy set to some degree, so its membership value can be between 0 and 1.

Fuzzy set theory, which was initially considered by practitioners and engineers to be of academic interest only, is now being recognized as capable of expressing the imprecision or fuzzy nature of real-life conditions better than, say, ordinary or classical set theory (Negata, 1979; Etschmaier, 1980; IFAC, 1984). As a result, the fuzzy set approach pioneered by Zadeh (1975) has been widely applied to decisionmaking (Bellman and Zadeh, 1970; Watson et al., 1979; Dubois and Prade, 1980; Kacprzyik, 1983; Xu, 1987) and specifically multicriterion decisionmaking (Roy, 1978; Carlsson, 1982; Siskos, 1982; Korhonen et al., 1987; Duckstein and Korhonen, 1987). Within the framework of this paper, the relationship of fuzzy set theory to probability theory has special importance (Dubois and Prade, 1986; Kandel and Byatt, 1979, 1980; Nakamias, 1978; Rao and Rashed, 1981; Zimmerman, 1978, 1985). Also, the use of fuzzy sets in safety analysis can be found in Terano et al. (1984) and Karwowski and Mital (1986). Dong et al. (1983) reviewed random and fuzzy models in engineering with application in earthquake assessment. Areas of civil engineering have been identified where fuzzy sets are useful and matched the general theory of fuzzy sets with the needs of civil engineering applications (Brown et al., 1984; Chameau and Gunaratne, 1984; Chameau et al., 1984; Gunaratne, 1984; Gunaratne et al., 1984; Yao and Brown, 1984). Fuzzy control of wastewater treatment processes is described in Tong et al. (1980). Hansel and Straube (1982) used fuzzy modeling to express soft information such as "high" or "low" water levels for hydrological forecasting. Alternative water quality control plans and reservoir operation have been studied in Duckstein et al. (1984) within a multicriterion framework. The decisionmaker's opinion about each of the criteria was encoded using degrees of fuzzy set membership. Chou (1987), Bardossy et al. (1987a, 1988), and Korhonen et al. (1987) present civil and industrial engineering applications of fuzzy sets.

Regional systems have been studied where environmental objectives such as flow, level, and quality of groundwater and surface water were expressed by fuzzy membership functions, and the best compromise policy was sought as a trade-off between non-fuzzy economic objectives and the fuzzy environmental objectives (Nachtnebel et al., 1982, 1986; Bogardi et al., 1983; Bardossy et al., 1984). However, the applications of fuzzy set theory to real-life problems did not entirely convince practitioners, partly because of the lack of guidelines for application. In addition, fuzzy set theory has often been offered as an alternative to statistical techniques, whereas it may be advantageous to combine fuzzy and statistical methods, which is endeavored in the present paper.

3.   FORMULATION OF RISK UNDER IMPRECISION AND UNCERTAINTY

Classical risk analysis (Equation 1) will be extended according to two principles:

1.   Use probabilities whenever a frequency-based estimation can be made, or alternatively, when subjective probabilities can be assessed.

2.   Use fuzzy sets to account for imprecision in nonprobabilistic cases; also when considering elements such as safety factor or the number of people affected.

These principles mean that the risk will be estimated as a fuzzy number (or qualitative measure) as defined in Appendix 1 using function (1) or fuzzy probabilities (possibilities) and fuzzy numbers:

$$\widehat{Risk} = \int_{d_0}^{+\infty} [\widehat{Resp}(\hat{k}(x)) \times \hat{g}(x)]\ dx \times Mort \qquad (3)$$

where $\hat{}$ means fuzzy number, fuzzy function, or fuzzy probability. Thus the response function values, $P(A|x)$, will be defined as fuzzy probabilities characterized by membership functions such as the one illustrated in Figure 1. Clearly, in the frequency domain the membership function may represent a nonfuzzy (crisp) probability while low dose extrapolation would involve wider membership functions--that is, higher imprecision.

The pdf of exposure, $\hat{g}(x)$, may be considered as a fuzzy probability function since a frequency basis to estimate the parameters of $\hat{g}(x)$ may not always be available, or imprecise measurements may necessitate the use of fuzzy geostatistics (Bardossy et al., 1987). The safety function $\hat{k}(x)$ is also represented by a membership function, reflecting the inherent imprecision of estimation. After elements of Equation 3 have been defined, fuzzy set mathematics, specifically the extension principle (Appendix 1), can be used to calculate the risk as a fuzzy number (Figure 2).

The following questions arise. Is this imprecise risk formulation more realistic than the classical, entirely probabilistic approach? How can one interpret the risk as a fuzzy number represented by its membership function? The problem is to compare fuzzy numbers--that is, ranking fuzzy sets. In fact, there are methods to rank fuzzy numbers (Dubois and Prade, 1980; Chen, 1985) and we will use such methods for ranking risk under uncertainty corresponding to various management options.

4.   RISK MANAGEMENT UNDER UNCERTAINTY

Risk estimates under stochastic and fuzzy uncertainty are traded off with economic performance indices, such as cost, by using MCDM techniques. The situation for nonfuzzy estimates of risk and cost is illustrated in Figure 3, where each point represents an alternative strategy and the existing situation is also assessed. Figure 3 represents a case where the existing risk of groundwater contamination is high but the related costs are relatively low. Alternatives 1 and 2 provide clear improvement since both the related risk and cost are less. In practice, such situations can be encountered; for example, better timing of fertilizer applications may not only reduce risk of nitrate contamination but also can reduce the amount of fertilizer needed. The other strategies identified in Figure 3 offer reduced risk only at a higher cost. Certainly one would seek the ideal solution (no risk, no cost), but the ideal point simply does not exist; thus a trade-off analysis is warranted.

The proposed methodology will be outlined for a distance-based trade-off technique (Goicoechea et al., 1982) which seeks to find the

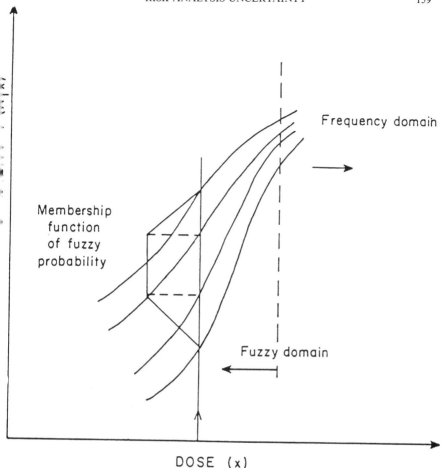

DOSE (x)

<u>Figure 1</u>. Low Dose Extrapolation Using Fuzzy Probability Concept

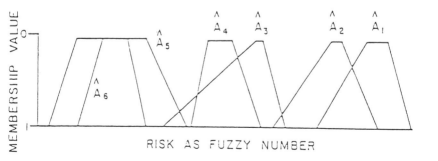

RISK AS FUZZY NUMBER

<u>Figure 2</u>. Fuzzy Numbers Ordered by Ranking Method of Dubois and Prade
(1980) as $\hat{A}_1 > \hat{A}_2 > \hat{A}_3 \approx \hat{A}_4 > \hat{A}_5 \approx \hat{A}_6$

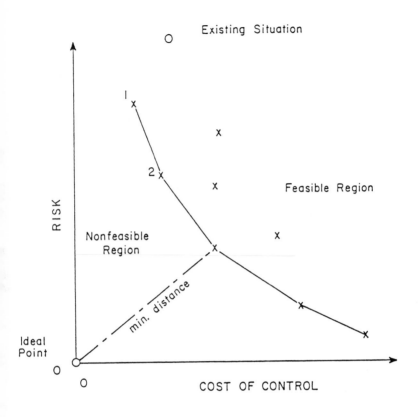

<u>Figure 3</u>. Elements of the Risk Management Without Considering Fuzzy
          Uncertainty

control strategy with minimum distance from an ideal point. In order to illustrate the general approach, an aquifer is considered with two water-supply systems. Control options $j = 1,\ldots,J$ incur different and imprecise costs considered as fuzzy numbers: $\hat{C}_{1j}$, $\hat{C}_{2j}$ for the two water supply systems, respectively.

On the other hand the health risk under uncertainty (Equation 3) can also be calculated for each control option: $\hat{R}_{1j}$, $\hat{R}_{2j}$. In such a situation, a two-level trade-off analysis may be necessary: in the first level between cost estimates and between risk estimates considering the two systems and resulting in a composite cost and composite risk; then in the second level, between composite cost and composite cost. Such a two-level trade-off analysis can be performed by composite programming (Bardossy et al., 1985). For the present purpose, composite programming is extended to the case of fuzzy criteria (risks and costs) using the following steps:

1. Costs $\hat{C}_j$ and risks $\hat{R}_j$ are normalized:

$$\hat{r}_j = \frac{R_{min} - \hat{R}_j}{R_{min} - R_{max}}$$

$$\tag{4}$$

$$\hat{c}_j = \frac{C_{min} - \hat{C}_j}{C_{min} - C_{max}}$$

where max and min represent the greatest and smallest values (risk and cost) for the control options (j) considered. The result of step 1 is the membership functions of normalized costs and risks for the two water-supply systems and the control options.

2. First-level trade-off analysis is performed and composite risk $\hat{cr}_j$ and $\hat{cc}_j$ are calculated for each control option $j = 1,\ldots,J$.

$$\hat{cr}_j = [\alpha_1(\hat{r}_{1j})^{p_r} + \alpha_2(\hat{r}_{2j})^{p_r}]^{1/p_r}$$

$$\hat{cc}_j = [\beta_1(\hat{c}_{1j})^{p_c} + \beta_2(\hat{c}_{2j})^{p_c}]^{1/p_c}$$

$$\tag{5}$$

where α and β are the weights expressing the relative importance between the two water-supply systems; p_r and p_c and the so-called balancing factors expressing the degree of substitution for risk and cost. If this balancing factor is great, the corresponding criteria may not compensate each other. Health risk may be such a criterion. If p is smaller, the degree of substitution is greater. For example, a lower cost at one place may compensate for a higher cost in another place.

The result of step 2 provides the membership functions for composite risk and costs.

3. Finally, the second level trade-off analysis is performed between composite risk and composite cost, leading to an overall indicator of risk management $\hat{G}_j$:

$$\hat{G}_j = (\gamma_1 \hat{c}r_j^2 + \gamma_2 \hat{c}c_j^2)^{1/2} \tag{6}$$

where γ_1 and γ_2 are the weights representing the relative importance between health risk and cost. The result of the second-level trade-off analysis is represented by the fuzzy numbers $\hat{G}_j$ given by the respective membership functions. Then a ranking among $\hat{G}_j$ would lead to a preferred risk management option.

Before an illustrative example is presented, three further remarks are necessary:

(a) The extension principle (Appendix A) is used to calculate the functions specified in the above three steps.

(b) The proposed approach is applicable even if the number of the considered water-supply systems or the number of trade-off analysis levels is greater than two.

(c) If the number of control options is too great, or the control options are continuous, mathematical programming methods should be used to select the best option. For the nonfuzzy case such methods have been used (Bogardi et al., 1984), but further research is needed in the present fuzzy case.

5. NUMERICAL EXAMPLE

A numerical example is presented to illustrate the main steps of the risk management methodology. The following table presents parameters of the membership functions for normalized risks $\hat{r}_j$ and costs $\hat{c}_j$ pertaining to two water supply systems and J = 6 control options. Triangular symmetric membership functions are considered in this example, where (a) is the value pertaining to the upper edge, $\mu = 1$ of the membership function and (b) is the half-width of the membership function.

Control Option	Normalized Risk				Normalized Cost			
	$\hat{r}_1$		$\hat{r}_2$		$\hat{c}_1$		$\hat{c}_2$	
j	a	b	a	b	a	b	a	b
1	0.70	0.07	0.4	0.08	0.60	0.05	0.80	0.07
2	0.60	0.10	0.48	0.05	0.40	0.07	0.73	0.09
3	0.90	0.06	0.3	0.06	0.35	0.03	0.62	0.13
4	0.53	0.05	0.55	0.04	0.52	0.07	0.93	0.09
5	0.13	0.05	0.85	0.11	0.72	0.02	0.88	0.04
6	0.73	0.07	0.85	0.12	0.32	0.04	0.48	0.05

The first-level trade-off analysis considers the two water supply systems for which risk and cost are assumed to be of equal importance from the aspects of both risk and cost. Consequently, $\alpha_1 = \alpha_2 = 0.5$; $\beta_1 = \beta_2 = 0.5$ in Equation 5. The balancing factor is $p_r = 9$, indicating that health risks can hardly be compensated with each other. On the other hand, $p_c = 1$; that is, the cost of control has a high degree of substitution. Using these parameters and (5) with the extension principle results in Figures 4 and 5. Figure 4 shows that the best risk reduction option is No. 6, while the worst is No. 5. Concerning costs (Figure 5), the best option is No. 5 and the worst is No. 6. The conflict situation is evident.

The second level trade-off analysis (Figure 6) between composite risk and cost assumes that health risk reduction and costs are of equal importance: $\gamma_1 = \gamma_2 = 0.5$ in the example. Note the difference between risk management without considering fuzzy uncertainty (Figure 3) and in its presence (Figure 6). The actual membership functions for the risk management indicators $\hat{G}_j$ are illustrated in Figure 7. Using the method of Dubois and Prade (1980) for ranking the fuzzy numbers $\hat{G}_j$, the following relations can be obtained:

$$\hat{G}_4 > \hat{G}_1 > \hat{G}_2 \approx \hat{G}_6 > \hat{G}_3 \approx \hat{G}_5$$

Thus the preferred risk management option, No. 4, provides a best compromise between health risk reduction and cost.

6. DISCUSSION AND CONCLUSIONS

Recent catastrophes (Chernobyl, Challenger, Rhine River pollutant, airplane crashes and near misses) have resulted in renewed public attention to safety issues. It seems that in these cases the classical approach of only considering probabilistic assessment of the safety was not really adequate: thus, the estimated probability of failure of a booster rocket varied from 0.015 to 0.00001. Similarly the probability of the failure of some structures (hazardous waste containment, for example) estimated by experts differs significantly from statistical data. These examples point out that there is a need to broaden our current concepts for assessing environmental risk. A possible way to incorporate our knowledge into mathematical models is the use of methods based on fuzzy sets. Recent studies (computers: Schmucker, 1984; existing structures: Hinkle, 1985) show that there may be a place for such techniques in the area of environmental risk.

Understanding the assessment of fuzzy set membership values and functions is important for applying models based on fuzzy set theory. Here two approaches have been taken: objective methods, usually based on the probabilistic nature of the defining characteristics of the set (Civanlar and Trussel, 1986; Dubois and Prade, 1986; Chameau and Santamarina, 1987) and subjective judgment methods based on psychophysical methods (Wallstein and Buddescu, 1983; Wallstein et al., 1985) or on having judges choose the parameters of suitable functions (Dickinson and Ferrell, 1985). The type of approach to be used depends on the type of fuzziness involved, e.g., the vagueness of the meaning of

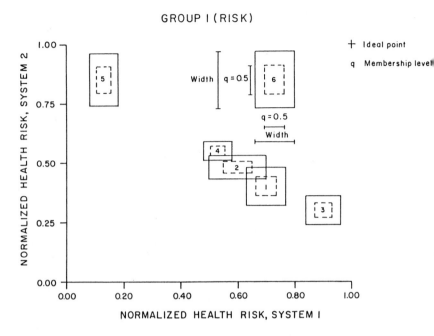

Figure 4. Results of First-Level Fuzzy Trade-Off Analysis: Risk

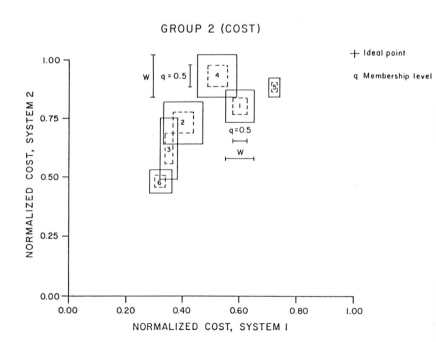

Figure 5. Result of First-Level Fuzzy Trade-Off Analysis: Costs

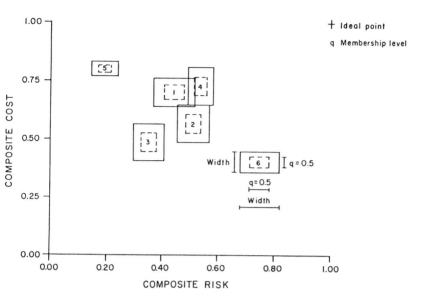

Figure 6. Results of Second-Level Fuzzy Trade-Off Analysis Between Composite Risk and Cost

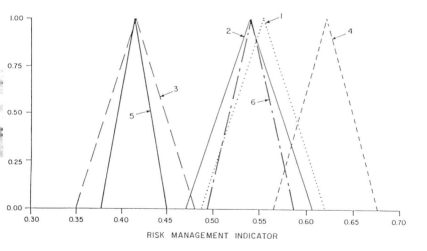

Figure 7. Risk Management Indicators Calculated as Fuzzy Numbers

a linguistic category such as "high risk" or the imprecision introduced in a variable due to use of an approximate model, and on the amount and type of information available about this vagueness.

The treatment of fuzzy numbers is different from the treatment of random variables. For instance, if one uses the extension principle for defining addition (Appendix A) then one can see that the fuzzy mean is not additive and is different from the mean of a random variable. Operations with fuzzy numbers are also different from operations with random variables; generally they are simpler. For example, the sum of two independent random variables requires the computation of a convolution integral, while the sum of fuzzy numbers involves only simple interval operations (Appendix A).

Results presented in this paper lead to the following conclusions.

1. Techniques of environmental risk analysis can be extended by a combined probabilistic and fuzzy set approach to model both stochastic and fuzzy uncertainty.

2. Fuzzy uncertainty or imprecision is the consequence of imperfect knowledge and information such as low dose extrapolation in dose-response relationships, and it can be encoded using fuzzy set theory.

3. The methodology can be specified for a system of a hazardous waste facility and an aquifer where a possible failure of the facility may cause contamination and lead to risk to public health.

4. Risk is interpreted as a fuzzy number and various ranking methods can be used to order such risk estimates.

5. Risk management is performed as a multicriterion decisionmaking (MCDM) analysis between risk estimates under uncertainty and economic indices such as the cost of control.

6. The MCDM technique composite programming can be extended to the case of fuzzy criteria.

REFERENCES

Bardossy, A., I. Bogardi, L. Duckstein, and H. P. Nachtnebel. 1984.
Fuzzy decisionmaking in regional water management. Session 11.8/A,
IFAC Ninth World Congress, Budapest, Hungary, July 1984.

Bardossy, A., I. Bogardi, and L. Duckstein. 1985. Composite programming
as an extention of compromise programming. In P. Serafini (ed.),
Mathematics of Multiple-Objective Optimalization. CISM, Udine,
Italy. Vienna: Springer-Verlag.

Bardossy, A., I. Bogardi, and W. E. Kelly. 1987a. Imprecise (fuzzy)
information in geostatistics. Invited paper, MGUS 1987 Conference
on New Advances in Geostatistics. Redwood City, California.

Bardossy, A., I. Bogardi, and W. E. Kelly. 1987b. Fuzzy regression for
electrical conductivity relationships. In Proceedings of the 1987
Annual Meeting, North American Fuzzy Information Processing
Society, Purdue University, Lafayette, Indiana.

Bardossy, A., I. Bogardi, L. Duckstein, and P. Nachnebel. 1988. Fuzzy
decisionmaking to regional conflicts between industry and the
environment. In C. W. Evans, W. Karwowsky, and R. M. Wilhelm
(eds.), Fuzzy Methodologies for Industrial and Systems Engineering.
Amsterdam: Elsevier. In press.

Bellman, R., and L. Zadeh. 1970. Decisionmaking in a fuzzy
environment. Management Science 17(4):141-164.

Benjamin, J. R., and C. A. Cornell. 1970. Probability, Statistics and
Decision for Civil Engineers. New York: McGraw-Hill.

Berger, J. O. 1985. Statistical Decision Theory and Bayesian Analysis.
2nd edition. New York: Springer-Verlag.

Bernier, J. 1987. Elements of Bayesian analysis of uncertainty in
hydrological reliability and risk models. In L. Duckstein and E.
J. Plate (eds.), Engineering Reliability and Risk in Water
Resources. Dordrecht, The Netherlands: Martinus Nijhoff.

Bogardi, I., L. Duckstein, and F. Szidarovszky. 1982. Bayesian
analysis of underground flooding. Water Resources Research
18(4):110-111.

Bogardi, I., L. Duckstein, and A. Bardossy. 1983. Regional management
of an aquifer under fuzzy environmental objectives Water Resources
Research 19(6):1394-1402.

Bogardi, I., L. Duckstein, and A. Bardossy. 1984. Trade-off between
cost and efficiency of pollution control. In Proceedings of the
Sixth International Conference on Multiple-Criteria Decisionmaking,
Cleveland, Ohio, June 1984.

Brown, C. B., H. Furuta, N. Shiraishi, and J. T. P. Yao. 1984. Civil engineering applications of fuzzy sets. In Proceedings of the First International Conference on Fuzzy Information Processing, Kauai, Hawaii, July 1984.

Carlsson, C. 1982. Tackling an MCDM problem with the help of some results from fuzzy set theory. European Journal of Operational Research 10:270-281.

Chameau, J. L., and M. Gunaratne. 1984. Performance evaluation in geotechnical engineering using fuzzy sets. Proceedings of the ASCE Specialty Conference on Probabilistic Mechanics and Structural Reliability, Berkeley, California.

Chameau, J. L., M. Gunaratne, and A. G. Altschaeffl. 1984. Type 2 fuzzy sets in engineering. In Proceedings of the First International Conference on Fuzzy Information Processing, Kauai, Hawaii, July 1984.

Chameau, F. L., and F. C. Santamarina. 1987. Methods for obtaining membership function. Working paper, Civil Engineering Department, Purdue University, Lafayette, Indiana.

Chen, S. H. 1985. Ranking fuzzy numbers with maximizing set and minimizing set. Fuzzy Sets and Systems 17:113-129.

Chou, K. C. 1987. Reliability of existing framed structures using fuzzy sets. In Proceedings of the Annual Meeting, North American Fuzzy Information Processing Society, Purdue University, Lafayette, Indiana.

Civanlar, M. R., and H. F. Trussel. 1986. Constructing membership functions using statistical data. Fuzzy Sets and Systems 18:1-13.

Davis, D. R., C. C. Kisiel, and L. Duckstein. 1972. Bayesian decision theory applied to design in hydrology. Water Resources Research 8(1):33-42.

DeGroot, M. A. 1970. Optimal Statistical Decision. New York: McGraw-Hill.

Dickinson, D., and W. R. Ferrell. 1985. Fuzzy set knowledge representation in a system to recommend management decisions. In L. Methlie (ed.), Proceedings of the IFIP Working Conference on Knowledge Representation for Decision Support Systems. Amsterdam: North-Holland.

Dong, W. M, A. C. Boissonade, and H. C. Shah. 1983. Review of models for processing information. In Proceedings of Applications of Statistics and Probability in Soil and Structural Engineering, Universita di Firenze, Florence, Italy.

Dubois, D., and H. Prade. 1980. Fuzzy Sets and Systems: Theory and Applications. New York: Academic Press.

Dubois, D., and H. Prade. 1986. Fuzzy sets and statistical data. European Journal of Operational Research 25:345-356.

Duckstein, L., J. Kempf, and J. Casti. 1984. Design and management of regional systems by fuzzy ratings and polyhedral dynamics. In P. Nijkamp, M. Despontin, and I. Spronk (eds.), Macro Economic Planning with Conflict Goals. Springer-Verlag.

Duckstein, L., and P. Korhonen. 1987. A visual and interactive fuzzy multicriterion approach to forest management. Working paper F-176, Helsinki School of Economics, Helsinki, Finland.

Etschmaier, M. M. 1980. Fuzzy controls for maintenance scheduling in transportation systems. Automatica 16:255-263.

Goicoechea, A., D. R. Hansen, and L. Duckstein. 1982. Multi-objective Decision Analysis with Engineering and Business Applications. New York: John Wiley & Sons.

Graham, J. D., and J. W. Vaupel. 1981. Value of a life: what difference does it make? Risk Analysis 1(1):89-95.

Gunaratne, M. 1984. The use of fuzzy sets mathematics in pavement evaluation and management. Report no. FHWZ/IN/JHRP-84/18, Purdue University, Lafayette, Indiana.

Gunaratne, M., J. L. Chameau, and A. G. Altschaeffl. 1984. An introduction to fuzzy sets in pavement evaluation. Presented at the Transportation Research Board Annual Meeting, Washington, D.C.

Haimes, Y. Y., and E. Z. Stakhiv (eds.). 1986. Risk-based Decisionmaking in Water Resources. New York: ASCE.

Hansel, N., and B. Straube. 1982. Fuzzy modelling for the forecast of the discharge and water level of large rivers. In Proceedings of IAHS Symposium, Exeter, U.K.

Hinkle, A. J. 1985. Linguistic assessment of fatigue damage in butt welds. In C. B. Brown, J. L. Chameua, R. Palmer, and J. T. O. Yao (eds.), Proceedings of NSF Workshop on Civil Engineering Applications of Fuzzy Sets.

Hipel, K. W. 1982. Fuzzy set techniques in decisionmaking. IFAC, Theory and Application on Digital Control. Pergamon Press.

IFAC. 1984. Imprecision in water resources. Session 11.8/A. In Proceedings of IFAC Ninth World Congress, Budapest, Hungary, July 1984.

Kacprzyik, J. 1983. Multi-stage Decisionmaking Under Fuzziness. Verlag TUV Rheinland.

Kandel, A., and W. J. Byatt. 1979. Fuzzy process. Fuzzy Sets and Systems 4:117-152.

Kandel, A., and W. J. Byatt. 1980. Fuzzy sets, fuzzy algebra and fuzzy statistics. Proceedings, IEEE SMC 66(12):1616-1639.

Karwowski, W., and A. Mital. 1986. Potential applications of fuzzy sets in industrial safety engineering. Fuzzy Sets and Systems 19:105-120.

Kaufmann, A., and M. M. Gupta. 1985. Introduction to Fuzzy Arithmetic: Theory and Applications. New York: Van Nostrand-Reinhold.

Korhonen, P., J. Wallenuis, and L. Duckstein. 1987. Multiple objective linear programming over a fuzzy feasible set. In C. W. Evans, W. Karwowski, and R. M. Wilhelm (eds.), Fuzzy Methodologies for Industrial Systems Engineering. Amsterdam: Elsevier. In press.

Nachtnebel, H. P., P. Hanish, and L. Duckstein. 1986. Multicriterion analysis of small hydropower plants under fuzzy objectives. Annals of Regional Science 86-11.

Nachtnebel, H. P., L. Duckstein, and I. Bogardi. 1982. Evaluation of conflicting regional water requirements: an Austrian case study. In Proceedings of International Conference of the IAHS, Exeter, U.K., July 1982.

Nakamias, S. 1978. Fuzzy variables. Fuzzy Sets and Systems 1:97-110.

Negata, C. V. 1979. On fuzzy systems. In Gupta, Ragade, and Yager (eds.), Advances in Fuzzy Set Theory and Applications, Amsterdam: North-Holland.

Rao, M. B., and A. Rashed. 1981. Some comments on fuzzy variables. Fuzzy Sets and Systems 6:285-292.

Roy, B. 1978. ELECTRE III, un algorithme de classements fonde sur une representation floue des preferences en presence de criteres multiples. Cahiers du Centre d'Etudes de Recherche Operationnelle 20:3-24.

Schmucker, K. J. 1984. Fuzzy sets, natural language computation, and risk analysis. Computer Science Press.

Sharefkin, J., M. Shechter, and A. Kneese. 1985. Impacts, costs and techniques for mitigation of contaminated groundwater. Water Resources Research 20(12):1771-1783.

Siskos, J. 1982. A way to deal with fuzzy preferences in multicriteria decision problems. European Journal of Operational Research 10:314-324.

Terano, T., S. Masui, Y. Muirayama, N. Aida, and N. Akiyama. 1984. Fuzzy fault tree and its application to petro-chemical plant accidents. Session 01.3/C, IFAC Ninth World Congress, Budapest, Hungary, July 1984.

Tong, R. M., M. B. Beck, and A. Lattens. 1980. Fuzzy control of the activated sludge waste-water treatment process. _Automatica_ 16:659-701.

Wallstein, T. S., and D. V. Buddescu. 1983. Encoding subjective probabilities: a psychological and psychometric review. _Management Science_ 29:151-173.

Watson, S. R., J. J. Weiss, and M. L. Donnell. 1979. Fuzzy decision analysis. _IEEE Transactions on Systems, Man, and Cybernetics_ 9(1).

Xu, Li D. 1987. Fuzzy multicriteria programming in economic systems analysis. In Proceedings of Annual Meeting, North American Fuzzy Information Processing Society, Purdue University, Lafayette, Indiana.

Yao, J. T. P., and C. B. Brown. 1984. Civil engineering applications of the theory of structural reliability. In D. Faulkner, M. Shinozuka, R. Fiebrandt, and I. Franck (eds.), _The Role of Design Inspection and Redundancy in Marine Structural Reliability_. Washington, D.C.: National Research Council, Committee on Marine Structures.

Zadeh, L. A. 1965. Fuzzy sets. _Information Control_ 8:338-353.

Zadeh, L. A. 1975. The concept of a linguistic variable and its application to approximate reasoning (part 3). _Information Science_ 9:43-80.

Zimmerman, H. J. 1978. Fuzzy programming and LP with several objective functions. _Fuzzy Sets and Systems_ 1:45-55.

Zimmerman, H. J. 1985. _Fuzzy Set Theory and its Applications_. Dordrecht, The Netherlands: Martinus Nijhoff.

APPENDIX A: FUZZY SETS, FUZZY NUMBERS, THE EXTENSION PRINCIPLE

This Appendix summarizes the basic definitions of fuzzy sets, fuzzy numbers, and fuzzy operations. Some of the definitions are not given in their most general forms but for convenience are restricted to the form used in this paper.

Definition: Let x be a set (universe). A is called a fuzzy subset of x if A is a set of ordered pairs:

$$A = \{(x, \mu(x)), x \in X\}$$

where $\mu_A(x)$ is the grade of membership of x in A. $\mu_A(x)$ takes its values in the interval $[0,1]$. The closer $\mu_A(x)$ is to 1 the more x belongs to A; the closer it is to 0 the less it belongs to A. If $[0,1]$ is replaced by $\{0,1\}$, then A can be regarded as a subset of X.

Definition: The q level set of the fuzzy subset A is the set of those elements which have at least q membership:

$$A_q = \{x; \mu_A(x) \geq q\}$$

Definition: A fuzzy subset M is called normal if there is at least one z such that $\mu_M(z) = 1$

Definition: A fuzzy subset M of the set of real numbers is called convex if for each $0 < \alpha < 1$ and each $x, y \in M$

$$\mu_M(\alpha x + (1 - \alpha)y) \geq \min(\mu_M(x), \mu_M(y)) \qquad (7)$$

Definition: A fuzzy subset M is called a fuzzy number if M is a normal convex fuzzy subset of the set of real numbers. The membership value of a real number reflects the "likeliness" of the occurrence of that number; the level sets (intervals in this case) reflect different sets of numbers with a given minimum likeliness (Kaufmann and Gupta, 1985). Any real number can be regarded as a fuzzy number, and is often called a crisp number in fuzzy mathematics.

Definition: Let $X_1, \ldots, X_n$ be sets and $A_1, \ldots, A_n$ be fuzzy subsets of $X_1, \ldots, X_n$ respectively. The Cartesian product of $A_1, \ldots, A_n$, that is $A_1 \times \ldots \times A_n$, is the fuzzy subset of $X_1 \times \ldots \times X_n$ with the membership function

$$\mu(x_1, \ldots, x_n) = \min\{\mu_{Ai}(x_i); i = 1, \ldots, n\}$$

The extension principle (Zadeh, 1965) is a method of extending point-to-point operations to fuzzy sets. It is the basic tool for the development of fuzzy arithmetic.

Definition: Let X and Y be two sets. Let f be a point-to-point mapping from X to Y

$$f: X \rightarrow Y \text{ for all } x \ X \in f(x) = y \in Y)$$

f can be extended to operate on fuzzy subsets of X in the following way:

Let A be a fuzzy subset of X with membership function μ_A, then

$$f(A) = B$$

where B is a fuzzy subset of Y with the membership function

$$\mu_B(y) = \sup\{\mu_A(x)\}$$

$$x \in X$$

$$f(x) = y \tag{8}$$

$\mu_B(y) = 0$ if there is no $x \in X$ such that $f(x) = y$

Applying this principle to functions like $f(x,y) = x + y$ or $f(x,y) = x.y$ we can define the basic operations on fuzzy sets.

The level set representation of fuzzy numbers can be used to describe the operations with fuzzy numbers. The application of the extension principle to addition, in the level set representation, can be written as (for detailed proof see Zimmermann [1985]):

$$\hat{A} + \hat{B} = \hat{C}$$

$$\hat{A}_q + \hat{B}_q = \hat{C}_q \qquad q \in (0,1] \tag{9}$$

where the sum of two intervals is:

$$[a,b] + [c,d] = [a + c, b + d].$$

A fuzzy number is multiplied by a real number as follows:

$$\hat{B} = \lambda \hat{A}$$

$\hat{B}_q = [c_q, d_q]$ and $\hat{A}_q = [a_q, b_q]$ then
if $\lambda > 0$ then $c_q = \lambda a_q$ and $d_q = \lambda b_q$
if $\lambda < 0$ then $c_q = \lambda b_q$ and $d_q = \lambda a_q$ \hfill (10)

Note that equations (9) and (10) are not definitions; they are derived from the extension principle and presented here to simplify later computations.

Definition: The mean of a fuzzy number is defined as:

$$M(\hat{A}) = \frac{\int_R x\mu_{\hat{A}}(x)dx}{\int_R \mu_{\hat{A}}(x)dx} \tag{11}$$

Definition: The width of a fuzzy number is defined as:

$$W(\hat{A}) = \sup\{x; \mu_A(x) > 0\} - \inf\{x; \mu_A(x) > 0\} \qquad (12)$$

One can see that if a fuzzy number needs to be discretized it is useful to discretize a fuzzy number with respect to its membership values. Further details and properties of fuzzy numbers can be found in Dubois and Prade [1980].

THE ECONOMICS OF WASTE MANAGEMENT IN AGRICULTURAL PRODUCTION SYSTEMS

Erik Lichtenberg and David Zilberman

ABSTRACT

Agricultural production is a growing cause for concern as a generator of health hazards via contamination of surface and ground waters by agricultural chemicals. Methodologies for formulating appropriate policy responses must take into account the key features of agricultural production, especially (1) heterogeneity among farms and farmers, (2) multiple-policy objectives and policy instruments, (3) uncertainty about estimates of risk assessment parameters, and (4) dynamic factors affecting both environmental hazards and the agricultural economy. We present models that incorporate one or more of the first three of these key features and discuss empirical applications demonstrating their importance in policy determination. We show, in particular, that analyses that fail to take them into account produce policy recommendations that are seriously flawed. We end with a discussion of dynamic factors likely to be important, with an illustration from agricultural drainage problems in the San Joaquin Valley, California.

1. INTRODUCTION

Agricultural production systems are major causes of environmental health hazards throughout the United States via contamination of surface or ground waters by agricultural chemicals (pesticides, fertilizers) in run-off and drainage. These agriculture-generated hazards are increasingly a cause of concern; consequently, regulatory efforts aimed at controlling agricultural water pollution are growing, as illustrated by the recent crisis resulting from the disposal of agricultural drainage water into the Kesterson Reservoir in California and by the recent promulgation of regulations on chemigation issued by the Environmental Protection Agency. These attempts to come to grips with agricultural pollution problems have also brought to the fore a lack of methodologies adequate for formulating and analyzing policies for controlling environmental health by-products of agricultural production. Thus, government agencies have actively sought the assistance of the scientific sector in developing methodologies and planning and policy-making decision procedures.

This paper (1) presents what we believe to be the key features of decision methodologies appropriate for these problems, (2) derives specification of models that incorporate these features (assuming static conditions and certainty), and (3) discusses applications of these

Erick Lichtenberg is on the faculty of the Department of Agricultural and Resources Economics, The University of Maryland, Beltsville, Maryland.

David Zilberman is a professor in the Department of Agricultural and Resource Economics at the University of California, Berkeley.

models to agricultural production in the United States. We then discuss incorporation of uncertainty considerations and finish with a brief discussion of extensions to dynamic situations.

2. KEY FEATURES

A comprehensive framework for control of agriculturally generated environmental health hazards must incorporate results and methodologies from many disciplines, including agronomy, life sciences, engineering, and behavioral sciences. The following are features we consider essential for economic modeling and analysis:

Bottom-up Derivation of Industry Responses to Regulation-Recognizing Heterogeneity among Farms and Prevalence of Technological Change in Agriculture. An understanding of the impacts of pollution control regulation on industry output levels, input uses, and pollutant generation is vital for policy analysis. The specific characteristics of the industry must be incorporated in modeling these response functions.

The literature on agricultural policy has identified several key characteristics of U.S. agriculture (Brandow, 1977; Schultz, 1965; Cochrane, 1958), of which the most important are that U.S. agricultural industries tend to (1) have competitive structure (many price- and policy-making units); (2) be heterogeneous (production units varying in size, environmental, technology, and human capital); and (3) go through frequent technological changes. Heterogeneity among farms is substantial with respect to production parameters and may be even greater when environmental considerations are introduced. Differences, say, in location and soil quality, may substantially affect water contamination by farms which are otherwise very similar. In modeling industry-wide response functions, we propose to start by constructing a decision model for a price- and policy-making competitive farm and to obtain decision rules (output, input use, and pollution) as functions of policy parameters and key variables that distinguish farms. The decision rules will be used to obtain the specific decisions for each type of farm in the industry, and the industry-wide response function will be obtained by aggregation over farms.

Many agricultural technologies are characterized by technological rigidities, expressed by relatively constant input/output ratios (Cochrane, 1958; the "asset fixity hypothesis" of Johnson, 1958). In these cases, use of programming models and the techniques developed by Johansen (1972) and Houthakker (1955-1956) is consistent with our approach to aggregation. But even in these cases, one must incorporate new technological options that new policies may render economical. We believe that while for some inputs production coefficients are quite rigid, the use of inputs such as water and fertilizers is responsive to price changes. Microlevel modeling is therefore non-trivial, and obtaining aggregation relationships may require ingenuity and effort.

It may also be necessary to incorporate other characteristics. When the polluting agricultural industry faces downward-sloping demand, it becomes necessary to model both demand and supply relations. The supply function can be constructed using the aggregation approach recommended here. The analysis in this case will predict the impact of environmental regulation on production, input uses, pollution levels, and prices. Note that in many cases, environmental regulations affect local agricultural industries, which are small relative to the national industry and which are pricetakers. Another consideration is government programs, which for many commodities set prices. Modeling at both the micro- and macrolevel should incorporate government program parameters and rules. When program participation is voluntary, the analysis of alternative antipollution policies should consider their impacts on participation as well. Finally, the behavior of farmers is affected substantially by financial considerations, and credit availability is often restricted. Credit and debt considerations may prevent adoption of new technologies and hence substantially affect the outcomes of environmental regulations.

Multiple Objectives and Multiple Policy Tools. Economic analysis of environmental policies traditionally involves efficiency grounds. For example, policies might be ranked according to benefit-cost ratios or total net benefits (the sum of producer, consumer, and government surpluses). The economic literature on regulation (Peltzman, 1976) and social choice (Buchanan and Tullock, 1971) has pointed out that policy makers are interested in many other policy criteria as well. There is thus a need to identify qualitatively and quantitatively the gainers and losers from policies and the impacts of policies on inflation, unemployment, government expenditures, etc. Distributional effects are especially important, as are impacts on solvency, since farmers' decisions are often driven by a need to stave off bankruptcy.

A similar line of argument suggests that a wide variety of antipollution policies--taxes and standards alike--should be considered. Even though a specific antipollution tax may be the most efficient, other taxes, subsidies, or forms of direct regulation may be superior on other grounds. Economic analysis should produce a matrix comparing the performance of alternative antipollution policies in several dimensions.

Uncertainty. There is typically considerable uncertainty about the impacts of antipollution policies in agriculture. One type of uncertainty concerns the economic impacts of the policies. Even more substantial uncertainty exists regarding the environmental impacts of the policies. These environmental health uncertainties may arise from the randomness of nature; differences of weather and precipitation patterns drastically affect water quality and the impact of policies directed toward controlling water quality. They may also spring from gaps in scientific knowledge, which necessitate the use of environmental health risk assessment models built on estimates that are subject to substantial error.

Many policy makers are averse to uncertainty and have grave hesitations about accepting the recommendations of scientific studies because of poor statistical reliability. Policy models should, therefore, provide indicators of the likelihoods of outcomes of importance. When there is substantial randomness about natural conditions, outcomes of policies should be compared at different states of nature. Where there is substantial scientific uncertainty, results should be presented with explicit specification of their reliability.

Incorporation of Dynamic Considerations. Many environmental problems are results of dynamic processes. Accumulation of residual materials in the ecosystems may reach levels that trigger profound impacts on certain species, causing environmental crises. Salinization of soils and groundwater is a continuous process with important time dimensions. Dynamic considerations are also important in production and economics. The diffusion processes of new technology are lengthy; it may also take substantial time from the moment a technology is introduced until it makes a substantial impact. Time is also essential for gathering environmental information and for developing and refining new technologies for ameliorating environmental hazards. For this reason, environmental policy modeling should take a dynamic approach; specifically, policies should be designed to evolve over time. Processes of learning, adoption, diffusion, and research and development should be considered. Dynamic analysis should be able to indicate both what to do and also when to do it.

3. MODELING IMPACTS OF ANTIPOLLUTION REGULATION ON AGRICULTURE--STATIC MODELS ASSUMING FULL CERTAINTY

This section introduces methodologies that possess our two key features and applications of the models which illustrate the importance of the features. The first model is conceptual and based on Hochman and Zilberman (1978). It assumes that the industry consists of J types of homogeneous production units; $j = 1, \ldots, J$ is the unit indicator. A plot of fixed size, a cow, and a fishing pond are examples of such production units. A fixed-proportions technology is assumed; output, variable input use, and pollution per production unit of type j are given by y_j, x_j, and z_j, respectively. To present the production unit distribution, let n_j denote the number of units of type j. Assume that the industry is price-making and farmers maximize profit; let output price be denoted by p and variable input price by w. Let the quasi-rent (revenue minus cost except production unit rent) of production unit type j be denoted by q_j.

Before the environmental policies are introduced, the group of farms that operate is denoted by the set

$$R^o\{j: q_j = p \, y_j - wx_j \geq 0\}$$

The farms that belong to the group R^o have non-negative quasi-rent; the rest of the industry will not operate. Let $s = 0, \ldots, S$ be a policy indicator; $s = 0$ is associated with no antipollution regulation. The group of farms that operate after policy s is introduced is denoted by R^s and includes all the farms which have non-negative quasi-rents with policy s. For example, if policy 1 is an antipollution tax denoted by V, the group of farms operating with this policy is denoted by

$$R' = \{j: q_j = p \, y_j - wx_j - vz_j \geq 0\}$$

The industry's supply, variable input demand, and pollution are

$$Y^S = \sum_{j \varepsilon R^S} y_j n_j \tag{1}$$

$$X^S = \sum_{j \varepsilon R^S} x_j n_j \tag{2}$$

$$Z^S = \sum_{j \varepsilon R^S} z_j n_j \tag{3}$$

Several other aggregate variables are important in policy analysis. Two are straightforward. They are

$$Q^S = \sum_{j \varepsilon R^S} q_j n_j \tag{4}$$

aggregate quasi-rent of the regulated industry and

$$N^S = \sum_{j \varepsilon R^S} n_j \tag{5}$$

the aggregate utilization of the industry's fixed asset as a result of policy s. Policy makers are also interested in the impacts of policies on the solvency of farms. Extra costs associated with compliance to environmental regulation may disrupt farmer's ability to service debt and hence cause bankruptcies. To measure this impact, another dimension of heterogeneity between production units is added. Let m_j denote the debt payment owed by the owner of production unit type j. To continue to hold a production unit, the quasi-rent should exceed the debt payment required ($q_j^S \geq m_j$). The aggregate measure of insolvency is the number of production units at which quasi-rent cannot cover this debt, and it is denoted by

$$D^S = \sum_{j \varepsilon B^S} n_j \tag{6}$$

$$B^S = \{j : m_j > q_j^S\} \tag{7}$$

Hochman and Zilberman (1978) used this conceptual framework to compare analytically the outcome of a pollution tax (s = 1) with a

pollution standard that sets an upper bound S on pollution per
production unit. For the standard s = 2 and the group of operating
farms in

$$D^2 = \{j: \ py_j - wx_j \geq 0, \ z_j \leq \overline{Z}\}$$

comparing a tax and a standard when both policies are designed to
restrict aggregate pollution to a target level $\overline{Z}$, Hochman and Zilberman
(1978) have shown that the tax attains the environment target level at
less cost (when cost is measured by value added last or net welfare
surplus, if demand is inelastic). However, the standard may be
preferable when other criteria are used. The loss of output and
increase in output price (when demand is inelastic) associated with the
use of standards are lower; the reduction in producers' profit is also
lower when standards are used. The distributional effects of the policy
within the industry vary. Standards may force very efficient but very
polluting production units to cease operation. These farms may continue
operation under taxes while other farms, much less efficient but less
polluting, will cease operation. Thus, if the policy maker considers
effects other than cost minimization, he/she may prefer standards
because of their lesser effect on output price or because they are more
desirable to producers.

Hochman, Zilberman, and Just (1977) and Moffit, Zilberman, and Just
(1978) extended the model to allow for pollution abatement and applied
it to the regulation dairy waste disposal in the Santa Ana River basin.
In this case, dairy waste contaminated groundwater through percolation
and run-off. The dairies were assumed to have a fixed proportion
technology, and pollution abatement was possible by shipping waste out
of the region. The policy makers set an overall target for salt deposit
to groundwater by the dairies. The analysis considered three policies:

  a. A standard proposed by the local Water Quality Control Board.
     This standard limits the cow-per-acre ratio identically for
     all dairies.

  b. Direct regulation restricting solid and liquid wastes
     separately. This standard was computed recognizing the
     heterogeneity of the industry.

  c. A tax on salts deposited to groundwater.

Policies (b) and (c) are close to the policies compared by the
conceptual analysis of Hochman and Zilberman (1978). The empirical
conclusions were:

  a. In heterogeneity matters, the proposed policy constructed
     without recognition of heterogeneity was found to be
     draconian. Up to 40 percent of the industry might be forced
     out of production, resulting in pollution generation much
     below the target level and welfare loss (reduction in the
     value of social surplus produced by industry) of 66 percent.
     The standard that recognized heterogeneity resulted in
     attaining the pollution goal with small losses in production
     and only a 16 percent welfare loss.

b. The welfare loss associated with the use of the well-designed (second) standard was only slightly (10 percent) higher than the welfare loss of the tax. Differences in output were also slight.

c. The loss to producers under the tax was substantially (more than 100 percent) greater than under the standard. At least in this example, producers have much to gain from a standard, while social loss from standards relative to tax is small.

The conceptual micro model presented above is rather simple and limited in use. Two principal limitations are the fixed proportion assumption and the lack of abatement technology. The micro model proposed by Caswell and Zilberman (1986) overcomes this limitation and is especially appropriate to cases where variable input application (water, pesticides, and/or fertilizers) is the source of the pollution problem. This model assumes constant return-to-scale technology, and output per unit is a function of the effective amount of input applied. Specifically, $y = f(e)$ where y is output and e is effective input use, i.e., the level of input actually utilized by the crop. The model distinguishes between x, the amount of variable input applied, and e, effective input use; the residual, $x - e$, is a source of pollution. In the case of irrigation, water that is not consumed by the crop may contaminate the environment through run-off or deep percolation. The relationship between effective and applied input is characterized as a function of locational characteristics (water-holding capacity) and application technology. Let $k = 0, \ldots, K$ be an indicator of application technology, $k = 0$ corresponding to not operate, $k = 1$ corresponding to use of the traditional technology, and $k > 1$ corresponding to adoption of new input-conserving technologies (in the case of water, $k = 1$ corresponds to application by flood irrigation while $k > 1$ corresponds to application with sprinklers or drip irrigation). Let a_j be a land-quality index measuring locational effectiveness in variable input use. For convenience, suppose that a_j assumes values from 0 to 1 and measures effective input to applied input ratio with traditional technology. Each modern technology has annual fixed costs denoted by C_k and it operates to improve input use effectiveness to $g^k(a_j)$ where $g^k(a_j) > g^{K-1}(a_j)$. Obviously, the relative gain in input use effectiveness is larger as the initial input use effectiveness is smaller and as k is larger.

The fraction of applied input that pollutes is a function of application technology and input use effectiveness and is denoted by $h^k(a_j)$. Obviously, $h^k(a_j) \leq 1 - g^k(a_j)$. Let b_j be a land fertility measure and output at location j is $b_j f[g^k(a_j) X_j]$.

For a farmer who pays a pollution tax v, the profit-maximization problem is

$$q_j = \max_k \; \Sigma \left(\max_x \{Pb_j f[g^k(a_j) \; x] - wx - vh^k(a_j) \; x - C^k, 0\} \right)$$

The farmer chooses application technology and the application level. His decisionmaking is sequential. First, the application level for each technology is derived and used to compute quasi-rent for each technology; the optimal k is then determined by comparing quasi-rents. Obviously, when all technologies yield negative quasi-rent, the production unit is not used.

The micro model yields an optimal technology for each j. The output, input, and pollution of this technology are used to obtain y_j, x_j, and z_j and allow using Equations 1 through 5 for aggregation.

Zilberman and Lichtenberg (1987) experimented with the micro model using data from the west side of the San Joaquin Valley, the area from which drainage water was supplied to Kesterson. They investigated the impacts of changes in water price, output price, drainage charge, and land fertility on water use and technology adoption in cotton production. They concluded:

a. Adoption of modern technologies becomes more likely as land fertility and output price increase. Higher water prices and drainage charges will encourage adoption of modern technologies but may also result in cessation of operation. When output prices are relatively good (80 cents per pound of cotton lint), profit maximization will dictate modification of the flood irrigation system to shorter runs or adoption of a sprinkler system when water prices and drainage charges are moderate. When water prices are high (above $80 per acre-foot) and drainage costs are substantial (above $100 per acre-foot), high irrigation efficiency and capital-intensive technologies such as drip irrigation become optimal. When cotton prices are low (50 cents per pound), operation is profitable only when prices of water and drainage are rather moderate, and only furrow and modified furrow irrigation technologies will be used.

b. Without the ability to switch technologies, the impact of higher water prices and drainage charges on water use and drainage generation are not substantial as long as farming is economical. Adoption of more effective irrigation technology is the source of most of the reduction in water use and drainage associated with higher water and drainage prices. Actually, adoption of drip irrigation may reduce drainage per acre by 90 percent.

The microlevel outcomes of the model of Equation 7 can be aggregated using Equations 1 through 7 to obtain regional policy impacts. A recent application derived impacts of proposed water policy standards for the San Joaquin River on the affected farmers (Hanemann, Lichtenberg, and Zilberman, 1988). In particular, the study investigated the impacts of cost increases associated with environmental regulations on utilized areas and solvency of farmers. It was found that when the cost increases are small (up to $20 per acre), there will be almost no impact on production and owners of less than 5 percent of the land will have financial difficulties. When cost increases are moderate ($50 to $60 per acre), less than 5 percent of the land will go out of production but close to 20 percent of the farmers will not be able to meet their debt. When the cost of the environmental policies are even higher ($80 to $90 per acre), about 7 percent of the land will not be in operation but the insolvency rate may reach 35 to 40 percent. The impacts on ownership and wealth seem to be much more substantial than the impacts on the industry's level of operation. Therefore, solvency impacts should be an important part of policy analysis.

4.   INCORPORATING UNCERTAINTY CONSIDERATIONS INTO POLICY MODELING

Policies are frequently discussed in an atmosphere of uncertainty. The randomness of nature and the unpredictability of economic factors are one source of uncertainty. Water quality regulations will have different effects in different years because precipitation, temperature, and other natural factors vary. The economic impacts of policies depend substantially on weather and product prices. Incomplete knowledge and insufficient information are other sources of uncertainty. Many parameters used for policy computations are statistical estimates subject to substantial error, and policy makers may have substantial doubt regarding their reliability. The uncertainty about policy outcomes is likely to affect policy choices. Weitzman (1974) demonstrated that the selection of a policy tool (whether price or a direct control) is dependent on the structure and reliability of the policy impacts' predictions.

Uncertainty considerations are especially important in methodologies that generate environmental health impacts in addition to economic impacts. Quantification of environmental health risks require modeling of a sequence of processes that generate risks. These processes include: (1) the generation of pollution (contamination), (2) the transformation and movement of the contaminant, (3) the exposure of vulnerable populations to the contaminants, and (4) the dose-response process--the reaction of the victim to the material he was exposed to. These processes are subject to substantial randomness (because of weather and human, topographic, and geographic variability) and there is typically considerable uncertainty regarding key parameter estimates.

Evidence suggests that policy makers are averse to uncertainty and are therefore prone to giving up some average gains to reduce the likelihood and magnitude of bad outcomes. Quantification of the uncertainty aversion behavior is thus important in the modeling effort. Here, we present the Lichtenberg and Zilberman (1988) model, which argues that "safety rules" provide a good foundation for environmental policy analysis under uncertainty.

In the Lichtenberg-Zilberman framework, environmental policies are set to maximize social gain subject to stochastic environmental quality constraint. The constraint sets a lower bound on a probability that a measure of environmental health risk will be below some target level.

For formal presentation of the model, let r be a measure of environmental health risk. In many studies, r is a probability of mortality or morbidity to a member of the population exposed by the environmental problem. The policy maker is uncertain about r because of insufficient knowledge and randomness in nature, so that r is a random variable. The distribution of r is affected by production choices of farmers as well as other activities (including preventive activities by potential victims). Government policies affect r indirectly through their impacts on the choices of farmers and victims. Let z be a vector of possible policies and let SC be the social welfare cost measure associated with consumption and production of marketable products affected by the regulation. We can view SC as the sum of consumers' and producers' suppliers in the affected markets and government revenues from pollution taxation minus social costs on pollution cleanup. Let α denote a statistical significance level used for government decisions. Using this notation, the social optimization problem becomes

$$\max_{z} SC$$

subject to $\Pr\{r \leq \bar{r} / \geq \alpha\}$ (8)

where $\bar{r}$ is the targeted risk level. The optimization in Equation 8 is an extension of Baumol and Oates' (1974) standard and taxes approach for uncertainty. Here, the policy makers set two parameters--a risk target $\bar{r}$ and reliability factor α--and the analysis obtains the social cost-efficient policies to meet these targets.

Where the model in Equation 8 is set up as a normative tool (selecting optimal policies), it is useful for positive purposes (explaining actual behavior) as well. Many environmental regulations are guided by legislation targeting the provision of sufficient protection for public health with an adequate margin of safety, which can be expressed in terms of a safety rule formulation. Also, natural scientists tend to use classical statistics in decisionmaking, specifically using statistical significance measures to measure reliability of estimates and containments of risks. The more conservative (in their aversion to uncertainty) the risk assessors are, the higher will be their α.

Repetition of the optimization problem varying $\bar{r}$ and α produces trade-off sets between economic efficiency and environmental risks for difference levels of statistical significance. These sets provide the decisionmakers with the economic costs associated with both imposing more strict limits on environmental health risks and increasing the probability that the risk standards will not be exceeded.

Several applications illustrate the methodology proposed here.

a.   Lichtenberg, Zilberman, and Bogen (forthcoming) estimated the trade-offs associated with water quality regulation (in this case, standards for DBCP contamination of drinking water) in Fresno County. More stringent standards reduce the risks of cancer at an increasing cost (due to the need to install filters or drill new wells in more areas). The risk estimates are uncertain because of randomness associated with exposure and uncertainty of dose response and contamination estimates.

Figure 1 depicts three trade-off curves--the one closest to the origin depicts trade-off between costs and mean level of risk, the second curve between costs and risk level which will not be exceeded at a 95 percent probability, and the third between costs and risk levels that will not be exceeded at a 99 percent probability. It is obvious that uncertainty and risk aversion matter. It costs about $80 million to assure that for a resident of Fresno County, the cancer risk of drinking water is below 10^{-6} with a 95 percent probability; and it will cost $100 million to assure that cancer risk will not exceed the same upper level with a 99 percent probability.

b. Lichtenberg and Zilberman (1987) studied alternative policies (taxes and standards) to regulate dairy run-off contamination of shellfish in San Francisco Bay. They derived a trade-off curve between cost of building run-off containment facilities and risks of food poisoning. The uncertainty in their model is due to weather. Their results show that marginal costs of safety are increasing (the costs of a marginal reduction in risk are greater for smaller initial risks). Moreover, the cost induced by higher reliability (higher probability that actual risk will not exceed target risk) increases as risk becomes smaller.

c. Hanemann, Lichtenberg, and Zilberman (1988) studied the economic impacts of alternative selenium standards for the San Joaquin Valley, explicitly recognizing weather variability. A given set of agricultural and conservation practices will result in different selenium concentrations in the river in different years. Therefore, in setting a policy, the regulator has to set an upper bound on selenium concentration and with what likelihood (under what conditions) it should be met. The study found that the cost to obtain a given standard may be more than twice as high if it is based on a bad year (and hence will be met about 95 percent of the time) than if based on a median year. Moreover, policies with standards required to hold 90 to 95 percent of the time are more likely to lead to higher levels of clean-up investment and adoption of modern irrigation technologies than standards not supposed to hold during bad years.

All of these applications demonstrate that uncertainties can be taken into consideration and are important in policy research.

5. MODELING DYNAMICS

Dynamic aspects have not been given adequate attention in formal models of agricultural pollution problems. This is a regrettable state of affairs since, as will be demonstrated below, dynamic considerations are essential for understanding some key elements of agricultural environmental health problems.

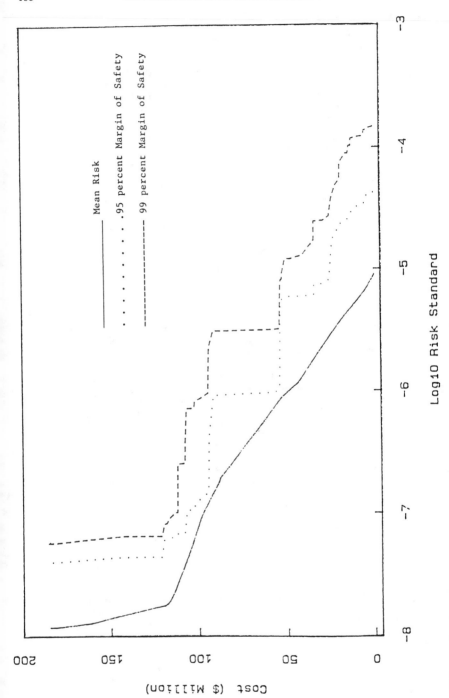

Some of the elements in which dynamic considerations play a key role include:

a. <u>The generation of environmental health risks</u>. In many situations, the health risks associated with agricultural pollution problems are dependent on accumulated toxin levels or accumulated levels of exposure rather than pollution generation at a given moment.

b. <u>Feasibility and costs of technologies</u>. Clean-up and abatement technologies play an important role in addressing pollution problems. Some of these technologies are in their infancy and require further research and development to make them practical and affordable. The costs of using them are likely to decline over time because of the processes of learning by <u>doing</u> (i.e., manufacturers of equipment get more proficient in production over time) and learning by <u>using</u> (i.e., the costs of use decline over time as users adapt these new products to their particular environments). Research and development activities can accelerate these cost reductions. Thus, the dynamics of learning and research and development are important for projecting abatement and clean-up costs over time and in using them for policy design.

c. <u>Input and output prices</u>. Some of the inputs that are important in generating or preventing pollution exhibit dynamic behaviors that affect their optimal use over time. For example, water prices in many locations are projected to rise because: (1) groundwater is an exhaustible resource with value, and extracting costs increase as stocks are being depleted; (2) the prices of surface water supplied by government projects are likely to rise when contracts are renegotiated in a few years; and (3) demand for water for non-agricultural uses is increasing, putting upward pressure on price. Similarly, final product prices are subject to important trends. Prices of many commodities have been falling over time, and others are likely to fall in the future. This trend may affect policy choices.

The dynamic nature of agricultural systems causing environmental health problems implies a need for time-varying policies, such as dynamic patterns of standards or taxes and timetables for constructing clean-up facilities or introducing conservation measures. In other words, it is crucial to prescribe not just what to do but also when to do it.

The drainage problem in California's San Joaquin Valley provides a good example. The root of this problem is the accumulation of percolating saline irrigation water because of an impenetrable clay layer. Over time, the presence of saline water in or near the crop root zone leads to reduced agricultural productivity. Installation of tile drainage seemed to be a solution for this problem, but it turned out to cause a new problem. The drainage water supplied to man-made wildlife preserves constrains naturally occurring toxins that damage wildlife. Two elements of a solution to this new problem are detoxification of drainage water and adoption of conservation technologies (drip, sprinkler) that reduce drainage volume. Knowledge about the hydrology of drainage water and the associated environmental risks is currently very limited, clean-up and water conservation technologies are in their infancy, and water prices in the region are likely to change at least in the 1990s. With so many dynamic elements, finding solutions to the drainage problem requires dynamic models capable of deriving dynamic paths for construction of drainage tile systems, for research and development, and for the extend of use of new clean-up and conservation activities, and for land-use, production, and water-use patterns. Appropriate policies (taxes and standards) are also likely to vary over time. While the work of specifying and analyzing such models is sure to be arduous, the gains in terms of additional flexibility of policy response and reduced disruption from regulatory intervention promise to make the effort worthwhile.

REFERENCES

Baumol, William J., and Wallace E. Oates. 1974. The Theory of Environmental Policy. Englewood Cliffs, New Jersey: Prentice-Hall.

Brandow, G. F. 1977. Policy for commercial agriculture. In L. Martin (ed.), A Survey of Agricultural Economics Literature 1. Minneapolis: University of Minnesota Press.

Buchanan, James M., and Gordon Tullock. 1971. Polluters' profits and the political response: direct control versus taxes. American Economic Review 65: 139-147.

Caswell, Margriet, and David Zilberman. 1986. The effects of well depth and land quality on the choice of irrigation technology. American Journal of Agricultural Economics 68: 798-811.

Cochrane, W. W. 1958. Farm Prices: Myth and Reality. Minneapolis: University of Minnesota Press.

Hanemann, Michael, Erik Lichtenberg, and David Zilberman. 1988. Conservation versus cleanup in agricultural drainage control. Department of Agricultural and Resource Economics, University of California at Berkeley, California.

Hochman, Eithan, and David Zilberman. 1978. Examination of environmental policies using production and pollution microparameter distributions. Econometrica 46(4): 739-760.

Hochman, Eithan, David Zilberman, and Richard E. Just. 1977. Two-goal regional environmental policy: the case of the Santa Ana River basin. Journal of Environmental Economics and Management 4(1): 25-39.

Houthakker, H. S. 1955-1956. The Pareto distribution of the Cobb-Douglas production function in activity analysis. Review of Economic Studies 23: 27-31.

Johansen, Leif. 1972. Production Functions: An Integration of Micro and Macro, Short Run, and Long Run Aspects. Amsterdam: North-Holland.

Johnson, G. L. 1958. Supply function -- some facts and notions. In E. O. Heady (ed.), Agricultural Adjustment Problems in a Growing Economy. Ames, Iowa: Iowa College Press.

Lichtenberg, Erik, and David Zilberman. 1987. Regulation of marine contamination under environmental uncertainty: shellfish contamination in California. _Journal of Marine Resource Economics_ 4: 211-225.

Lichtenberg, Erik, and David Zilberman. 1988. Efficient regulation of environmental health risks. _Quarterly Journal of Economics_ (forthcoming).

Lichtenberg, Erik, David Zilberman, and Kenneth T. Bogen. 1988. Regulating environmental health risks under uncertainty: groundwater contamination in California. _Journal of Environmental Economics and Management_ (forthcoming).

Moffitt, L. Joe, David Zilberman, and Richard E. Just. 1978. A 'putty clay' approach to aggregation of production/pollution possibilities: an application in dairy waste control. _American Journal of Agricultural Economics_ 60(3): 452-459.

Peltzman, Sam. 1976. Toward a more general theory of regulation. _The Journal of Law & Economics_ 19: 711-760.

Schultz, Theodore W. 1965. _Economic Crisis in World Agriculture_. Ann Arbor, Michigan: University of Michigan Press.

Weitzman, J. L. 1974. Price vs. quantities. _Review of Economic Studies_ 41: 477-491.

Zilberman, David, and Erik Lichtenberg. 1988. Modeling the effects of environmental politics on the adoption of irrigation technologies. Department of Agricultural and Resource Economics, University of California at Berkeley, California.

INSTITUTIONAL RISK AND UNCERTAINTY IN IRRIGATION WATER MANAGEMENT

L. Douglas James and Margriet F. Caswell

ABSTRACT

Recent cases of irrigation-induced water quality problems have caused doubts about whether the benefits of irrigated agriculture justify the costs. Irrigation can alter the hydrologic balance and the environmental characteristics of an area. What is often forgotten, however, is that the technical and institutional aspects of water management are inexorably tied. Irrigation management systems often lack the flexibility to respond adequately to major technical problems, and may even impede desirable change. This paper describes some technical and institutional problems inherent in irrigation water management in arid regions, and offers suggestions for research on applying risk assessment techniques to institutional changes.

1. INTRODUCTION

Civilization and irrigated agriculture began together. In the ancient world, technology and institutions for watering crops developed simultaneously to foster economic growth. Over the last 140 years, new technologies and new supporting institutions have developed in the American West. Today, growing problems are spreading doubts about whether the benefits from continuing American irrigated agriculture justify the costs.

The technical problem limiting irrigation is that the properties that give water value (Newell, 1972) also cause deterioration with each use. The "universal solvent" carries salts and other chemicals downstream to be joined by multiple loadings that eventually make the water unusable. The water that makes plants grow leaves salts that inhibit future crop growth. It is estimated that over 50 million hectares of farmland worldwide have severely salinized soils, and that the crop losses from these lands are close to $3 billion per year (Bobchenko, 1986). The challenge is to leach salts and other contaminants from these fields and transport them to the sea without harming downstream uses.

Water is essential to life and central to ecological systems, and deteriorating water quality (rather than insufficient quantity) has become the principal factor limiting their productivity. In the past, the environment had a large capacity to absorb and neutralize the residuals of production. Now, the assimilative capacity of the

L. Douglas James is the Director of the Utah Water Research Laboratory, College of Engineering, Utah State University, Logan, Utah.

Margriet F. Caswell is an assistant professor in the Department of Economics, University of California, Santa Barbara, California.

environment is increasingly extended. The cumulative impacts and synergistic effects associated with the many uses of the environment compound the problem. The general level of water quality has declined severely in many areas, so that the marginal impact of irrigation contamination may be very high (even though the absolute quantity may be small).

The institutional problem is that irrigation management systems lack the flexibility to respond adequately to the major technical problems, which are growing in severity and complexity over time. Throughout the following discussion, the term "institution" will be used to describe the complex legal, economic, social, and administrative infrastructures associated with irrigation water management. The ideal irrigation institution would help farmers protect their fields, reallocate marginal irrigation water to higher valued uses, and keep agricultural drainage from harming downstream uses. Downstream impacts have become more troublesome as environmental resources have become more highly valued. As high-quality natural environments have become increasingly scarce, more people (and a wider range of people) have recognized the importance of these resources to the quality of human life. The increasing relative value of high quality natural environments has altered the constituency for water management institutions. In its prime, agriculture had the political strength to overpower opponents (Kneese and Brown, 1981), gain rights to most of the water, and protect those rights thereafter. As water shifts toward urban and environmental uses, a political assessment will be made on whether Western water institutions are protecting continuing societal interests or are vestigial anachronisms impeding desirable change (El-Ashrey and Gibbons, 1986).

Risk analysis has been used for many years to assess such technical aspects of water resource management as floods, droughts, and dam safety. New developments incorporating risk and uncertainty into the decisionmaking calculus have given planners and policy makers a valuable new tool. However, its application has not been extended to the study of the risks associated with evolving institutions. The technical and institutional aspects of water management are inexorably tied. This paper will describe some of the potential interfacing technical and institutional problems inherent in irrigation water management in arid regions, and will offer suggestions for research on applying risk assessment techniques to institutional changes. "Risk" and "uncertainty" will be used interchangeably here since we recognize the use of subjective probabilities within the analysis.

2. IRRIGATION INSTITUTIONS IN THE AMERICAN WEST

The long history of American irrigation institutions has not prepared them well to meet the present need. As the American frontier moved westward into a drier and more variable climate, imaginative leaders adapted ancient irrigation technology to encourage the settlement of vast tracts of land. While seeking federal money, they made the argument that a small initial investment would lead to immense long-run benefits. The selling phraseology changed "irrigation" to "reclamation" and told of making waste lands blossom. This mirrored the general belief that an undeveloped resource was wasted. Water was valued for its consumptive uses, and society did not recognize the importance of instream uses. Institutions developed to serve each water use (e.g., agricultural, municipal, recreational, etc.) without an apparent acknowledgment of their interrelatedness.

As Congress financed more projects and irrigation technology advanced, visions grew. The small was found good, the large was dreamed to be better, and large projects were authorized in every arid state. Deserts became green and the Western economy prospered. Supporters spoke of economic benefits, dietary diversity, regional development, increased tax revenues, and the social values of the family farm. As the number of jobs increased in the agricultural sector, irrigation gained the support of nearly every political faction. Due to such support, irrigation works currently deliver virtually the full flow of some major rivers.

The irrigation institutions which developed during this time did an unparalleled job of promotion, finance, and construction. They saw irrigation as a vision, successful projects as monuments of success, and results that were wholly good. They failed to prepare the public for the changing roles of water in society and the technical failures that were bound to come.

3.   PROBLEMS WITH HYDROLOGIC BALANCES

In order to understand the need for modifying institutions, it is necessary to study the technical problems in achieving a steady state of water, sediment, salt, and toxic specie movements through agricultural lands. Many of these fluxes are just beginning to be understood. The design, operation, and maintenance of existing water management systems lacked the knowledge required for long-term productivity. To optimize crop growth and keep their soil productive, farmers need to hold soil water content within a fixed range, limit deep percolation, and control return flows. System managers must encourage individual water management decisions in harmony with public needs.

Irrigation diversions and return flows convert river flow patterns from floods and droughts of fresh water to smaller and more uniform flows of more water. The inflows of water, sediment, salt, and toxics are increased while the transport capacities of the outflows are reduced. The impact, however, transcends the farm with external effects that threaten many sectors of society. Irrigation that falls short of a steady state alters the hydrologic balance and the environmental characteristics of an area. Eventually four recurring problems must be faced:

•   Water Sinks: Efficient storage, conveyance, and application of irrigation water to grow crops is difficult. It has been noted that only 20 to 35 percent of project irrigation water actually reaches crops (Golubev and Biswas, 1985). The losses have two serious consequences: They add to the cost of delivered water; and they raise the water table and cause waterlogging. The additional cost of distributing more water and reduced productivity diminishes the financial resources available for dealing with the problem.

• Sediment Sinks: Sediment erodes from uplands, fills reservoirs and canals, and adds to the cost of maintaining the system (Wolman, 1967). Financial resources are further strained by this process.

• Salt Sinks: About 20 percent of the irrigated fields in the West have been rendered less productive by salts left by evapotranspiration (Miller et al., 1985). Reservoir evaporation concentrates salinity in incoming irrigation water. In addition, deep percolation from irrigated areas flows through salt bearing formations and adds salt loading downstream. Salt sinks develop in fields, streams, and reservoirs as water flows through the system.

• Toxic Sinks: Some soils and geologic formations have high toxic mineral contents and can be considered as natural geologic hazards. Leaching waters dissolve these toxics and then accumulate above impervious layers in newly formed water sinks downstream. Irrigation can mobilize these contaminants, accelerate their movements to these toxic sinks, and expose a variety of ecological systems to dangerous contamination.

4.   WILDLIFE AND HUMAN HEALTH IMPACTS

Toxic substances leached from hazard areas and waters contaminated with pesticides accumulate in these sinks and have potential health impacts. The danger to aquatic habitats has been suspected for some time. In 1966, Glacken expressed alarm that irrigation could have long-run impacts that might be devastating to natural environments. Concentrations of toxic substances bioaccumulate in wildlife exposed to the sinks and eventually pose a serious threat to ecosystem viability and human health. The selenium contamination of the Kesterson Wildlife Refuge will probably be followed by many more cases.

The sinks result from complex interactions among hydrologic processes and chemical and biological changes that are poorly understood. Impact estimation is an art more than a science. A major research effort is needed to quantify the risks associated with an irrigation project. Their institutional impacts are virtual unknowns.

5.   POTENTIAL INSTITUTIONAL PROBLEMS

Irrigation management must shift, from expounding the benefits of irrigation in order to obtain money for new construction to countering the forces of deterioration with more effective operations and maintenance. History raises doubts that this shift can be achieved; society has done much better at construction than at mitigating the pervasive forces of deterioration. It is our contention that risk assessment can and should be applied to the failure of irrigation institutions, much as it has been applied to dams.

Unlike a dam failure, an irrigation failure occurs slowly and it not seen as a catastrophe. The irrigation projects built in the American West have seldom produced their expected long-term benefits. The full impact of the four types of hydrologic sinks was not adequately understood, although salt and waterlogging damage to crops caused by inadequate drainage was recognized.

Irrigation institutions began by assigning farmers all of the risks (i.e., lost profits) inherent in accepting irrigation water. Drains were provided to carry away excess water, but system management saw no compelling reason to influence farm water management as long as the water was not used wastefully. It is now known that farm water management may significantly affect others and that these effects vary considerably over short distances on irrigated lands. Since individual decisions have external effects, there is a role for centralized management. There are several reasons why the irrigation institutions in place today may not be adequate to deal with the technical problems that must be addressed.

• Institutional Inertia: The institutional alliances and political coalitions that supported major irrigation projects often made commitments that did not allow for changing conditions (either physical or social). Authorization and water rights legislation often fixed water use with no provision for the transfer of water outside designated service areas or even trade between uses. Irrigation institutions in the American West have changed over the last 140 years, but their evolution has been slow. Once successful institutions do not easily recognize changing times. Effective response may have to be very rapid, and current institutions may lack the expertise and flexibility to move quickly in unfamiliar arenas. The original operating rules may no longer be valid, but they may be difficult to adjust in the face of legal and cost-sharing agreements.

• Conflicting Jurisdictions: In 1903, the Chief of Irrigation Investigations (Mead, 1903) stated that "The paramount need of the West is relief from some of the evils of the haphazard development of the past." Eighty years later, Bowden et al. (1982) said, "The major weaknesses today in California water law and management are primarily the result of this haphazard development of the allocation system." The government established an effective bureaucracy for project building but was not able to integrate irrigation with water resources management for other purposes. Each aspect of water use has its own supporting institutions, and their jurisdictions often overlap. Several laws and regulations may apply to a particular circumstance and they may be contradictory. Since land use and water management decisions are inexorably tied, coordination among institutions which deal with these resources is needed. Without coordination, no one feels responsible, and the problems compound without causing action.

• Financial Burden: Irrigation projects become an increasing financial burden over time. All of the relatively inexpensive storage sites were developed first, so the marginal costs of developing irrigation water have increased dramatically. Water supply systems to irrigate new lands have immediate benefits that are easy to see. The maintenance costs of these systems (especially as they begin to age) are less apparent and easily delayed. Drainage facilities, although usually an integral part of the original system design, are often not built until waterlogging and salinity problems become too costly to ignore. The postponed construction costs are usually substantially higher than the original projection. Raising aging project standards to meet new regulatory requirements (e.g., water quality standards) would compound the burden. Currently, mounting expenses are discouraging further federal funding for irrigation projects. These costs could be covered by raising water charges. However, water prices to irrigators have often been set to meet political goals (e.g., regional development) rather than to repay costs or to reflect the social value of the resource (Clark, 1967). The artificially low price of water has added to the financial burden imposed on taxpayers by many irrigation projects, and the national political climate is becoming increasingly unfavorable for continued subsidies.

• Inappropriate Scale: Designing an institutional structure at an effective scale is not easy. The more centralized the management system (or rule-making entity), the less likely the institution will be sensitive to local conditions or flexible enough to accommodate change. There is also a fear that the large information systems required to operate a centralized system threaten basic human freedoms. However, some centralized control is important. If the management scale is too small, the external effects will not be addressed. Rights to water and environmental quality are poorly specified, and incentives for farmers to consider the external effects of their irrigation decisions are lacking.

• Simplistic Operation: Irrigation institutions are ideally designed to (1) deliver water; (2) keep sediment, salt, and toxic flows from concentrating in sinks; and (3) coordinate diversions with instream water uses. All of these needs vary in space and time, and, as a generalization, the emphasis of the institutions has had to change from (1) to (2) to (3) over time. Whereas stream diversion may initially have been envisioned as a "plumbing" problem, it has evolved into a complex balancing of the costs and multi-dimensioned benefits of many water uses.

A multidisciplinary approach is needed to adequately assess problems and devise responses, and the administrative infrastructure often lacks the necessary expertise. Ongoing monitoring and anticipatory research will be necessary parts of these organizations. In the past, government agencies have been consistently unreceptive to research developing effective management institutions. White (1974) expressed deep anxiety about the capacity of bureaucracies to examine the social aspects of water development objectively. Maher (1974) ties the failure of an organization to utilize research to its poor adaptive capacity.

There is a strong need for irrigation institutions to recognize that they operate within a broad context. Few problems exist in isolation, either in a physical or a social sense. The technical problems inherent in irrigation systems are directly influenced by federal agricultural policies, export quotas, pollution regulations, etc. Attempts to narrowly define the sphere of influence affecting any problem will severly diminish the probability of finding a solution.

• Equity Considerations: The economies of scale in capital-intensive irrigation systems favor larger farm units. Therefore, publicly financed irrigation and salinity control projects have contributed toward the concentration of wealth among fewer farmers. Reisner (1986) opines that Congress has "authorized probably $1 billion worth of engineered solutions to the Colorado salinity problem in order that a few hundred upstream farmers could go on irrigating and poisoning the river." It may no longer be politically feasible to initiate such projects or to even continue operation of some existing ones.

6.   THE FUTURE OF IRRIGATED AGRICULTURE

    The future of irrigated agriculture in the American West lies at the interface of agricultural production, water resources planning, and political decisionmaking. Trouble looms on all three fronts.

• The National Water Commission (1973) concluded that irrigation investment is generally less cost-effective than agricultural production in rain-fed areas. Exceptions may be found in specialty crops grown only in irrigated areas (Gardner et al., 1982) and in crops having major roles in world trade. This raises the question of whether the benefits from sustaining irrigated agriculture outweigh the environmental costs.

• In the river basins with the largest irrigated areas (California, Great Basin, Colorado, and Rio Grande), the available water is nearly fully developed. The continued rapid municipal and industrial growth will compete more vigorously with the agricultural sector for a fixed supply of water.

• Although agriculture is still an important part of the economic base of the West, other sectors are gaining prominence. The agricultural political alliances which supported irrigation projects in the past no longer hold sway. There has been a fundamental change in the way that society values the environment: "[T]he emphasis is shifting from production of commodities to preservation and restoration of the quality of life" (Seckler, 1971). To many, agriculture no longer represents the "American way of life"--especially because of the concentration of wealth in fewer agri-businesses and the decline of the "family farm."

    The area of irrigated land is going to decline. An important issue is whether the change will be marginal or whether federal facilities will be abandoned or turned over to local interests. Large-scale changes are seldom optimal, and dispersed management responsibility may not be productive. Bennett (1974) describes how the benefits of irrigation in ancient Egypt peaked under a central political authority and then deteriorated as responsibility was dispersed to a feudal system, which was less able to provide needed maintenance. Even if irrigation systems are abandoned, large costs may be required to disassemble projects to protect public health and safety and environmental quality.

The maximum economic benefit would result from the efficient management of existing systems and orderly withdrawal of irrigation to supply new needs in marginal water-short regions. Current irrigation institutions may not be adequate to accomplish this.

On the other hand, the required institutional change may be even more difficult. There are risks inherent in changing irrigation institutions, and the long-range purpose of a change must be clearly defined. As Ciriacy-Wantrup (1967) stated, "The function of water institutions is not to maximize economic efficiency for particular conditions and points in time but to structure decisionmaking on the lower level under various and constantly changing conditions," where the "lower level" refers to the direct "control of inputs, outputs, and other quantitative characteristics of the water resource system." Institutions evolved within the context of their times and the way that they will respond to changes may be predictable. The economic theory of Public Choice may offer new insights into institutional behavior. Policies proposed to solve future technical problems should be assessed with respect to the physical and social responses to that policy.

7.   ROLE OF RISK ASSESSMENT

Risk assessment techniques have been developed for the study of physical failures. Research could define the probabilities, locations, and consequences of sinks being formed. The results could be used to guide cost-effective preventative and corrective programs.

The more difficult problem is assessment of the interactions between physical and institutional components. In most studies, the risk of institutional failure is ignored. In past assessments, all of the controllable variables were technical in nature, while all of the institutional parameters were specified as state variable--hence, uncontrollable (Seckler and Hartman, 1971). Single estimates (or at best, high and low values) are assigned, and the "most likely" outcome is computed. The lack of consideration for institutional variability masks outcomes and makes results misleading:

> For example, two projects may have the same benefit-cost ratio based on the most likely values of each component, but one project may be more "risky" in the sense that it is much more vulnerable to possible adverse conditions. (Rausser and Dean, 1971)

Sensitivity analysis can offer a range of outcomes, but likelihood estimates require information on the stochastic variability of the institutional parameters. Research of this sort is sorely needed, not only with respect to the problems of irrigation management, but also on a wide variety of institutional responses to drought, floods, dam failures, etc. Institutional risk offers great promise as a profitable area of research.

8.   SUMMARY

The above analysis probes management options for securing the long-term productivity of agriculture within the larger contexts of economic and environmental needs in the American West. Desirable options protect beneficial irrigation systems by solving their technical problems and improving their performance. In other cases, they shift land and water to more productive uses. Studies are needed to determine what to do in particular cases.

Variable response is the name of the game. Economic needs change over time. Past infrastructure may be either abandoned or adapted. A rigid policy to preserve every irrigated acre or even every irrigation project is the path to collapse. Adaptations can sustain productivity. However, these will require a more thorough scientific understanding of the physical-chemical biological processes governing water flow and pollutant transport and storage in local sinks than did initial project construction. Fortunately, irrigation management is turning more to adaptive compromise and using technological advances in conflict resolution.

Irrigation institutions became popular by building visible monuments with large regional benefits. However, these institutions must now shift emphasis from construction to operations, from political lobbying to technical sophistication, and from monopolizing to sharing water supplies. Special attention must be given to how this can best be done, and to the risks inherent in such an effort. During this time of change, both agriculture and the environment will suffer catastrophes. Objective analyses employing risk assessment techniques must be supplied to strong political leadership to preserve public confidence, keep the government from abandoning valuable capital investments, and protect the quality of life. Objective risk assessment can be used to preserve the good in irrigation and eliminate the nonproductive.

ACKNOWLEDGMENTS

Special thanks are given to colleagues on the Committee for Irrigation-Induced Water Quality Problems, Water Science and Technology Board, National Research Council, for stimulating the thinking.

REFERENCES

Bennett, J. W. 1974. Anthropological contributions to the cultural ecology and management of water resources. In L. D. James (ed.), Man and Water: The Social Sciences in Management of Water Resources. Lexington, Kentucky: University Press of Kentucky.

Bobchenko, V. I. 1986. Salinization control of irrigated land. Water International 11(4):150-156.

Bowden, G. D., S. W. Edmunds, and N. C. Hundley. 1982. Institutions: customs, laws and organizations. In E. A. Engelbert (ed.), Competition for California Water. Berkeley, California: University of California Press.

Ciriacy-Wantrup, S. V. 1967. Water policy and economic optimizing: some conceptual problems in water research. American Economic Review 57(2):179-189.

Clark, C. 1967. The Economics of Irrigation. Oxford, England: Pergamon Press.

El-Ashry, M. R., and D. C. Gibbons. 1986. Troubled Waters: New Policies for Managing Water in the American West. Washington, D.C.: World Resources Institute.

Gardner, B. D., R. H. Coppock, C. D. Lynn, D. W. Rains, R. S. Loomis, and J. H. Snyder. 1982. In E. A. Engelbert (ed.), Competition for California Water. Berkeley, California: University of California Press.

Glacken, C. J. 1966. Reflections on the man-nature theme as an object for study. In F. F. Darling and J. P. Milton (eds.), Future Environments of North America. New York: Natural History Press.

Kneese, A. V., and F. L. Brown. 1981. The Southwest under Stress: National Resource Development Issues in a Regional Setting. Baltimore, Maryland: Resources for the Future, Inc. Johns Hopkins University Press.

Maher, T. 1974. Epilogue: recommendations from a social science viewpoint. In L. D. James (ed.), Man and Water: The Social Sciences in Management of Water Resources. Lexington, Kentucky: University Press of Kentucky.

Mead, E. 1903. Irrigation Institutions: A Discussion of the Economic and Legal Questions Created by the Growth of Irrigated Agriculture in the West. New York: Macmillan. Reprinted, 1972, by Arno Press (New York).

Miller, T. O., G. D. Weatherford, and J. E. Thorson. 1986. The Salty Colorado. Napa, California: John Muir Institute.

National Water Commission. 1973. <u>New Directions in U.S. Water Policy: Summary, Conclusions, and Recommendations</u>. Washington, D.C.: U.S. Government Printing Office.

Newell, F. 1903. <u>Water Resources: Present and Future Uses</u>. New York: Macmillan. Reprinted, 1972, by Arno Press (New York).

Rausser, G. C., and G. W. Dean. 1971. Uncertainty and decisionmaking in water resources. In D. Seckler (ed.), <u>California Water: A Study in Resource Management</u>. Berkeley, California: University of California Press.

Reisner, M. 1986. <u>Cadillac Dessert: The American West and Its Disappearing Water</u>. New York: Viking/Penguin.

Seckler, D., and L. M. Hartman. 1971. On the political economy of water resources evaluation. In D. Seckler (ed.), <u>California Water: A Study in Resource Management</u>. Berkeley, California: University of California Press.

White, G. F. 1974. Role of geography in water resources management. In L. D. James (ed.), <u>Man and Water: The Social Sciences in Management of Water Resources</u>. Lexington, Kentucky: University Press of Kentucky.

FUTURE CLIMATIC CHANGE AND ENERGY SYSTEM PLANNING: ARE RISK ASSESSMENT
METHODS APPLICABLE?

Stephen H. Schneider

ABSTRACT

Nearly all energy systems entail some significant forms of health,
environmental, or socio-political risks. But there is considerable
difference among various energy systems in the magnitude, timing, and
nature of their associated risks. It is this difference that allows a
degree of choice, or policy, with regard to selecting energy
alternatives. The National Research Council's Committee on Nuclear and
Alternative Energy Systems compiled a detailed analysis of risks
associated with different types of energy use. A condensed table of
that analysis is presented.

The issue of assessing risks is also discussed, stressing
complications involved in making valid quantitative risk assessments,
and the worth of risk assessment as a public policy-making device.
Among questions raised: Can risks and benefits be measured in the same
units? Can we, for example, compare the risk of climatic change with
the risk of accidental injury or death? How much do we value the
future? Can quantitative risk assessment methods be applied to such
complex problems? The example of greenhouse effect climatic change is
developed in some detail to illustrate these issues. Since risks will
be proportional to the total energy produced and consumed by society, it
is important that global population be kept in check so that total
energy usage can be minimized, while per capita consumption is not too
low.

1.   INTRODUCTION

There is a widespread perception that whereas professional risk
assessors think in terms of expected values, the public and politicians
fear catastrophes, regardless of their placement on some
probability-times-consequences scale. Indeed "risk" has been
classically defined as:

$$\text{risk} \frac{\text{consequence}}{\text{unit time}} = \text{frequency} \frac{\text{events}}{\text{unit time}} \times \text{magnitude} \frac{\text{consequence}}{\text{event}}$$

But this classical probability-times-consequence measure is too
simplistic. For example, some people may not play a game whose
probability-times-consequence is the same as another, when the former
game has an extremely high consequence and a low probability of

Stephen H. Schneider is with the National Center for Atmospheric
Research, Boulder, Colorado.

The National Center for Atmospheric Research is sponsored by the
National Science Foundation.

achieving it. Others, however, might have the opposite reaction. Alternatively, we could express risk more generally than probability, P, and consequence, c, in the form

$$risk = Pc + f(P,c,x)$$

where f can be some function of P,c and other relevant variables, x, where x could be a measure of the fairness of the risk, its nearness to today, how catastrophic it is, etc. It could also be a nonlinear function of P or c.

A person or society which follows the basic rule of simply multiplying probability by consequence to get an expected value, and then acts accordingly, is said to be risk-neutral (for them, f = 0). Alternatively, one who estimates a risk value higher than P × c is said to be risk-averse (f>0). Or, a person who estimates risk to be less than P × c is risk-prone (f<0). The only thing general that can be said is that different people and different societies act differently in regard to different risks, and by no means are their actions (i.e., their behavior with regard to f) consistent. For example, societies are often unwilling to spend heavily to prevent the statistical loss of life in, say, highway accidents, whereas the same society might spend very heavily to maintain the life of a known threatened or injured person. In any case, about the best we can do is to identify as quantitatively as possible both the probability and the consequence of some activity so that the decisionmaking process can choose explicitly the extent to which our society will be risk-neutral, risk-prone, or risk-averse.

To place the climatic risks of increased greenhouse gas build-up in the atmosphere of formal quantitative risk analyses is no mean task. For example, although some of the potential consequences of greenhouse gas build-up can be roughly defined (e.g., decreased snow pack in the Sierra Nevada or Rocky Mountains), assignment of probabilities for any scenario can, at best, be grounded in the intuition of climate experts. Furthermore, there always is a good probability (say 0.1 to 0.5 -- intuitively determined, of course) that some consequence or sequence of consequences we cannot now define will occur as real climatic changes actually evolve. To make this issue clearer, I will summarize the nature of the carbon dioxide/greenhouse effect issue. After that, we can put this issue in the perspective of the overall problem of energy system risk analysis, and finally consider the issue of whether climate change can be addressed productively by quantitative methods of risk analyses.

2. SCIENTIFIC ISSUES SURROUNDING THE "GREENHOUSE EFFECT"

We'll focus on the CO_2 greenhouse issue for the sake of simplicity, although similar series of stages apply to other atmospheric trace "greenhouse" gases as well. Survey assessments of this issue can be found in Bolin et al. (1987), Carbon Dioxide Assessment Committee (1983), or Schneider and Londer (1984).

Behavioral Assumptions

At the very foundation of the problem, behavioral assumptions must be made in order to project the future use of fossil fuels (or deforestation, since this too can affect the amount of CO_2 in the atmosphere). The essence of this issue, then, is not chemistry or physics or biology, but social science. It depends upon projections of human population, the per capita consumption fossil fuel, deforestation rates, reforestation activities, and perhaps even countermeasures to deal with the extra carbon dioxide in the air. These projections depend on issues such as the likelihood that alternative energy systems or conservation measures will be available, their price, and their social acceptability. Furthermore, trade in fuel carbon (for example, a large scale transfer from coal-rich to coal-poor nations) will depend not only on the energy requirements and the available alternatives but also on the economic health of the potential importing nations. This in turn will depend upon whether those nations have adequate capital resources to spend on energy rather than other precious strategic commodities, such as food or fertilizer, as well as some other strategic commodities such as weaponry. In order to project the future we can make scenarios (such as seen in Figure 1) that show alternative CO_2 futures based on assumed rates of growth in the use of fossil fuels. Most typical projections are in the one-to-two percent annual growth range, implying a doubling of CO_2 in the 21st century. CO_2 concentration has already increased by some 25 percent this century. (Other-than-greenhouse gases like CFC's, methane, nitrogen oxides, etc., taken together could be as important as CO_2 in the future greenhouse effect, but these have complicated biogeochemical interactions; thus we'll focus on CO_2 alone for the present).

The Carbon Cycle

Once we have some plausible set of scenarios for how much CO_2 will be injected into the atmosphere, we next need to determine what interacting biogeochemical processes control the global distribution and stocks of carbon. This involves uptake by green plants (since CO_2 is the basis of photosynthesis, more CO_2 in the air means faster rates of photosynthesis), changes in the amount of forested area, what is planted, and how climate change affects natural ecosystems. As a concrete example of how large natural climatic change (from the Ice Age to interglacial formations, for instance) affects natural ecosystems (the North American forests), we can look at Figure 2. This set of "snapshots" of forest status is obtained by counting forest pollen grains in lake beds and soils.

During the past 15,000 years, the Ice Age conditions gave way to our present interglacial period of warmth. This represented some 5°C global warming, with 10-20°C warming locally near where the ice sheets used to be. The boreal forests of Canada were hugging the rim of the great glacier in the U.S. Northeast some 10,000 years ago, while presently abundant hardwood species were then hanging on in small

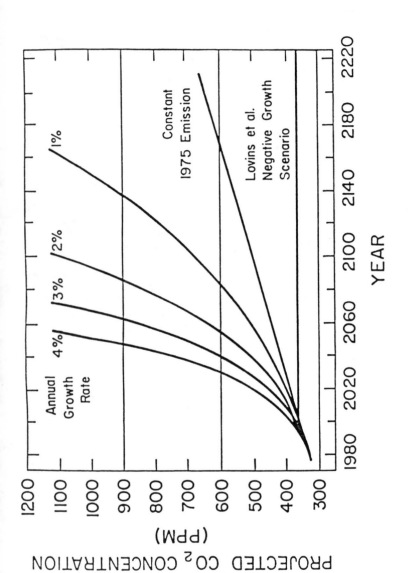

Figure 1. Various CO_2 scenarios based on specified sets of energy growth assumptions (from Lovins et al., Least-Cost Energy: Solving the CO_2 Problem, Andover: Brick House, 1981)

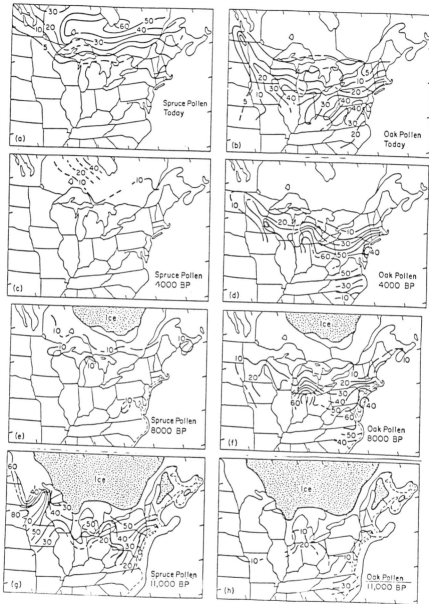

Figure 2. The relative fraction (in percent) of fossil pollen from various species of trees is shown on the maps. These pollen patterns can be correlated to climatic conditions. These maps show, for example, that at the end of the last Pleistocene Ice Age 11,000 years ago much of the northeastern United States was covered with spruce forest, which is found today in central Canada near Hudson Bay. (From Bernabo and Webb III, 1977, Quaternary Research 8:64-95)

refugia, largely in the South. The natural rate of forest movement that can be inferred from Figure 2 is some one km movement per year, in response to an average temperature change of some 1-2°C per thousand years. If climate were to change much more rapidly than this, then the forests would not be in equilibrium with the climate--i.e., they could not keep up with the fast change and would go through a period of transient adjustment in which many hard-to-predict changes in species distribution or productivity would very likely occur. Also, since the slow removal of CO_2 from the atmosphere is largely accomplished through chemical processes in the oceans which take decades to centuries, the rate at which climate change modifies mixing processes in the ocean also needs to be taken into account. There is considerable uncertainty about just how much CO_2 will remain in the air, but most present estimates put the so-called "airborne fraction" at about 40 percent $\pm$ 25 percent, which suggests that over the time-frame of a century at least, something like half the CO_2 we inject will remain and may exacerbate the greenhouse effect.

Global Climatic Response

Once we have projected how much CO_2 (and other trace greenhouse gases) may be in the air over the next century or so, we have to estimate what its climatic meaning is. The greenhouse effect, despite all the controversy that surrounds the term, is really not at all a scientifically controversial subject. In fact, it is one of the best, most well-established scientific theories in the atmospheric sciences! For example, with its very, very dense carbon dioxide atmosphere, Venus has ovenlike temperatures at its surface. Mars, with its very thin carbon dioxide atmosphere, has temperatures comparable to our polar winters. The explanation of the Venus hothouse and the Martian deep-freeze is really quite clear and straightforward: the greenhouse effect (e.g., see Kasting, Toon, and Pollack, 1988). The greenhouse effect works because some gases and particles in an atmosphere preferentially allow sunlight to filter through to the surface of the planet relative to the amount of radiant energy that the atmosphere allows to escape back up through the atmosphere to space. This latter kind of energy, so-called terrestrial infrared energy, is affected by the amount of "greenhouse" material in the atmosphere (see Figure 3). Therefore, if there is an increase in the amount of greenhouse gases, there is an increase in the planets' surface temperature because the amount of heat trapped in the lowest part of the atmosphere has increased. While that part is not controversial, what is controversial about the greenhouse effect is exactly how much the earth's surface temperature will rise given a certain increase in a trace greenhouse gas such as CO_2. Complications arise due to processes known as feedback mechanisms. For example, if the warming due to added CO_2 were to cause a temperature increase on Earth, the warming would likely melt some of the snow and ice that now exists. Thus, the white surface covered by the melted snow and ice would be replaced with darker blue ocean or brown soil: surface conditions which would absorb more sunlight than

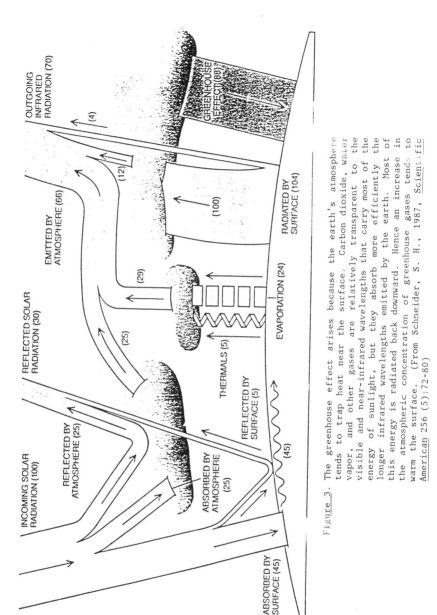

Figure 3. The greenhouse effect arises because the earth's atmosphere tends to trap heat near the surface. Carbon dioxide, water vapor, and other gases are relatively transparent to the visible and near-infrared wavelengths that carry most of the energy of sunlight, but they absorb more efficiently the longer infrared wavelengths emitted by the earth. Most of this energy is radiated back downward. Hence an increase in the atmospheric concentration of greenhouse gases tends to warm the surface. (From Schneider, S. H., 1987, Scientific American 256 (5):72-80)

the previous white snow or ice. Thus, the initial warming would create a darker planet, which would absorb more energy, thereby creating a larger final warming. However, this is only one of a number of possible feedback mechanisms. And because many of them are interacting simultaneously in the climatic system, it is extremely difficult to estimate quantitatively how many degrees warming the climate will undergo.

Unfortunately, since there is no time over Earth's history that we can turn to when the CO_2 amounts in the atmosphere were, say, twice what they are now, and at the same time have reliable, quantitative knowledge of what the Earth's climate was then, we cannot directly verify our quantitative predictions of greenhouse warming on the basis of purely historical analogues. Instead, we must base our estimates on natural analogues and climatic models (e.g., see Schneider, 1987). These are not laboratory models, since no physical experiments could remotely approach the complexity of the real world. Instead we try to simulate the present Earth climate by building mathematical models, in which the known basic physical laws are applied to the atmosphere, oceans, and glaciers, and the equations that represent these laws are solved with the best computers available. Then, we simply change the computer program's expression of the effective amount of greenhouse gases, repeat our calculation, and compare it to the so-called "control" calculation for the present Earth. Many such models have been built over the past few decades and are in rough agreement that if CO_2 were to double, then the Earth's surface temperature would warm up somewhere between one and five degrees Celsius. Just for a point of comparison, the world average surface temperature difference between our present climate and the Ice Age of 18,000 years ago is about five degrees Celsius, colder then than now. Thus, global temperature change of more than a degree or two really is a very substantial alteration.

Regional Climatic Response

To estimate the societal importance of climatic changes, however, it is not so much global average temperature we need to study but the regional distribution of evolving patterns of climatic change. Will it be drier in Iowa in 2010, too hot in India, wetter in Africa, more humid in New York, or flooded in Venice? Unfortunately, to predict the fine-scale regional responses of variables such as temperature and rainfall requires climatic models of greater complexity and expense than are currently available. It's not that we haven't made preliminary calculations of these variables. Rather, to be honest, it would be hard to get a consensus of knowledgeable atmospheric scientists to agree that the regional predictions of state-of-the-art models are reliable. Nevertheless, there is at least some plausible suggestion that the following coherent regional features might well occur over the next fifty years or so:

- wetter sub-tropical monsoonal rain belts

- longer growing seasons in high latitudes

- wetter springtimes in high and mid-latitudes

- drier midsummer conditions (see Figure 4 in some mid-latitude areas--a potentially serious problem for the future agriculture and water supply in major grain-producing nations)

- increased probability of extreme heat waves (with possible health consequences for people and animals in already warm climates and concomitant reduced probability of extreme cold snaps) (See Table 1)

- increased sea levels by as much as a meter over the next hundred years.

It must be stressed again that considerable uncertainty remains in these predicted regional features, even though many plausible scenarios have been investigated. The principal reasons for the uncertainty are two-fold: the crude treatment in climatic models of hydrological processes and the usual neglect of the effects of the deep oceans. The latter would respond slowly--over many decades to centuries--to climatic warming at the surface, but would also act differentially (that is, not-uniformly in space and over time). That means that the oceans, like the forests, would be out of equilibrium with the atmosphere if greenhouse gases increase as rapidly as typically projected and if climatic warming were to occur as fast as a few degrees over a century. This, recall, is some ten times faster than the natural average rate of change of temperature seen from the end of the last Ice Age to the present time. Furthermore, if the oceans are out of equilibrium with the atmosphere, then it will be hard to assign much credibility to specific regional forecasts like that of Figure 4 until fully coupled atmosphere/ocean models are tested and applied. This is a formidable scientific and computational task.

Environmental Impact of CO_2 Scenarios

Given a set of scenarios for regional climatic change, we must next estimate the impacts on the environment and society. Most important are the direct effects on crop yields and water supplies. Also of concern is the potential for altering the range or numbers of pests that affect plants or diseases that threaten animals or human health. Also of interest are the effects on unmanaged ecosystems. For example there is major concern among ecologists that the destruction rate of tropical forests due to human expansion is eroding the genetic diversity of the planet. That is, since the tropical forests are in a sense major libraries for the bulk of living genetic materials on Earth, the world is losing some of its irreplaceable biological resources through overdevelopment. The connection of this already formidable environmental problem to climatic change becomes clear when one recognizes that substantial future changes to tropical rainfall have been suggested by climatic models, which means that reserves (or refugia) presently set aside and designed as compromise solutions to preserve genetic resources into the future may not be as effective as presently planned if rapidly evolving climatic change will cause conditions in these refugia sufficiently different from the present. Simply, they may not sustain even those species that they are designed to protect.

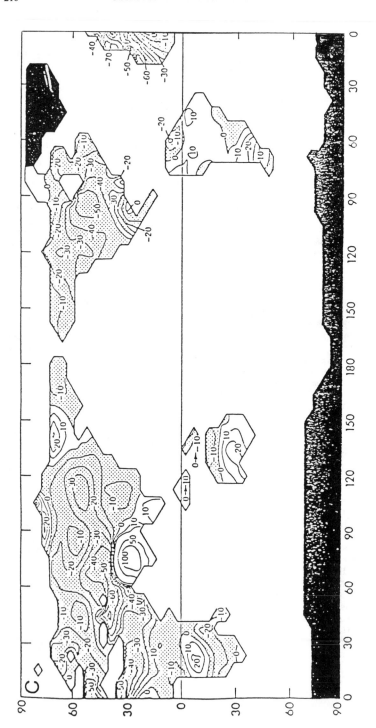

Figure 4. CO_2-induced change in soil moisture expressed as a percentage of soil moisture obtained from the normal CO_2 experiment. (From Manabe and Wetherald, 1986, Science 232:626-628)

Table 1

IF ONLY TEMPERATURE GOES UP BY 3°F, THEN ODDS FOR JULY HEAT WAVES* GO UP:

	ODDS NOW	ODDS IF +3°F
WASHINGTON D.C. (95°F)	17%	47%
DES MOINES (95°F)	6%	21%
DALLAS (100°F)	38%	68%

Source: Mearns, L. O., R. W. Katz and S. H. Schneider, 1984, Changes in the probabilities of extreme high temperature events with changes in global mean temperature, J. Clim and Applied Meteorl. 23: 1601-1613.

* "Heat wave" means 5 or more days in a row with max temp above 95/100°F. threshold

Economic, Social, and Political Impacts

The estimation of the distribution of economic winners and losers, given a scenario of climatic change, involves more than simply looking at the total dollars lost and gained--were it possible somehow to make such a calculation credibly! It also requires looking at these important equity questions: "Who wins and who loses?" and "How might the losers be compensated and the winners taxed?" For example, if the Corn Belt in the United States were to "move" north by several hundred kilometers from a warming, then a billion dollars a year lost in Iowa farms could well eventually become Minnesota's billion dollar gain. Although some macroeconomists viewing this hypothetical problem from the perspective of the United States as a whole might see no net loses here, considerable social consternation would be generated by such a shift in climatic resources, particularly since the cause was economic activities (i.e., CO_2-producing) that directed differential costs and benefits to

various groups. Moreover, even the perception that the economic activities of one nation could create climatic changes that would be detrimental to another has the potential for disrupting international relations--as is already occurring in the acid rain case. In essence, what greenhouse gas-induced environmental changes create is an issue I like to call the problem of "redistributive justice."

Policy Responses

The last stage in dealing with the greenhouse effect--or other atmospheric and even some non-atmospheric problems--concerns the question of appropriate policy responses. Three classes of actions could be considered. First is mitigation: purposeful interventions in the environment to minimize the potential effects (for example, deliberately spreading dust in the stratosphere to reflect the sunlight to cool the climate as a counter-measure to the inadvertent CO_2

warming). This "geo-engineering" solution suffers from the immediate and obvious flaw that if there is admitted uncertainty associated with predicting the inadvertent consequences of human activities, then likewise substantial uncertainty surrounds any deliberate climatic modification. Thus, it is quite possible that the inadvertent change might be overestimated by our computer models and the advertent change underestimated, in which case our intervention would be a "cure worse than the disease." The prospect for international tensions resulting from such deliberate environmental modifications is so staggering, and our legal instruments to deal with this so immature, that it is hard to imagine acceptance of any substantial mitigation strategies for the foreseeable future.

The second kind of policy action, one that tends to be favored by most economists, is simply adaptation. Adaptive strategies simply let society adjust to environmental changes without attempting to mitigate or prevent the changes in advance. We could adapt to climate change, for example, by planting alternative crop strains that would be more widely adapted to a whole range of plausible climatic futures. Of course, if we don't know what is coming or we haven't developed or tested the seeds yet, we may very well suffer substantial losses during the transition to the new climate. But such adaptations are often recommended because of the uncertain nature of the redistributive character of future climatic change.

Finally, the most active policy category is <u>prevention</u>, which could take the form of sulfer scrubbers in the case of acid rain, abandonment of the use of chlorofluorocarbons and other potential ozone-reducing gases, or a reduction in the amount of fossil fuel used around the world. The latter policies, often advocated by environmentalists, are controversial because they involve, in some cases, substantial immediate investments as a hedge against large future environmental change, change which--it must honestly be admitted--cannot be precisely predicted. The sorts of preventive policies that can be considered are increasing the efficiency of energy end use (in a word, conservation), the development of alternative energy systems that are not fossil fuel-based, or, in the most far-reaching proposal I've seen, a "law of the air." This was proposed in 1976 by anthropologist Margaret Mead and climatologist William Kellogg (Mead, 1976). They suggest that various nations would be assigned polluting rights to keep CO_2 emissions below some agreed global standard.

In summary, then, it is safe to conclude that a strong consensus of knowledgeable atmospheric scientists agrees that a warming of the climate through the greenhouse effect is a highly plausible future condition. The consensus would hold that very rapid climatic changes will cause both ecological and physical systems to go out of equilibrium--a transient condition that makes detailed predictions tenuous. The consensus begins to crumble over detailed assessments of the timing and distribution of potential effects, and thoroughly falls apart over the question of whether present information is sufficient to generate a societal response stronger than more scientific research on the problems--self-serving advice which we scientists, myself included, somehow always manage to recommend.

3.   CLIMATIC CHANGE IN THE PERSPECTIVE OF OVERALL ENERGY SYSTEM RISKS

Since the single largest component of the greenhouse effect is the emission of carbon dioxide gas through fossil fuel use, it is obvious that the climatic risks of fossil fuel use need to be weighed against the risks of alternative energy systems and balanced against the overall benefits of energy end usage. This is precisely what was attempted a decade ago by the Risk/Impact Panel of the NAS/NRC Committee of Nuclear and Alternative Energy Systems (CONAES). Although that panel never issued a formal refereed book, as a member of the panel I published three tables that we produced (Schneider, 1979). Through that article, I had hoped to insure some public exposure for the panel's labor. They are reproduced here as Tables 2, 3, and 4. What is obvious from these tables is the presence of many non-quantified entries, such as "aesthetic impacts" or "arms proliferation" or "governmental interventions." On the other hand, some risks are more easily quantifiable, such as "accidental deaths and injuries" or "acid rain" or "capital costs." The obvious problem for risk analysts, politicians, or

TABLE 2. SOME HEALTH RISKS FROM ENERGY GENERATION AND END-USE

Stage of the fuel cycle	Nuclear	Fossil Fuels — Coal	Fossil Fuels — Oil	Fossil Fuels — Gas	Hydroelectric	Geothermal	Solar
Extraction	Lung cancer; Accidental deaths and injuries	Accidental deaths and injuries; Lung disease; Water contaminants	Accidental deaths and injuries	Accidental deaths and injuries	Accidents in construction	Accidents in construction	Accidents in construction and equipment shipping; Toxic by-products of construction materials
Processing	Cancer and mutation from tailings	Lung disease	Lung disease				
Transportation	Theft; Accidents	Accidents	Accidents	Death or injury from explosion			
Generation	Cancer and mutation; Accidents; Sabotage	Bronchitis and exacerbation of chronic disease; Possible risk of cancer and mutation; Water contaminants; Acute respiratory disease, increased mortality; Annoyance reactions; Altered physiological functions			Accidents from dam failures	Possible toxic effects from emissions (H₂S)	Possible toxic effects (e.g. from metals); Accidents
Transmission	Accidents; possible ozone and electromagnetic radiation changes						
Waste disposal	Cancer and mutation	Cancer; lung disease, mutation				Unknown	
End-use	Deaths and injuries from automobile and other transportation; accidents from other sources; electrocution; fires; accidents from solar rooftop maintenance						

Source: CONAES Risk/Impact Panel.

TABLE 3. SOME ENVIRONMENTAL RISKS FROM ENERGY GENERATION AND END-USE

Stage of the fuel cycle	Nuclear	Coal	Oil	Gas	Hydroelectric	Geothermal	Solar
Extraction	Scarred land	Scarred land Damaged and decreased water supplies Subsidence Excess rubble	Water pollution from seepage and spills Subsidence	Subsidence			
Processing	Mine tailings	Effluents Water shortage	Air and water pollution				
Transportation		Coal dust	Pollution and wildlife harm from spills				
Generation	Thermal effects Water shortage	Acid rain, water pollution particles			Loss of rivers Silting Estuarine imbalance Water loss	Subsidence Brine discharge Thermal effects H₂S Water pollution Water shortage	Local temperature effects Land use Water shortage Loss of shade
		Climate changes from CO₂; thermal effects; water shortages NOₓ effects					
Transmission		Damage to wildlife, forests, and aesthetics from transmission lines					
Waste disposal	Radioactivity	Excess rubble Trace metals				Pollution	
End-use		Ecological damage from high energy use (photochemical smog) Water shortages Habitat loss					

Source: CONAES Risk/Impact Panel.

TABLE 4: SOME SOCIOPOLITICAL RISKS FROM ENERGY GENERATION AND END-USE

Stage of the fuel cycle	Electric			Fossil electric			Geothermal
	Nuclear	Solar thermal	Hydroelectric	Coal	Oil	Gas	
Extraction	Land-use impacts withdrawal from other uses; legal impacts; land rights; access aesthetic	Land-use impacts aesthetic legal impacts boomtowns Siting impacts	Aesthetic (loss of free-flowing rivers)*; Boomtown effects; Siting conflicts*; Land use*; Risks of accidents/sabotage*	Same as nuclear* (more wide-spread)*; Boomtown effects*; Water shortages*; Reclamation costs*; Regional/local inequities	Spill impacts* aesthetic recreational; foreign policy impacts (wars, embargoes); Land subsidence		Siting, land-use impacts; Land subsidence; Noise
Processing	Safeguards* civil liberties diversion				Refinery siting* tank farm siting		
Transportation	Safeguards civil liberties diversion			Accidents		Pipeline and LNG tankers; availability, safety conflicts	
Generation	Safeguards civil liberties diversion Siting fiscal impacts institutional impacts Risk of accidents	Siting impacts		Regional equity disputes (mine site vs remote conversion)* Infrastructure investment			
Transmission			Land use and aesthetic impacts; power lines System vulnerability	Capital intensity of control technologies*			
Waste disposal	Opposition of siting Long-term care/intergenera-tional impacts						
End-use			Local political conflicts Equity impacts Federal preemption	Second-order health impacts Conflicts over safety, economics, equity Intergenerational impacts (e.g. CO_2 effects on climate)			
Whole-system risks	Arms proliferation* Political conflicts over safety, economics, equity (local, national, international)* Federal preemption Second-order health impacts			Capital requirements; inflated demand from inefficient energy utilization; greatest dissociation of costs and benefits; regional disputes; CO_2 and climate control*			
	Inflated demand from inefficient energy utilization; greatest dissociation of costs and benefits; regional disputes*						

continued

Comparative risk assessment of energy systems

Stage of the fuel cycle	Fossil direct use			Nonelectric	
	Coal	Oil	Gas	Solar heating and cooling	Conservation
Extraction	Same as fossil electric*			Legal impacts Sun rights Land use impacts scattering*	Capital costs materials redesign of industrial processes** Institutional impact** Curtailment allocation (equity) problems* Lifestyle impacts* First-cost equity impacts Civil liberties Second-order health risks
Processing	Same as fossil electric				
Transportation	Traffic generation				
Generation	Same as fossil electric			High initial costs Operation and maintenance risks	
Transmission				Governmental interventions	
Waste disposal	Capital intensity of control technologies*				
End-use			Transportation impacts*		
Whole-system risks	Second-order health impacts* Regional disputes Allocation problems Intergenerational impacts (e.g. CO₂ effects on climate)			Back-up requirements* Regional imbalances	

Source: CONAES Risk/Impact Panel.
Note: * Means a particularly important risk.

citizens alike is how to weigh this complex tangle of partially quantifiable and unquantifiable entries, each entry having some fuzzy, often intuitive probability attached.

Even for the relatively "narrow" problem of climatic risks from alternative energy systems, complexity and uncertainties soon dominate the analysis. Consider, for instance, the problem of estimating an energy system-environmental and societal impacts matrix S. This matrix could be disaggregated into a product of an energy systems-emission matrix, E, an emission-climatic effects matrix C, and a climatic effects-environmental and societal impacts matrix I:

$$S = E \times C \times I.$$

The matrix E would have energy systems entries such a solar, hydro, coal, nuclear, etc. for the matrix rows, and categories such as CO_2, dust, water vapor, heat, radionuclides in the emissions columns. The matrix C would then have all the emissions columns of E as its rows, and climatic effects such as temperature, drought frequency, sea level rise, and acid deposition, etc., as matrix C's columns. The final matrix, I, repeats as its rows the climatic effects columns of C and adds impact entries such as crop yield changes, costs of coastal flooding, heat stress on animals, species extinction, forest fire frequency, etc., as its columns. To determine S, the effect of changes in the mix of energy supply systems on environmental and societal impacts, one could simply (in principle only, or course) multiply the three matrices ($E \times C \times I$) for the answer. But since it is obvious that many of the entries are associated with major uncertainties, the exercise is essentially useless in practice. For example, if only one of the matrix elements in the first matrix E were highly uncertain, by the time the three matrices are multiplied, the uncertainty would potentially have spread to all elements of S (e.g., see Schneider and Temkin, 1979).

Where does all this lead with respect to the problem of treating climatic change in the context of quantitative risk methods? Clearly, it is impossible to hope for a well-quantified assessment of comparative climatic risk assessment across various energy options, let alone a fully comprehensive risk assessment that includes all factors on Tables 2 to 4. Nevertheless, we can and do possess the ability to quantify many subelements of the complex whole. I believe it would be just as foolish to dismiss these well-quantifiable elements as to use them as a sole basis for policy choice. Well-quantified aspects are to be welcomed, as they help to place decisionmaking on a firmer factual basis, but they cannot serve as a whole-system risk assessment. There is simply no methodological substitution for human judgment.

Despite all this uncertainty, there are still some general conclusions that I believe can be drawn from this complex problem. Indeed, I draw these conclusions after the CONAES experience and reproduce them here because I think they are still applicable (see Schneider, 1979).

4.   SUMMARY AND CONCLUSIONS

Despite a tremendous range of uncertainties and the diverse character of both risks and benefits associated with alternative energy systems, a few general conclusions seem to emerge from available studies:

1.   No energy system is without risk, and often the risk is not recognized until the system has been used for some time.

2.   Current methods of risk analysis are embryonic. Improvements in all basic areas of risk analysis--monitoring, assessment, and evaluation--are badly needed.

3.   Regardless of the real risk or benefit of an energy system, or even the best calculated values of risks or benefits, it is the public's perception of these that will determine the political acceptability of an energy option. It is thus imperative that the public and its leaders be familiar with the basics of risk analysis and that issues of fact and value are as clearly and distinctly stated as possible.

4.   The risks and impacts of energy generation and usage of health, on the environment, and on society are unequally distributed in space, time, and severity. Those who receive benefits are often not the same as the ones who suffer risk. One chief objective of risk analysis and rational decisionmaking in choosing a mix of alternative energy systems should be an attempt to minimize the inequities inherent in the distribution of risks and benefits from energy usage. Coal and nuclear power appear now to be systems which tend to "export their externalities" more than systems of solar, geothermal, or wind power, whose impacts are more likely to be localized in space and time.

5.   In view of the increasing number of studies of the side-effects of energy use, it is likely that assessed risks from all sources will increase in the future.

6.   In view of the very large uncertainties, particularly regarding coal and nuclear risks, It seems prudent to keep several energy supply options open rather than to rely on a single major source of expansion. This sort of diversity provides a hedge against unforeseen adversities, such as the discovery of heretofore unsuspected risks or the verification of risks now only suspected (e.g., climate changes from CO_2 emissions.)

7.   In the meantime, a shift away from coal towards natural gas would cut the CO_2 emissions substantially, as well as reduce acid rain, local air pollution crises, and other environmental or health risks of massive coal use.

8. Given the prospect of long-term resource depletion of oil and gas, the dangers of expanding coal and nuclear too rapidly, and the long lead times necessary to develop new technologies such as large-scale solar, conservation must be a major part of any future energy program. Fuel not consumed does not produce cancers, does not destroy native habitat, does not change the radiation balance of the earth, and does not deplete non-renewable resources. Of course, the degree of conservation must be carefully chosen to minimize damaging economic or health impacts from "too much" or inappropriate conservation measures.

Finally, a last conclusion that can be drawn about potential risks from energy systems is a simple and obvious one: risks will be proportional to the total energy produced and consumed by society. Therefore, since per capita consumption of energy can be associated, at least for low levels of usage, with improved standards of living, it is essential that global population levels remain as low as possible. Since total energy usage is simply the per capita consumption times the total number of people, steady-state impact of energy usage, for a given supply technology, will be proportional to the steady-state population size of the earth and the per capita levels of consumption.   This suggests that if more energy is to be used as a means of improving standards of living, then parallel steps are also urgently needed to see to it that the benefits of such expanded energy usage do not foster such a large population that the ultimate health, environmental, and social impacts of energy generation and usage will begin to outweigh the very benefits for which the expansion was originally justified.

REFERENCES

Bolin, B., B. R. Doos, J. Jager, and R. A. Warrick (eds.). 1987. The Greenhouse Effect, Climatic Change, and Ecosystems. SCOPE 29. Chichester, U.K.: John Wiley.

Carbon Dioxide Assessment Committee. 1983. Changing Climate. Washington, D.C.: National Academy Press.

Kasting, James F., O. B. Toon, and J. B. Pollack. 1988. How climate evolved on the terrestrial planets. Scientific American 258(2): 90-97.

Mead, M. 1976. Society and the atmospheric environment. Preface in W. W. Kellogg and M. Mead, (eds.), The Atmosphere: Endangered and Endangering. Fogarty International Center Proceedings No. 39. Washington, D.C.: DHEW Publications.

Schneider, S. H. 1979. Comparative risk assessment of energy systems. Energy 4:919-931.

Schneider, S. H. 1987. Climate modeling. Scientific American 256(5): 72-80.

Schneider, S. H. and R. Londer. 1984. The Coevolution of Climate and Life. San Francisco: Sierra Club Books.

Schneider, S. H. and R. L. Temkin. 1978. Climatic changes and human affairs. In Climatic Change. J. Gribbin (ed.). Cambridge, U.K.: Cambridge University Press.

RISK-COST EVALUATION OF COASTAL PROTECTION PROJECTS WITH SEA LEVEL RISE AND CLIMATE CHANGE

David A. Moser and Eugene Z. Stakhiv

ABSTRACT

Federal coastal projection projects can incorporate forecasts of sea level rise and storm frequency changes due to climate change through the application of risk and uncertainty analysis techniques. The incorporation of these forecasts is not a trivial matter, but well within the probabilistic analyses currently employed to estimate project benefits and costs for coastal projects. When the effects of sea level rise and climate change occur in the future, in-place structural projects of a larger scale than that warranted under the current sea level and storm frequencies would offer greater benefits than those designed for the current conditions. In addition, sea level rise and climate change which increase recurring project maintenance costs tend to favor structural-type projects. At the present time, however, there is considerable disagreement on the amount of sea level rise and the impact of climate change on storm frequency. More importantly, the adverse impacts of storm damages occur too far into the future, given the nature of discounting and the level of the federal discount rate, to have much influence on the economically efficient type and scale of project recommended today. There is likely to be a greater reliance on nonstructural, land use management solutions that require state and local regulatory controls. These controls, however, are generally outside the purview of the federal agencies that have some responsibilities for coastal zone management.

1. INTRODUCTION

The impending threat of climate change and sea level risk has brought calls from various sectors for governmental institutions to prepare for this creeping natural hazard. This institutional preparation includes factoring in the relevant consequences into the respective regulatory, planning, and design procedures. The U. S. Army Corps of Engineers is one such federal agency that is responsible for various aspects of a diverse water resources and shoreline protection program whose analyses of the economic efficiency of projects would be affected by climate change and sea level rise. The evaluation principles, procedures, and decision rules the Corps must abide by in

---

David A. Moser and Eugene Z. Stakhiv are with the U. S. Army Corps of Engineers, Water Resources Support Center, Institute for Water Resources, Casey Building, Fort Belvoir, Virginia.

planning coastal protection projects will be highlighted in an attempt to show how current evaluation approaches could incorporate such long evolving natural hazards as sea level rise. The Corps recognizes that its activities are likely to be affected by the hydrologic, meteorologic, and oceanographic consequences of global warming and expected climate changes that are outlined in Figure 1. While the Corps may appear to be relatively unresponsive to an increasingly important potential hazard, it has been delving into the ramifications of these changes for some time. Most of the Corps' incremental modifications so far have dealt with its planning and design procedures which address the manner in which the Corps computes physical (hydrologic and hydraulic) changes and forecasts economic conditions. A primary change has been in the explicit introduction of risk analysis for the selection of an appropriate design basis for alternative plans and component projects involving natural hazard extremes and the mitigation of their social and economic consequences.

It is no longer a question of whether the substantial increases in carbon dioxide and chlorofluorocarbons will cause significant climate changes, resulting in weather changes and sea level rise. Instead, the focus has shifted to when the consequences will become apparent and how society should respond to these changes through adaptation to or mitigation of these consequences. In terms of hydrologic consequences, some areas around the globe will not experience much change, while others will become either drier or wetter. Sea level rise is a phenomenon associated with climate change. The inevitable warming of the atmosphere, especially in the polar latitudes, will accelerate snowpack and glacial melting, whose effect would be felt uniformly worldwide.

Weather changes associated with global warming could imply increased variability and intensity of individual coastal storm events. This phenomenon combined with sea level rise would further exacerbate the present conditions of beach erosion and property damages in coastal areas. Sea level rise alone, even with the present weather regime, will logically cause the landward retreat of the shoreline, following the Bruun rule. Figure 2 shows the relationship between sea level rise, shore retreat, and foreshore profile changes. Any change in the intensity and frequency of storm events, the direction of which is presently largely speculative, will either accentuate or diminish the physical consequences of sea level rise. An additional factor in the proper selection of strategies for societal adaptation to sea level rise and storm frequency is their rates of change. Thus, the immediacy of the consequences to shorelines and coastal development will influence the choice of action.

The fundamental question that climate change and sea level rise poses for society is how to effectively cope with the changes that appear irreversible. Many federal, state, and local institutions are currently debating the possible strategies and specific measures for anticipating the most severe consequences and adapting to the inevitable

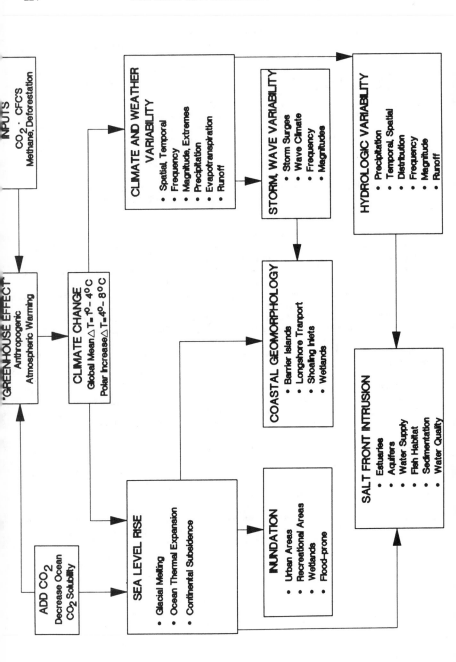

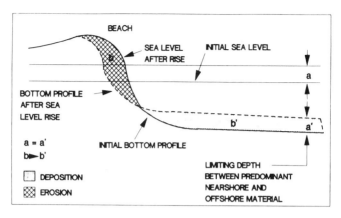

SOURCE: Adapted from M.L. Schwartz (1987).

Figure 2. Bruun Rule

Figure 3. Project Benefits at Time 0

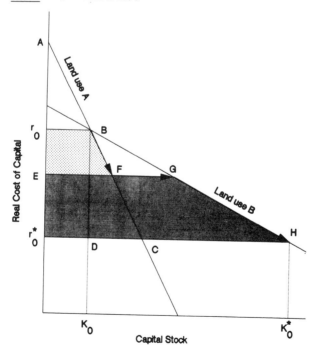

r_0 = Real Cost of Capital without project at Time 0

r_0^* = Real Cost of Capital with Project at Time 0

☐ + ■ = Benefits with Project at Time 0

changes. The Corps of Engineers, as one of these institutions, can effectively deal only with protective measures. This paper will show how the Corps' economic evaluation principles and decision rules influence the choice of a particular shore protection measure. There are many other effective alternative management measures, residing within the responsibilities of the states and local communities, that should be strongly considered in adapting to sea level rise. The range of public measures to mitigate the potential hazards to life and property from sea level rise and climate change will be the same as those available today under "normal" conditions. The probable difference will be that the emphasis on alternative management strategies will change to reflect the reality that the baseline condition is changing. Thus, it is likely that shore protection strategies would shift from protective measures such as groins, bulkheads, sea walls, and beach nourishment to land use modification measures. These adaptive strategies would consist of limiting investments in and subsidies to hazard-prone areas through regulation and disinvestment strategies such as transferable development rights and the use of financial incentives and tax deductions. A major component of such adaptive strategies would be a greater emphasis on emergency warning and evacuation preparedness planning for coastal areas at risk.

The primary point is that adaptive strategies may be economically more efficient and more socially effective than assuming a "King Canute" posture in the face of the inevitable. These adaptive strategies, however, must rely in large part on the state and local regulatory and land use control authorities. The Corps of Engineers does not have the responsibility for such decisions nor does the Corps feel that it is appropriate for a federal agency to exercise such authority. Yet public interest groups and scientific institutions are calling for the Corps to respond to the threat of sea level rise. The following analysis will demonstrate the philosophy, rationale, and procedures by which the Corps would incorporate climate change and sea level rise in shore protection project evaluation.

2.   AN OVERVIEW OF ECONOMIC EVALUATION PRINCIPLES

The federal government has a long history of planning coastal protection projects. The rationale for these projects follows that of the inland flood control program: the risks due to natural hazards inhibit more intensive use of land in these areas by offsetting their location advantage for recreational, residential, or commercial use. By providing protection against the hazard, efficiency gains can be achieved, resulting in an increase in the national output of goods and services. There are also additional regional and local economic gains that result from the transfer of economic activity from some other location. These income transfers are not included because they are not net increases in natural output. The identification and measurement of the national efficiency gains follows benefit-cost analysis (BCA) procedures developed, to a significant extent, to evaluate the national economic implications of federal investments in what are inherently local water resources projects. These procedures are codified in the Economic and Environmental Principles and Guidelines for Water and

Related Land Resources Implementation Studies (P&G), (WRC, 1983).

The types of approaches to mitigating coastal storm damages include:

1. structural: sea walls, revetments, breakwaters, and surge barriers,
2. maintenance: beach nourishment and beach and dune stabilization, and
3. no federal action.

The P&G requires the evaluation of the "no federal action" alternative to provide a baseline, thus quantitatively defining the "without project" future condition. This alternative does not mean "do nothing." Rather, it defines the baseline that would properly include state, local, or individual protective/mitigative measures that would likely occur in the future without a federal project. Each alternative is then evaluated and the benefits determined by comparison to the "without project" baseline. The economic benefit standard of "willingness to pay" (WTP) incorporates, but is not restricted to, storm damages prevented to the current coastal development. Because of the random nature of the intensity and occurrence of storms, the analyst must adopt an evaluation framework that allows for a comparison of the project costs, which are normally deemed more certain, with the relatively uncertain storm damages prevented by the project. P&G recommends the adoption of a risk-neutral evaluation viewpoint, so that the benefits from alternatives are based in part on the reduction in the expected value of future storm damage. Coastal protection projects involve spending money today to gain the predicted benefits in the future. In addition, many types of projects, particularly the beach nourishment maintenance type, require the commitment to future spending to maintain the project. Another important feature of federal BCA is that the current and future dollar costs and benefits must be commensurated or compared in a common unit of measurement; typically in terms of their present values or the average annual equivalent of the present values. Therefore, the level of the interest rate used to determine the present values, the discount rate, influences the economic feasibility of alternative projects. As is well known, large discount rates reduce the influence of future benefits and costs on present values. High interest rates generally favor the selection of projects with low first costs but relatively high planned future maintenance expenditures over those with high first costs but low future maintenance expenditures.

The standard for the identification and measurement of benefits from a water resources investment project is individual WTP. For projects such as flood control and coastal protection, this value can be generated by a reduction in the cost to a current land use activity or the increase in net income possible at a given site. A coastal protection project generates these values by reducing the risk of storm damage to coastal development. Conceptually, the risk from storms can be viewed as incurring a cost to development, i.e. capital investment, at hazardous locations. Thus, the cost per unit of capital, invested at risky locations, is higher than at risk-free locations. Therefore, economic theory predicts that the risk of storm damage results in less intensive development and lower land values compared to the same location with a lower risk or otherwise equivalent risk-free locations.

From an economic standpoint, the risk component of the marginal cost of capital is composed of the expected value of the per unit storm damages plus a premium for risk. This risk premium results from the attitudes or preferences to the individual decisionmaker toward risk. If the individual is risk-averse, the risk premium is positive, indicating that capital must earn a return to cover not only expected storm damages but also to compensate the capital owner for the risk. It is convenient for explanatory purposes to assume that capital owners are risk-neutral, making the risk premium equal to zero. Assuming risk aversion would needlessly complicate the analysis and require including the effects on benefit estimation of increasing, constant, or decreasing risk aversion. Conceptually, the following analysis of the benefits from coastal protection investment applies to both risk-neutral and constantly risk-average capital owners, although the numerical value of benefits would differ slightly.

3.   NATURAL SOURCES OF RISK AND UNCERTAINTY

Storms produce damage to property in coastal areas from several causes. Many storms contain high winds that can cause damages to structures in the immediate area of the shoreline as well as far inland. For the purposes of this discussion, wind damage will be ignored. A storm typically produces a storm surge raising the water surface elevation above the mean high tide level. This surge may be sufficient, even in absence of waves, to produce flooding in low-lying areas from standing water and sheet flow. In addition to the surge, storms also produce larger waves. One common measurement of the size of waves from a storm is the significant wave height. This is measured as the average of the largest one-third of all the waves over some time period. Property subject to direct wave attack can suffer extensive damage to the structure and contents as well as foundation erosion, threatening the stability of the entire structure. The combination of storm surge and waves produces damages that can be identified and classified in three zones in the coastal area. In the immediate shore area, waves are the most significant cause of damage. Where the coastal area is economically well developed with many structures, these structures dissipate much of the wave energy. Immediately inland from the shore, property is subject to damage resulting from water flowing over the natural dunes or other barriers at the shoreline from waves and storm surge. This "weir flow" produces damages similar to those produced by riverine flooding. Inland from the zone of weir flow, damages result from "sheet flow." This is low velocity flow from the weir flow zone as the water spreads out and ponds in the low-lying areas. Obviously, the boundaries between these zones are not fixed, but depend on the severity of the storm. Generally, the more severe the storm, the further inland the boundaries of the damage zones; the inland incursion being directly proportional to the average slope of the land surface.

In addition, the extent of damages in each of these zones, particularly those caused by waves, increases with the duration of the storm.

Storms also produce at least temporary physical changes at the land-water boundary by eroding the natural beach and dunes that serve to buffer and protect the shore from the effects of storms. Increased wave energy during storms erodes the beach and carries the sand offshore. At the same time, the storm surge pushes the zone of direct wave attack higher up the beach and can subject the dune to direct wave action. Some combinations of storm surge, significant wave height, and storm duration can result in breaching of the dune. Thus, the beneficial storm protection effects of the beach and dune are at least temporarily lost following particularly large storms. Therefore, a subsequent storm of even moderate intensity can result in larger property damage until the beach and dune are replenished. Coastal development has also subjected the natural protective system of the beach and dune to stress and direct destruction. In some areas, the dune has been used for building sites, while in others the dune has been breached to provide scenic views and easier access to the beach. In addition, structures such as groins have been built at some locations to trap sand, stabilizing beaches, but impeding the predominant littoral drift, often resulting in sand starvation to down-drift beaches.

In terms of this presentation, the evaluation principles and procedures of coastal protection projects parallel those risk analysis concepts that are more commonly used in the evaluation of inland flood control projects. In reality this approach is not yet practiced by the Corps, although the uncertainty of sea level risk and climate change argues for its implementation. Several components of the evaluation are stochastic, so that the evaluation can be computationally complicated. For instance, the damages from storms are dependent on characteristics described in probabilistic terms, such as intensity, duration, wind direction, and diurnal tide level. Since these characteristics in turn influence the level of storm surge and significant wave height, these two direct factors in storm damage are also stochastic. Note that the uncertain sea level rise is not yet addressed.

4. THE EVALUATION FRAMEWORK

We begin by defining the variable <u>sea state</u> such that it is measured as the sum of the level of storm surge plus the significant wave height or:

$$H = S + h_s \qquad\qquad (1)$$

where

H = sea state in feet above mean sea level (ft. msl.),

S = storm surge (ft. msl.),

h_s = significant wave height in feet.

The level of storm surge is a function of the storm characteristics, so that the annual probability of storm surge exceeding some level $\overline{S}$ depends on the annual probability of storms that can generate a surge of $\overline{S}$ or greater.

The distribution of wave heights from a storm is not independent of the level of storm surge (Bakker and Vrijling, 1981). One can consider the storm surge to shift the probability density function for significant wave heights. Therefore, since

$$f_{h_s \mid S}(h_s, S) = \frac{f_{h_s, S}(h_s, S)}{f_S(S)} \tag{2}$$

and

$$f_{H \mid S}(H, S) = f_{h_s \mid S}(H-S, S) \tag{3}$$

so that

$$f_{H, S}(S, S) = f_{h_s, S}(H-S, S)$$

then the distribution of sea states is

$$f_H(H) = \int f_{H, S}(H-S, S) \, dS. \tag{4}$$

The $f_H(H)$ distribution may be estimated for a particular location from the historical record or derived using estimates of $f_S(S)$ and $f_{h_s \mid S}(h_s, S)$. The $f_H(H)$ distribution can then be used along with a local sea state-storm damage function to estimate expected annual storm damages.

The final component for incorporating classical risk analysis techniques within the benefit evaluation framework for storm protection, specified by the P&G, is to estimate the "with" and "without" project future economic development and land values. Recall that a project produces benefits based on the location advantage of the site. With the project, landowners realize increases in economic rental values of land at protected locations. Because this increase in rental value is location-based, resulting from a reduction in the external costs imposed by storms, it represents a national economic development (NED) benefit as required under P&G. It is this type of economic benefit that is compared to project costs to determine the economic feasibility of any proposed federal project.

Viewing storms as incurring a risky cost to capital invested at hazardous sites, Figure 2 is useful in schematically portraying the basis for the measurement of NED benefits from a project at a particular site. The curve labeled Land Use A represents the demand for capital at the site by land use A. Land use A can be the current pattern of land use or the land use that emerges "without" a federal coastal protection project. This demand curve is derived considering land as a fixed factor of production and by subtracting the cost of labor and intermediate goods from the total revenue generated at the site. Thus the area under D_A measures the maximum potential amount available to pay the owners of capital and labor for their services. Concentrating for the time being on the current land use, the "without project" cost of capital at the site at some initial time (time 0) is r_0, resulting in optimum economic development indicated by the capital stock level K_0. Recall that in the risk-neutral case presented here, r_0 includes the equilibrium market cost of capital plus the expected annual storm damages per unit of capital.

Now assume that the project reduces the cost of capital at the site at time 0 to r_0^*. In Figure 2, the project reduces the expected annual storm damages to existing development by an amount equal to the area $r_0^* r_0 BD$. Total benefits to the existing land use activity include this amount plus the area DBC, which represents the net income from the optimal increase in site use intensity. The sum of these two areas represents the increase in residual return to land and is equivalent to the increase in the rental value of land in its current use with the project.

Note that the project may allow an economically efficient change in land use. Consider an alternative demand denoted by Land Use B. From Figure 2, land use A outbids land use B for the use of the site "without" the project. Note also that land use B can outbid land use A when the cost of capital at the site falls below E. Given the lower site cost of capital with the project, r_0^*, land use for the site is predicted to be different with the project. This complication implies that project benefits are no longer measured primarily as the reduction in expected storm damage. Benefits, however, are still the increase in the rental value of the site with the project. From Figure 2, project benefits are equal to area $r_0^* r_0 BD$ plus DBC plus CFGH. This last area represents the amount by which land use B could outbid land use A for the use of the site with the project.

The reduction in the expected cost of capital and the benefits produced depends on the type and scale of the protection project. Even where two alternative projects have the same scale, as defined by the design level of storm protection (e.g., 100-year storm or probable maximum hurricane), the impact on the cost of capital will differ depending on the magnitude of residual losses from storms that exceed the level of protection. Consider two types of projects: one a structural type, such as a sea wall, and the other a maintenance type, such as beach and dune restoration, stabilization, and periodic nourishment. For a given level of protection, the sea wall is likely to result in different residual storm losses compared to beach and dune restoration, stabilization, and nourishment.

Given the preceding considerations, risk analysis concepts can be applied in the economic evaluation of alternative plans to identify the type and scale of project that maximizes NED benefits net of costs. This is a key conceptual point in risk analysis: the net benefits decision rule for selecting the economically optimal project simultaneously selects the degree of protection and level of residual risk bearing. In terms of Figure 2, each scale of project, measured by the reduction in capital cost, will yield a unique level of expected benefits. Thus, by varying the scale of project for each type, a benefit function can be derived for each type of project. For increase in the level of protection, benefits increase, but eventually at a decreasing rate. If for no other reason, this would result from the fact that increases in project scale offer protection against storms that are less likely to occur. Thus, the contribution to the reduction in expected capital cost for each increment in level of protection becomes smaller as the level of protection increases beyond some level. The benefit curves would also reflect any change in the elasticity of demand for capital as the cost of declined with the project.

In addition to NED benefits, a second major consideration in applying benefit-cost analysis concepts to determining the choice of projects and level of protection is the stream of future project costs. The appropriate costs used in the analysis should provide a measure of all the opportunity costs incurred to produce the project outputs. These NED costs may differ from the expenses of constructing and maintaining the project. For coastal protection projects, expenses would include the first costs of project construction, any periodic maintenance costs, and future rehabilitation costs. In addition, the project may incur environmental or other non-market costs whose monetary value can be imputed. The nature of the stream of future costs depends on the type of project. For instance, a structural-type project typically has high first costs and high future rehabilitation cost but low future periodic maintenance costs. On the other hand, a maintenance type project is composed of relatively low first costs but larger recurring future maintenance costs.

Figure 3 represents the cumulative total project costs over time typical of a structural project and of a maintenance project. Each of the time streams of costs that are the bases for the curves in Figure 3 must be converted into present value terms using the prevailing federal discount rate. Note that the stream of future costs for both types of projects, but especially the maintenance, must be defined in probabilistic terms. The realized amount and timing of maintenance and rehabilitation expenditures depends on the number and severity of storms experienced at the project site in the future. Thus, the expected future cost stream represented in Figure 3 is based on the estimated probability density function for sea states.

Once the alternative formulated plans are evaluated in economic terms, the expected net benefits can be calculated. Following the project selection criteria in P&G, the recommended plan should be the one which "reasonably maximizes" net NED benefits. This plan is defined as the NED plan. Deviations from the NED plan can be recommended to incorporate risk and uncertainty considerations in addition to the explicit risk analysis used in the economic evaluation. These could be considerations for human health and safety or non-monetized environmental concerns.

5. CLIMATE CHANGE AND SEA LEVEL RISE

Thus far the evaluation and selection of federal coastal protection investments has assumed stationarity of climate and mean sea level. The underlying physical parameters and relationships yielding the historically observed distribution of sea states have been assumed to be constant. Sea level rise can be included in the evaluation of planning alternatives as a mean-increasing, variance-preserving shift in the probability density function of storm surge, incorporated in Equations 2 and 4. This results in an increase in the site cost of capital "with" and "without" each alternative. The change in benefits depends on the response of quantity demanded of capital to the increase in site cost determined by the slope of the demand curve for capital. Assume that a rise in sea level over time increases expected storm damage. In Figure 4, which is based on Figure 2, this change is represented by the rise in the cost of capital from r_0^* to r_0^T "with" the project and the equivalent increase from r_0 to r_T "without" the project. Note that the rise in sea level will likely have different impacts on the site cost of capital for different types of planning alternatives. For instance, with a sea wall, a rise in sea level will result in more frequent wave overwash but the structure will retain much of its damage prevention capability. In contrast, higher sea levels will result in more frequent breaching of the dune during storms and loss of much of the damage prevention effectiveness. Thus, the rise in capital costs with sea level rise will likely be greater with beach and dune restoration than with a sea wall. This difference in project effectiveness suggests that the incorporation of higher future sea levels in project evaluation will favor the recommendation of larger, structural type projects over maintenance type. Two additional considerations temper the conclusion that risk analysis will benefit structural projects, however. First, building higher levels of protection than are economically efficient given the current mean sea level means that current net benefits are sacrificed. The larger levels of protection are economically efficient only at higher mean sea levels, which may or may not occur in the future. Second, since the increase in net benefits for a larger scale project occur in the future, the discounting process necessary to determine the present values of benefits and costs will reduce the influence of these

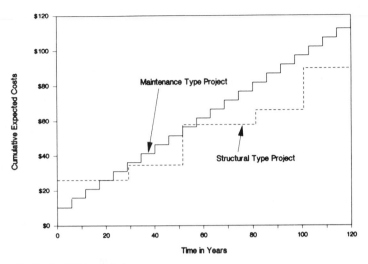

Figure 4. Cumulative Expected Costs

Figure 5. Project Benefits at Time T

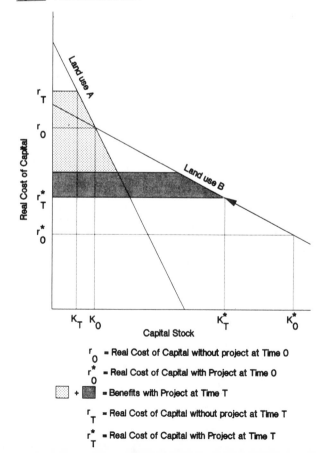

r_0 = Real Cost of Capital without project at Time 0

r_0^* = Real Cost of Capital with Project at Time 0

▨ + ▦ = Benefits with Project at Time T

r_T = Real Cost of Capital without project at Time T

r_T^* = Real Cost of Capital with Project at Time T

future benefits on the determination of the appropriate project scale.

Most forecasts for sea level rise suggest that it is not an immediate problem for coastal development. In addition, there is a wide variation in the estimates of the rate of sea level rise. For instance, a recent National Research Council report (NRC, 1987) notes that relative sea level rise is composed of two components: (1) the localized land subsidence or uplift, and (2) a worldwide rise in mean sea level. Thus, the total relative sea level rise is:

$$T(t) = L(t) + E(t) \qquad (5)$$

where

   $T(t)$ = total relative sea level rise at a particular location,

   $L(t)$ = the local component of relative sea level rise due to land subsidence or uplift, and

   $E(t)$ = eustatic component of relative sea level risk.

The NRC report adopted equations of the following form to forecast total relative sea level rise:

$$L(t) = (M/1000) \bullet t \qquad (6)$$

$$E(t) = (0.0012 + b \bullet t) \bullet t \qquad (7)$$

where

   M = the local subsidence or uplift rate in mm/year, and

   b = the eustatic component of relative sea level risk by the year 2100 in m/year^2.

Therefore,

$$T(t) = (0.0012 + M/1000) \bullet t + b \bullet t^2 \qquad (8)$$

The value of M is fairly well established for many coastal locations: the value of b, however, is subject to wide forecast differences. Table 1 shows the estimates of the total relative sea level rise at Hampton, Virginia and Grand Isle, Louisiana for the three scenarios adopted in the NRC report. The variability in the predicted sea level rise offers a case for the application of sensitivity analysis in the evaluation of project scale. In addition, the disagreement over the eustatic component of relative sea level rise argues for projects whose scale can be staged to account for sea level rise as it occurs.

The problem of choosing the economically optimal level of protection can be generalized to one of maximizing the present value of expected net benefits. This can be written mathematically as:

Table 1. Total Relative Sea Level Rise Forecasts in Feet

	Hampton, VA			Grand Isle, LA		
t	I	II	III	I	II	III
1985	0.0	0.0	0.0	0.0	0.0	0.0
1990	0.1	0.1	0.1	0.2	0.2	0.2
1995	0.2	0.2	0.2	0.3	0.4	0.4
2000	0.2	0.3	0.3	0.5	0.5	0.6
2005	0.3	0.4	0.4	0.7	0.8	0.8
2010	0.4	0.5	0.6	0.9	1.0	1.0
2015	0.5	0.6	0.7	1.1	1.2	1.3
2020	0.6	0.8	0.9	1.3	1.4	1.6
2025	0.7	0.9	1.1	1.5	1.7	1.9
2030	0.8	1.1	1.3	1.7	1.9	2.2
2035	0.9	1.3	1.6	1.9	2.2	2.5
2040	1.1	1.4	1.8	2.1	2.5	2.9
2045	1.2	1.6	2.1	2.3	2.8	3.2
2050	1.3	1.8	2.4	2.5	3.1	3.6
2055	1.4	2.1	2.7	2.8	3.4	4.0
2060	1.6	2.3	3.0	3.0	3.7	4.4
2065	1.7	2.5	3.3	3.2	4.0	4.9
2070	1.9	2.8	3.7	3.5	4.4	5.3
2075	2.0	3.0	4.1	3.7	4.8	5.8
2080	2.2	3.3	4.4	4.0	5.1	6.3
2085	2.3	3.6	4.9	4.2	5.5	6.8
2090	2.5	3.9	5.3	4.5	5.9	7.3
2095	2.7	4.2	5.7	4.8	6.3	7.8
2100	2.8	4.5	6.2	5.0	6.7	8.4

Scenario Eustatic Component by 2100 Scenario		b
I	0.5	0.000028
II	1.0	0.000066
III	1.5	0.000105

$T = (0.0012 + M/1000) \cdot t + b \cdot t^2$
M = rate of subsidence or uplift in mm/yr
  = 3.1 for Hampton, VA
  = 8.9 for Grand Isle, LA
b = the rate of change in rate of growth in eustatic sea level rise for scenarios I, II, and III.

SOURCE: Based on NRC (1987).

$$\text{Maximize} \Bigg\{ \left[\frac{B_t(D,T(t)) - C_t(D,T(t))}{(1 + r)^t} \right] \Bigg\} \tag{9}$$

where

D = design level of protection, and

r = the discount rate.

The hypothesized signs for the components of the solution to (9) are:

$$\frac{\partial B_t}{\partial D} > 0 \; ; \qquad \frac{\partial^2 B_t}{\partial D^2} < 0 \; ; \qquad \frac{\partial B_t}{\partial T} < 0 \; ; \qquad \frac{\partial^2 B_t}{\partial D \partial T} > 0$$

$$\frac{\partial C_t}{\partial D} > 0 \; ; \qquad \frac{\partial^2 C_t}{\partial D^2} > 0 \; ; \qquad \frac{\partial C_t}{\partial T} > 0 \; ; \qquad \frac{\partial^2 C_t}{\partial D \partial T} > 0$$

The impact of sea level rise on expected project benefits and choice of design is displayed in Figure 5. This figure shows the cumulative expected net benefits of two different scaled projects as a function of time. Project A provides the economically efficient level of protection ignoring sea level rise. Project B provides a higher level of protection than A in year 0. With sea level rise, the level of protection for both A and B declines over time, but it declines at a faster rate for A, so that B provides the efficient level of protection by some year T given sea level rise scenario III. The shaded area for A indicates the range of cumulative net benefits given the low (scenario I) and the high (scenario III) sea level rise estimates from the NRC report. Figure 5 shows that building for a future level of protection to incorporate sea level rise generates higher future net benefits but at the expense of lower near-term net benefits. Whether this trade-off makes economic sense requires the comparison of present values between projects A and B.

One way of presenting the economic trade-offs between design project scales that have different time streams of future net benefits is to determine the break-even discount rate for the projects. The break-even discount rate is the interest rate that equates the present values of two streams of future net benefits. Using the net benefit streams for project A, scenario III, and project B, scenario III, shown in Figure 5, the present value of net benefits as a function of the discount rate is shown in Figure 6 for A and B. Notice that for the numerical values in Figure 5, the present value of future net benefits for project B exceeds the present value of future net benefits for project A if the discount rate is less than approximately 3.2 percent. This compares to the 1987 federal discount rate of 8-7/8 percent used for project evaluation. In general, because sea level rise and its effects occur relatively far in the future, incorporating even a high

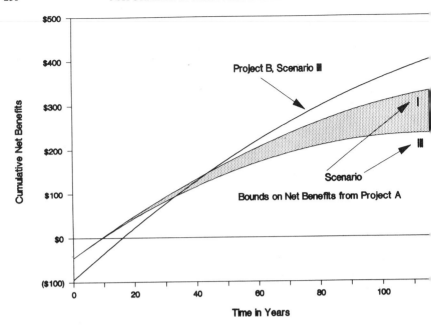

Figure 6. Cumulative Expected Net Benefits

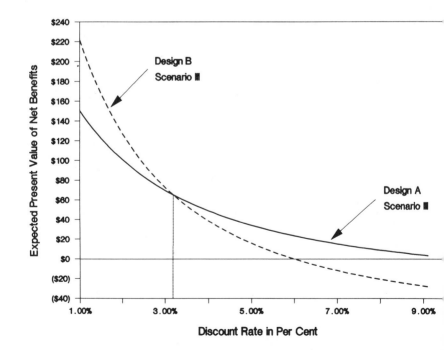

Figure 7. Expected Present Value of Net Benefits as a Function of Discount Rate

forecast of future sea levels in the evaluation of project scale will have little impact on the economically efficient project design. Additionally, the uncertain prospect of the amount of sea level rise may support projects that are more flexible and that can easily incorporate staging of project increments as sea levels change.

Similar to the above analysis, project evaluation could incorporate the effects of forecasted climate change, expressed as a change in the frequency of storm events, through the calculation of expected values and sensitivity analysis. One hypothesis about the effect of climate change is that in many locations the frequency of severe storms will increase over time. This will increase the site cost of capital and reduce the benefits from a coastal protection project of a given scale compared to the situation without the climate change. In addition, since recurring maintenance expenditures depend primarily on the frequency of storms, climate change that increases storm frequency will shorten the time period between these expenditures. This would tend to favor structural-type projects since they have lower maintenance costs. Again, perhaps an overriding consideration for federal projects is the impact of discounting on these future costs and their influence on project type and scale. Thus, even though climate change may result in a dramatic increase in total lifetime project costs, most of the increase occurs beyond the first 15 to 20 years of project life, which are the most influential on the present value of net benefits. Therefore, discounting will mean that these future costs have little impact on plan selection.

REFERENCES

Bakker, W. T., and J. K. Vrijling. 1981. Probabilistic design of sea defences. In Proceedings of the Seventeenth Coastal Engineering Conference, Vol II. New York: ASCE.

Bruun, Per. 1962. Sea level rise as a cause of shore erosion. Journal of Waterways and Harbors Division 1:116-130.

National Research Council. 1987. Responding to Changes in Sea Level: Engineering Implications. Washington, D. C.: National Academy Press.

Schartz, M. L. 1967. The Bruun theory of sea level rise as a cause of shore erosion. Journal of Geology 75:76-92.

U. S. Water Resources Council. 1983. Economic and Environmental Principles and Guidelines for Water and Related Land Resources Implementation Studies. Washington, D. C.: U.S. Government Printing Office.

APPLICATION OF EXTREME VALUE THEORY TO GREAT LAKES RISE

Jonathan W. Bulkley and Robert H. Barkholz

ABSTRACT

The record high level of Lakes Michigan-Huron in the Great Lakes observed in 1986 contributed to the public demand for increased control measures to limit lake level changes for the Great Lakes. At the present time only Lake Superior and Lake Ontario are controlled. Lakes Michigan-Huron, considered a single lake from a hydraulic perspective, and Lake Erie are free-flowing lakes. The application of extreme value theory to the lake level record for Lakes Michigan-Huron provides insight into the cumulative probability of occurrence $F^*(X^*)$ where X^* is the highest monthly water level recorded for each year of record. The return period $T(X^*)$ for specified lake levels is calculated as well. While the total length of record is 125 years, adjustments must be made to reflect that the flow regime draining Lakes Michigan-Huron was significantly altered with major dredging which took place from 1908-1925. The results of the extreme value analysis indicate that the record high lake level observed in 1986 has a return period of 41 years and a $F^*(X^*)$ of 0.9755.

1. INTRODUCTION

This paper addresses the issue of water level in the Great Lakes as an extreme value problem. The water levels in the Great Lakes fluctuate as a consequence of a variety of natural factors. Precipitation and evapotranspiration are the two most critical natural variables with an impact upon long-term lake levels in the Great Lakes system. The historical record for lake levels in the Great Lakes extends back to 1860. However, as a result of physical changes which have taken place in certain of the connecting channels, the useful length of record is less than the full 127 years. Also, the levels of Lake Superior and Lake Ontario are controlled. Accordingly, the application of extreme value theory to the Lakes themselves will be limited to those lakes which are not controlled. Furthermore, because of the nature of the connection between Lake Michigan and Lake Huron at the Straits of Mackinaw, these two lakes are considered as one lake from a hydraulic perspective.

This paper will first present certain basic information on the Great Lakes system as a whole. This will provide the reader with an appreciation of the physical size and complexity of the Great Lakes system. Next, a brief discussion will be provided to introduce the

Jonathan W. Bulkley is Professor of Natural Resources and Professor of Civil Engineering, The University of Michigan, Ann Arbor, Michigan.

Robert H. Barkholz is a graduate student in Environmental and Water Resources Engineering, The University of Michigan, Ann Arbor, Michigan.

concepts of increased diversions into and out of the Great Lakes. This topic has become an issue of concern in the last seven years as a consequence of two primary forces.

First, there has been continuing concern over the problems of adequate water for agricultural purposes in the Ogallala High Plains region of this country. Since the end of World War II, irrigation agriculture has become a major source of economic strength in this region of the country. The water for this irrigation is drawn from groundwater. The depletion of the groundwater resources in this region poses an issue for the transport of other waters to maintain the irrigated agriculture economy which has become so important in the area.

The second force operating to focus attention upon future diversions from the Great Lakes is climate change in the region. There will be a brief discussion regarding the preliminary results of modelling the carbon dioxide-induced climatic change in the Great Lakes Region. This pattern of global warming, which presents the possibility of sea level rise, reverses itself in the Great Lakes. The initial results suggest that while the precipitation may increase in certain portions of the watershed, the overall impact upon lake levels will cause a decrease. This fall in lake level is expected to result from the increase in evapotranspiration as a consequence of the increase in ambient air temperature.

The application of extreme value theory will be made to the lake levels observed in Lake Michigan-Lake Huron. This application will be made for the entire period of record, and for a shorter period of record, which reflects the changes which have taken place as a consequence of the major dredging in the St. Clair River, which serves as the outflow from Lake Michigan-Lake Huron. The current status of the lake levels will be reported and the importance of extreme value theory for future conditions will be discussed.

Finally, in accordance with the request made of all speakers at the conference, the following working definition of "risk" is offered (McCormick, 1981):

$$\text{Risk} = \left\{ \frac{\text{Events}}{\text{Unit time}} \cdot \frac{\text{Consequences}}{\text{Event}} \right\}$$

Another possibility would modify the above definition as follows:

$$\text{Risk}_k = \left\{ \frac{\text{Events}}{\text{Unit Time}} \cdot \left(\frac{\text{Consequences}}{\text{Event}} \right)^k \right\}$$

where $k > 1$ is used to amplify the importance of events with large damages.

2. PHYSICAL SYSTEM

The information presented in Figure 1 provides a graphic layout of the five Great Lakes. It should be noted that the elevation drop from

Lake Superior to the Atlantic Ocean is 600 feet. Lakes Superior and Ontario are controlled. Lakes Michigan-Huron and Lake Erie are not controlled and their levels fluctuate according to inflow, outflow, precipitation, and evaporation. All of the flow figures shown in Figure 1 are in thousands of cubic feet per second. Accordingly, note that the outflow from Lakes Michigan-Huron averages 187,000 cfs. For comparison purposes, this outflow through the St. Clair River, Lake St. Clair, the Detroit River, and into Lake Erie is roughly equivalent to 135 million acre feet per year. Also, Figure 1 indicates that the very large surface areas of Lakes Superior and Michigan-Huron, coupled with the precipitation in the region, provide the greatest single water source input to these lakes on an annual basis. As a consequence, changes in either precipitation and/or air temperature will have significant impacts upon resulting lake levels (Quinn and Croley, 1983).

Table 1 provides certain data on the physical size of the Great Lakes system. This paper is concentrating upon Lakes Michigan-Huron, since this lake level is not controlled and since the high lake levels in Michigan-Huron and Lake Erie have been experienced in the recent past. These high lake levels have generated much interest among shoreline property owners who want to find means to lower these levels and control them in the future. In terms of the physical characteristics of Lakes Michigan-Huron, it is important to note that the total volume of water contained in the system exceeds 2000 cubic miles. The maximum depth of Lake Michigan is 923 feet; the maximum depth of Lake Huron is 750 feet. Even though the flow leaving these lakes is 187,000 cfs, the mean residence times for the two lakes is very long. For Lake Michigan, it is on the order of 106 years; for Lake Huron, it is on the order of 21 years.

Table 1

Physical Characteristics[*]

	LAKE MICHIGAN	LAKE HURON
Water Surface	22,300 sq. mi.	23,000 sq. mi.
Land Drainage	45,600 sq. mi.	51,700 sq. mi.
Maximum Depth	923 ft.	750 ft.
Volume	1,180 cubic miles	850 cubic miles
Mean Outflow	52,000 cfs	187,000 cfs
Mean Residence Time	106 years	21 years

Lake Level (Michigan-Huron considered as one lake: 1900-1983)

Average: 578.25 feet (above sea level)
Maximum: 581.04 feet
Minimum: 575.35 feet

Note: 1986 Maximum Lake Level (Michigan-Huron): 582.00 feet

[*](IFYGL; 1981 U.S. Army Corps of Engineers, Detroit District, 1985)

to Each of the Great Lakes*

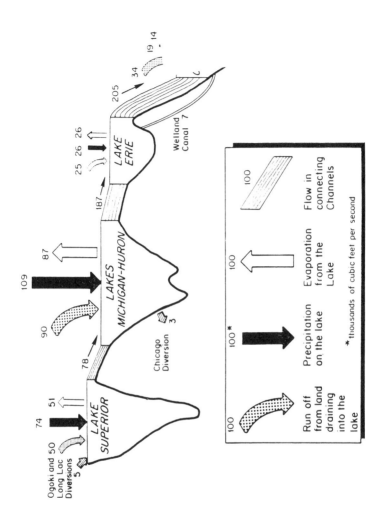

LAKE SUPERIOR

Ogoki and 50
Long Lac
Diversions
5

74
51

90
109
87

LAKES
MICHIGAN-HURON

78
187

Chicago
Diversion
3

LAKE
ERIE

25 26 26

205
34
19 14

Welland
Canal 7

Legend:

100	Flow in connecting Channels
100	Evaporation from the Lake
100*	Precipitation on the lake
100	Run off from land draining into the lake

*thousands of cubic feet per second

*U.S. ARMY CORPS OF ENGINEERS (DETROIT DISTRICT), 1985

3. GREAT LAKES DIVERSIONS

At the present time, there are five diversions into or out from the Great Lakes. The Ogoki and Long Lac diversions are into Lake Superior. There, combined flow is 5600 cfs into Lake Superior. The Chicago diversion takes 3200 cfs from Lake Michigan at Chicago. This water does not return to Lake Michigan. Rather, it enters the Mississippi River drainage via the sanitary canal and the Illinois River. The Welland Canal diverts 9,200 cfs from Lake Erie to Lake Ontario. It is the navigation link that permits vessels to operate between Lake Erie and Lake Ontario. Furthermore, a significant quantity of hydroelectric power is generated by waters flowing from Lake Erie to Lake Ontario via the Welland Canal. The fifth diversion is from the Niagara River to supply the New York State Barge Canal. The diversion is 700 cfs. The diversion takes place below the Niagara River hydraulic control section and, as a result, there is no effect on Lake Erie levels. Also, there is no effect on Lake Ontario levels, since most of the water from the barge canal flows into Lake Ontario. The overall net impact of all of these current diversions into/out from the Great Lakes is to lower the level of two lakes and to raise the level of two lakes. Lake Superior has been raised by 0.07 foot (0.84 in.); Lake Ontario has been raised by 0.08 foot (0.96 in.); Lakes Michigan-Huron have been lowered by 0.02 foot (0.24 in.); and Lake Erie has been lowered by 0.33 foot (3.96 in.). Accordingly, for the purpose of this paper (namely, the level of Lakes Michigan-Huron as an extreme value problem), it is clear that the existing diversions into/out from the lakes have had a minimal impact on the net change in the level of Lakes Michigan-Huron (DeCooke et al., 1984).

There has been considerable speculation regarding the possible use of Great Lakes water to supply portions of the Western states and Canadian provinces for agricultural and other purposes. The North America Water and Power Alliance (NAWPA), as conceived in 1964, proposed to bring water from Alaska and Canada for delivery to the Great Basin, Lake Superior, Texas, and Mexico. NAWPA included plans to divert a flow of 152,000 cfs for the purpose of power generation and agricultural development (NWC, 1973). The Great Recycling and Northern Development (GRAND) Canal concept proposed to convert James Bay into a fresh water lake and divert the fresh water into Lake Huron. The maximum diversion into Lake Huron and hence into the Lakes Michigan-Huron as well as Lake Erie and Lake Ontario would be 70,000 cfs. This proposal called for the diversion out of the lakes of 35,000 cfs, to be used primarily for agricultural purposes in both Canada and the United States (Futures in Water, 1984).

In 1982, a Task Force on Water Resources for the State of Michigan requested that a special study be undertaken to identify possible uses of waters if diverted from the Great Lakes at some time in the future. The study did not propose that such diversions should take place. Rather it was conducted to identify how the water might be used and to make some preliminary calculations regarding the costs and impacts of such diversions on the Great Lakes. The final paper from this study estimated the impact on lake levels for up to three 10,000 cfs diversions for water to be provided to the Missouri River and the Mississippi River as compensation for water being held back in these two watersheds for agricultural purposes. Depending upon the location and number of these diversions, the impacts on the Great Lakes would be as follows (DeCooke et al., 1984):

LAKE	REDUCTION IN LAKE LEVEL[*]
Lake Superior	-1.00 in. to -9.36 in.
Lake Michigan-Huron	-2.00 in. to -18.00 in.
Lake Erie	-6.00 in. to -17.00 in.
Lake Ontario	-6.00 in. to -18.00 in.

It should be noted that the costs to build the conveyance systems and the power plants to divert these volumes of water is estimated to exceed $29 billion. These capital costs alone would require water charges ranging from $45/acre foot to $350/acre foot. Finally, as a consequence of the large volume of the lakes in relation to these diversions, it would take nearly 15 years for the lakes to come to a new steady-state condition, i.e. to establish a new long-term average lake level reduced from the present long-term average as shown above.

Figure 2 shows a plot of lake levels in the Great Lakes for the time period from 1950 to 1982. As previous noted, Lakes Superior and Ontario are controlled for navigation and power production. Lakes Michigan-Huron, Lake St. Clair, and Lake Erie are not controlled. It should be observed that relatively high levels were observed in 1952 and 1973. Low lake levels were observed in 1964. Finally, the patterns of lake levels in Lakes Michigan-Huron, Lake St. Clair, and Lake Erie are very similar to one another since these lakes are not controlled and the flow proceeds through them, as shown in Figure 1.

4.   CLIMATE CHANGE IN THE GREAT LAKES REGION

The issue of climate change in the Great Lakes Region is a subject of sharp interest. One preliminary study (Cohen, 1986) indicates that carbon dioxide in the atmosphere in the region has increased 20% since 1886. Furthermore, using two climate models, the effect of induced warming in the Great Lakes Region has been estimated to reduce the net basin supply of water in the system. Furthermore, as a consequence of the warming, it is anticipated that there will be an additional consumptive use within the basin itself, which will further act to reduce the net basin supply (Quinn, 1985). The reduction in net basin supply will mean a reduction in lake levels in the Great Lakes system.

---

[*] Three diversions of 10,000 cfs each would be required to fully meet the four irrigation routes (A,B,C,D) identified in the Ogallala High Plains Study (HPA, 1982). The maximum reduction in lake level occurs with all three diversions in place.

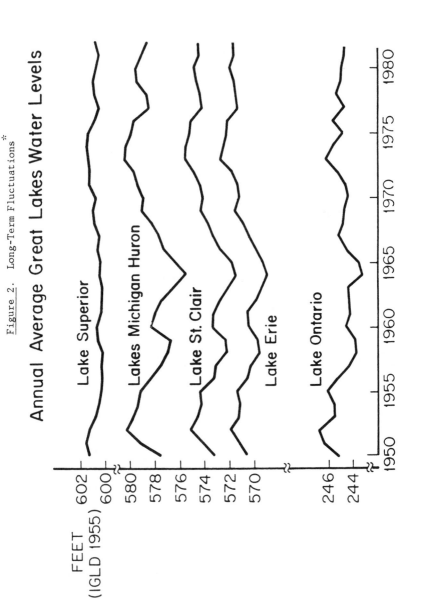

Figure 2. Long-Term Fluctuations *

Annual Average Great Lakes Water Levels

* U.S. ARMY CORPS OF ENGINEERS (DETROIT DISTRICT), 1985.

The preliminary studies indicate that the net basin supply will be reduced from 10,000 cfs to 54,000 cfs as a consequence of weather impacts alone. The net basin supply will be further reduced by 26,000 cfs as a consequence of increased consumptive use within the Great Lakes Region. The overall impact on lake levels of the climatic change plus the increased consumptive use will come from the reduction of net basin supply of from 36,000 cfs to 80,000 cfs. Flow reductions of this magnitude may reduce lake levels from 2.0 feet to 5.0 feet.

In summary, there are five diversions into/out of the Great Lakes at the present time. The overall net impact of these present diversions has been insignificant in terms of changes in long-term lake levels. It is conceivable that future diversions to the Missouri and Mississippi River basins could result in long-term lowering of the mean lake levels. Also, it is conceivable that as a consequence of global warming patterns, the levels of the Great Lakes could be reduced by two feet or more. Finally, it is very unlikely that the GRAND Canal project will be implemented in the foreseeable future as a consequence of a recent statement of Federal Water Policy in Canada (Federal Water Policy, November 1987). The record-high lake levels observed in 1986 prompted legislation at both the state and federal levels in the United States, calling for increased diversions from the Great Lakes in order to reduce the high lake levels. It is clear that the magnitude of diversions necessary to reduce lake levels would need to be on the scale of 30,000 cfs or more. Once such diversions are initiated, it is difficult to contemplate how one could reduce or terminate such diversions. Finally, the climate change models suggest that lake levels will be reduced by two or more feet as a consequence of global warming. Accordingly, to further reduce lake levels through increased diversions may be ill-advised at this time. All of this interest in lake levels and actions being studied to reduce lake levels leads to the application of extreme value theory to estimate the return period for the high lake levels observed in 1986.

5. APPLICATION OF EXTREME VALUE THEORY TO LAKE LEVELS

Extreme value distributions describe the probability of occurrence $F^*(X_*)$ of either the maximum or minimum x_* when a large number of independent events are sampled from an initial distribution. The asterisk is used to emphasize that the distribution $F^*(X_*)$ differs from the parent distribution $F(X)$, and x_* differs from x. By restricting attention to cases in which the number of events in the sample is large, one need consider only the asymptotic distributions for maximum or minimum values.

As applied to the water level of Lakes Michigan-Huron, x_* will refer to the highest monthly water level for the year m. The distribution which applies to sizes of floods over a large number of years will be used to evaluate the extreme high lake levels. The distribution is the Type I asymptotic distribution (Gumbel, 1958).

The cumulative probability of occurrence for such a distribution is defined as

$$F^*(X_*) - \exp(-e^{-y}),$$

where

$$y = \alpha(X_* - u),$$

and X_* = the extreme value for which one seeks the probability of occurrence,

$\alpha = \dfrac{1.28255}{\sigma}$ where σ is the sample standard deviation

and

u = the characteristic largest value. (If one has N samples, each composed of n observations taken from the same distribution, and if one chooses the N largest values, about 36.8% of them will be below the characteristic value.) u is defined by

$$u = m - .577\alpha^{-1}$$

where

$$m = \text{the mean value of } X_*.$$

The return period $T(X_*)$ of the extremes of magnitude at least X is given as

$$T(X_*) = (1-F^*(X_*))^{-1}.$$

The application of these extreme value distributions to the lake levels for Michigan-Huron will be presented for two cases. The first case will be for the entire period of record (1860-1985) in order to calculate the $F^*(X_*)$ and the $T(X_*)$ for the all-time record high level of 582 feet observed in Lakes Michigan-Huron in October 1986. It is recognized that the observed lake levels are not independent events. Because of the large volume of the lakes, changes in lake level are auto-correlated. A high lake level last year will influence the lake level this year. However, for this preliminary application, this factor is not being considered. Future studies will be undertaken to examine ways to handle the auto-correlation of lake levels. Secondly, it is also recognized that man-made changes through channel dredging have modified the flow characteristics of the connecting channels which drain Lakes Michigan-Huron (International Great Lakes Levels Board, 1973). It is the recognition of this fact that resulted in the second case. This second case utilizes the lake level data from 1926--1985 (59 years of record), since this is the time after the major channel modifications were made in the St. Clair River, Lake St. Clair, and the Detroit River--the connecting channels which drain Lake Huron.

Case 1 (Complete Record, 1960-1985)

$m = 579.3508$ ($n = 126$)

$\sigma^2 = 1.7572$

$u = 578.7544$

Results

X_*	$F^*(X_*)$	$T(X_*)$
(Lake level)		(Years)
580.0	0.7411	3.9
582.0	0.9577	23.6
583.0	0.9837	61.3
584.0	0.9938	160.5
585.0	0.9976	420.0

It is noted that according to these results, the observed lake level of 582.0 feet in Lakes Michigan-Huron has a return period of 24 years. However, this actual level has only been observed once in the entire period of record-keeping since 1860. Furthermore, in examining the data, it is noted that during the 19th century and early 20th century, the lake level for Michigan-Huron appeared to be generally higher than for the period following 1920. A further examination determined that in the period from 1908-1925, dredging in the connecting channels draining Lake Huron had been undertaken for navigation improvements. This navigation improvement increased the outflow from Lake Huron with a subsequent impact to reduce the lake level. Accordingly, a second case was examined from 1926 to 1985.

Case 2 (1926-1985: following modifications for navigation)

$m = 578.7415$ ($n = 59$)

$\sigma^2 = 1.7936$

$u = 578.1390$

Results

X_*	$F^*(X_*)$	$T(X_*)$
(Lake level)		(Years)
580.0	0.8451	12.9
581.0	0.9375	16.0
582.0	0.9755	41.0
583.0	0.9905	106.0
584.0	0.9964	274.0
585.0	0.9986	714.0

The results obtained in Case 2 with the data limited to the years following completion of the major navigation improvements offer more credible outcomes. For example, the return period for a lake level of 581.0 is 16 years. In the 59 years of record, there are three observations of the maximum monthly lake level for a given year at 581.0 feet. Furthermore, the maximum observed level of 582.0 feet has an estimated return period of 41 years, which is reasonable given a record length of 59 years. Finally, it should be noted that the return periods for lake levels greater than 582.0 feet for Lakes Michigan-Huron increase significantly at levels above 583.0 feet. While the maximum observed level to date (582.0 feet) has a return period of 41 years, the return period of a level of 583.0 feet is estimated at 106 years.

6. SUMMARY

This paper has presented a number of issues associated with the application of extreme value techniques to the problem of lake levels in the Michigan-Huron system of the Great Lakes. Care must be taken in the application of these techniques to consider the fact that over time, the natural conditions may have been altered by human intervention. Such intervention may change the basis for the data used for extreme value analysis. The issue of independence still needs to be addressed with specific regard to lake levels in the Great Lakes. Nevertheless, the results obtained from the Case 2 analysis, which limits the data to the time period following the major connecting channel modifications, suggest that extreme value applications may be very useful in providing insights into the probabilities of occurrence and the return periods associated with high lake levels. It should be noted that the same framework could be used to examine low lake levels as well.

The paper also presented the background and current status of diversions both into and out of the Great Lakes at the present time. With the record-high lake levels observed in 1986, many suggestions were made to increase diversions out of the Great Lakes in order to reduce the high lake levels. The paper presented information on the economic cost implications of building such diversion structures. Furthermore, it presented information on the fact that it would take 12 to 15 years for such man-made diversions to have their full impact on lake level reduction. Also, the preliminary results of climatic change modelling for the Great Lakes region indicates that major lake level reductions may occur within the next century as a consequence of global warming.

Finally, it should be reported that the actual level of Lakes Michigan-Huron has fallen since October 1986. At that time the level was nearly 582.0 feet or nearly 3.40 feet above the long-term average for October. At the end of October 1987, the level of Michigan-Huron was 579.2 feet--a decrease in lake level of more than 2.70 feet. It is now only 0.80 feet above its long-term average for the month of October. This drop in lake level has had an impact on the damage caused to shoreline property. For example, 60 homes in the state of Michigan fell into the Great Lakes in 1986. In 1987, fewer than ten homes have been lost to the lakes.

REFERENCES

Bulkley, J. W., S. J. Wright, and D. Wright. 1984. Preliminary study of the diversion of 10,000 cfs from Lake Superior to the Missouri River basin. Journal of Hydrology 68:461-472.

Cohen, S. J. 1986. Impacts of CO2 induced climatic change on water resources in the Great Lakes basin. Climatic Change 8:135-153.

Croley, T. E., II. 1986. Understanding recent high Great Lakes water levels. Contribution No. 499, Great Lakes Environmental Research Laboratory.

DeCooke, B. G., J. W. Bulkley, and S. J. Wright. 1984. Great Lakes diversions: preliminary assessment of economic impacts. Presented at conference on Our Great Lakes: Resources for Growth with Quality. Michigan State University, East Lansing, Michigan, March 24, 1984.

Environment Canada. 1987. Federal water policy. Ottawa, Canada.

Futures in Water. 1984. In Proceedings of the Ontario Water Resources Conference. Toronto, Ontario, June 12-14, 1984.

Gumbel, E. J. 1958. Statistics of Extremes. New York: Columbia University Press.

High Plains Associates. 1982. Six-state High Plains Ogallala aquifer regional resources study. Cambridge, Massachusetts: Arthur D. Little, Inc.

National Oceanic and Atmospheric Administration, Great Lakes Environmental Research Laboratory. 1981. The International Field Year for the Great Lakes. Ann Arbor, Michigan: U.S. Department of Commerce.

International Great Lakes Levels Board. 1973. Regulation of Great Lakes water levels. Report to the International Joint Commission.

McCormick, N. J. 1981. Reliability and Risk Analysis. Florida: Academic Press.

National Water Commission. 1973. Water policies for the future. Final report to the President and the U.S. Congress. Washington, D.C.: U.S. Government Printing Office.

Quinn, F. H. 1985. Implications of interbasin diversions, consumptive use, and the greenhouse effect on future Great Lakes water management. In Preprint Volume of the Sixth Conference on Hydrometry.

Quinn, F. H., and T. E. Croley. 1983. Climatic basin water balance models for Great Lakes forecasting and simulation. In Proceedings of the Fifth Conference on Hydrometeorology, American Meteorological Society, Boston, Massachusetts.

U.S. Army Corps of Engineers (Detroit District). 1985. Great Lakes: water level facts.

WARNING SYSTEMS IN RISK MANAGEMENT: THE BENEFITS OF MONITORING

M. Elisabeth Paté-Cornell

ABSTRACT

A method is presented here that allows probabilistic evaluation and optimization of warning systems, and comparison of their performance and cost-effectiveness with those of other means of risk management. The model includes an assessment (1) of the signals, and (2) of human response, given the memory that people have kept of the quality of previous alerts. The trade-off between the rate of false alerts and the length of the lead time is studied to account for the long-term effects of "crying wolf" and the effectiveness of emergency actions. An explicit formulation of the system's benefits, including inputs from a signal model, a response model, and a consequence model, is given to allow optimization of the warning threshold and of the system's sensitivity. An application to the problem of monitoring of dams is presented. The emphasis is on the study of the uncertainties regarding the observation of precursors of the different failure modes, the lead time that they provide, and the possibility of using this lead time either for preventing the collapse of the dam or for evacuation.

1. INTRODUCTION

Risk management for a critical engineering facility often involves monitoring and therefore requires the design and operation of a warning system. This is true, for example, for dams or nuclear power plants. When designing a warning system, three types of decisions have to be made: technical, strategic, and tactical. Technical decisions involve choosing the phenomena to be monitored and the observation system. For example, in a nuclear power plant, what characteristics (temperature, pressure) of a subsystem are to be observed and by what means? Strategic decisions include the choice of the signal or set of signals that will constitute an alert, and the kind of response to be considered or recommended when a signal occurs. For example, for a fire alarm, this can be the choice of a particular density threshold of smoke particles in the air that will trigger an alarm calling for immediate evacuation. Tactical decisions involve response decisions to be made at the time of an alert on the basis of chosen strategies and the specific circumstances of the warning (lead time, additional information, local constraints). The goal of this paper is to present and illustrate a method of probabilistic analysis of warning systems that can be used as a support for technical and strategic decisions of this type.

Most warning systems can issue false alerts or miss the event altogether. If the system is too sensitive, it will issue too many false alerts and people will stop responding. If it is not sensitive

---
M. Elisabeth Paté-Cornell is an associate professor in the Department of Industrial Engineering and Engineering Management, Stanford University, Stanford, California.

enough, it may not provide any warning at all, or the alert may occur with insufficient lead time to allow for appropriate protective measures. Therefore, in the choice of an alert threshold, a trade-off often exists between Type I errors (defined here as failure to warn) and Type II errors (defined here as false alerts). Stochastic methods permit the analysis and optimization of warning systems by assessment of the probabilities and consequences of all types of scenarios (i.e., true alert, false alert, missed prediction) (Paté, 1986). The model described in this paper involves the characteristics not only of the technical system but also of the human response to alerts under different types of circumstances. In particular, the method accounts for the effects of "crying wolf."

    The focus of this paper is on technical and strategic decisions. First, a general Bayesian analysis of the value of warning information is described using as an illustration the problem of monitoring dams. The question is to what degree one can expect to reduce potential losses due to catastrophic failures, given a particular warning system as described by the nature and quality of the signals. The simple model presented in this first part does not involve the memory effect or the possibility of reduced response due to a history of false alerts. Secondly, a general theory of warning systems is briefly presented. It involves a stochastic analysis of the signals and a study of human response according to the type of risk involved and the lead time given by the warnings. According to the type of risk, the emphasis is put on different aspects of human response. For frequent events (for example, fire alerts in a student dormitory), the phenomenon of interest is the memory of the system's past performance and the "cry-wolf" effect. For rare events (for example, accidents in nuclear power plants), the emphasis is on the willingness to respond given the perception of the risk and the sociological context of the warning (Paté and Claudio, 1987). Thirdly, several illustrations of the method are briefly described, each of them emphasizing different aspects of the model: fire alarms in a student dormitory, camera monitoring in an oil refinery, warning systems for nuclear power plants, and the link between missile launch policy and the reliability of the command and control system for nuclear weapons.

2.   SIMPLE BAYESIAN ANALYSIS: MONITORING OF DAMS

    In the course of developing a method to introduce risk costs into a cost-benefit analysis for new dams (Baecher, Paté, and de Neufville, 1980; Paté, 1984; Paté and Tagaras, 1986), it became clear that careful monitoring of dams can substantially reduce the failure losses but not eliminate the risk altogether. The Bayesian model presented here permits computation of the fraction of these potential losses that can be avoided, given the characteristics of the signals corresponding to the different failure modes.

Elements of the Model

    The efficiency of a warning system depends on several factors (Paté, 1986):

- the probability that these signals are detected and properly interpreted, given that they appear
- the lead time of these signals
- the actions that can be taken during the lead time, including actions that can prevent failure altogether and actions that can save part of the population if failure cannot be prevented
- the expected proportion of the population that can be evacuated, given the lead time.

It is assumed that property damage cannot be significantly reduced by warnings, except by actions that actually prevent a failure that would otherwise have occurred.

Our Bayesian analysis of the effects of monitoring dams therefore includes the following random events and random variables:

- Potential failure occurrence (annual probability)
- Failure modes
- Signals occurrence
- Lead time
- Potential avoidance of dam failure
- Proportion of people saved

The results of the analysis are (1) the probability that a failure that would have occurred otherwise is avoided by appropriate actions following a warning, (2) the expected proportion of property damage that is avoided by preventing the collapse of the dam when lead time permits, and (3) the expected proportion of casualties avoided by evacuation following a warning. The risk costs can then be computed, accounting for the effects of the considered warning system.

Figure 1 represents a decision tree for the question of whether or not to adopt a warning system of given characteristics for a hypothetical earth dam. The performance of the warning system is assessed by the difference in risk costs with and without the considered monitoring system. What has not been included in this evaluation is the actual cost of false alerts (not only financial but also psychological) and the fact that false alerts may erode people's confidence in subsequent warnings.

Failure Modes and Signals

Failure Modes

Statistical information on failure causes can be found in the literature (see, for example, USCOLD, 1975). It should be noted that the actual cause of failure is not always clear and that there may be more than one cause involved in a given accident. Statistics about failure modes including external stimuli are the following:

- Overtopping: 20% of cases.
- Internal erosion (e.g., piping or seepage): 40% of cases.
- Embankment slide: 30% of cases.
- Other external causes, such as acts of war or earthquakes: 10% of cases.

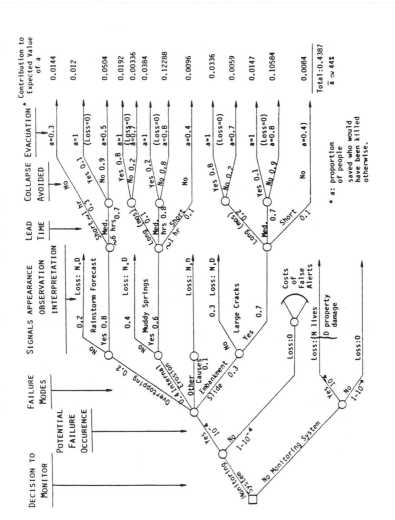

Figure 1. Decision Tree for the Monitoring of an Earth Dam

As a first approximation and by default, it can be considered that these percentages provide the probability of each possible failure mode, conditional on failure occurrence.

Signals

Three types of signals, each related to one failure mode, are considered here. They reveal a potentially serious situation that may require evacuation. These signals are: (1) the appearance of large cracks on the embankments, (2) the appearance of springs carrying soil particles at the foot of the dam, and (3) the forecast of an extreme storm.

The appearance of large cracks may be a precursor of an embankment slide. The lead time can range from less than an hour to months, most often between a few hours and a few days. To prevent failure in this case, immediate actions can be taken, such as drawing down the reservoir, plugging the cracks to prevent inflow of surface water and, depending on conditions, stabilizing the slope by placing material downstream. Immediate evacuation is appropriate if a slide develops that removes the crest of dam and water seeps out of the slide zone. Immediate evacuation is also appropriate if a slide develops during a severe storm, which reduces freeboard.

The appearance of muddy springs along the downstream slope or on the flat ground below may be an indicator of piping, and therefore a precursor of failure by internal erosion. The key issue is to detect whether or not the spring water is actually carrying away soil particles, in which case there is an imminent hazard. Internal erosion may worsen with time: as the phenomenon progresses, the gradient (change of "total head") that causes particle movement increases. The lead time is in the order of a few days, although signals can appear earlier (e.g., Teton Dam). The problem is to properly interpret the spring signal, because a spring of water that does not contain soil particles is much more frequent and of much less concern. Clearly, a large and growing muddy spring, carrying away a lot of material, requires immediate evacuation. Immediate actions to be taken include, as in all cases, lowering the water level in the reservoir. The next possible step is to construct an outside filter that could retain the soil while letting the water escape. Later, one can construct drains or improve the core by injection of chemicals (colloidal chemicals, cements, or clay).

Extreme rainstorms can lead to overtopping. Their forecast can be a reason for evacuation when the spillway capacity is much smaller than required by the magnitude of the storm. Immediate protective measures include removing as much water as possible from the reservoir and evacuating the downstream area. Lead times are short, typically up to six hours.

Other external mechanisms are difficult to detect in advance. If and when earthquakes can be predicted, it will be possible to evacuate the zone downstream from a dam, when an earthquake with a larger magnitude than the design magnitude of the dam is about to happen. Unfortunately, such a system is not yet developed and cannot be relied upon to eliminate the risk of human loss from dam failure in earthquakes. A landslide in the reservoir can also cause overtopping and be observed with a short lead time. Finally, the failure of a dam upstream can also lead to the failure of a dam downstream, again leaving a very short lead time for evacuation.

Analysis for an Earth Dam

The case studied in Figure 1 (Paté, 1984) concerns a hypothetical earth-filled embankment dam and assumes regular monitoring. Assumptions have been made about the performance of the system and the efficiency of the measures taken, including an evacuation model. The result, for each possible scenario, is its contribution to the expected proportion of avoided casualties. It is obtained by computing the joint probability of the different sequences of events and multiplying this probability by the corresponding proportion of inhabitants that are protected, either by avoiding the failure or by evacuation. In the cases presented here, it is in the order of one-half.

Property damage is avoided to the extent that failure is avoided by appropriate actions following a warning. This benefit is computed by adding the probabilities of all scenarios in which the dam is saved. Probabilities and results of the evacuation model are indicated on Figure 1. The costs due to false alerts have not been included because their expected value appears to be negligible in the economic analysis of the dam. A false alert, however, could also cause a subsequent real alert to be ignored. A simplification made in this model has been to assume that consecutive alerts are independent.

The model, as set in Figure 1, leads to the following results:

- There is a 0.12 probability that a failure that would have occurred otherwise will be avoided because of timely warning and appropriate actions.

- The number of casualties can be reduced by 44% of what it would be without warning system.

- The property damage that would occur otherwise can be reduced by 12% (the probability of avoided failure).

In this example, direct assumptions were made about response and evacuation. Evacuation models have been developed for facilities such as nuclear power plants and will not be presented here. Response models are more complex. The "cry-wolf effect" can have a significant impact on the results. The following theory was developed to illustrate the trade-off between lead time and frequency of false alerts.

3. THEORY OF WARNING SYSTEMS: THE CRY-WOLF EFFECT

Expected Number of Avoided Casualties

Assume that a fire alarm in a student dormitory has to be set (or located), i.e., a level of sensitivity has to be chosen. A trade-off has to be found between a high sensitivity and therefore a short lead time for response (or no lead time at all) (see Figure 2). Optimization in this case can be performed as a maximization of the number of people saved within reasonable cost constraints. The dominant effect that determines the response level is the memory of past performance of the warning system.

By combination of these different submodels, one can compute the expected value of the number of people saved. The notations are the following:

$EV(\Delta L)$:   expected value of the number of people saved per time unit (result)

ϕ: threshold level or other sensitivity parameter

$\lambda_1(\phi)$:   rate of irrelevant false alerts that are not linked to the occurrence of events (e.g., an insect in a fire alarm)

$\lambda_2(\phi)$:   rate of alerts (true or false) due to actual exceedence of the threshold (e.g., a real fire or simply cigarette smoke under a fire alarm)

N_E: loss model (expected number of deaths or injuries per event, assuming no warning)

i: index of different patterns of past true (T) and false (F) alerts (in chronological order) that are assumed to constitute the historical memory affecting the response (e.g., TFFT, the most recent event being the last one in the sequence)

$\{K\}$:   set of alert patterns ending in a true alert (e.g., FFFT)

α_i: proportion of people who are willing to respond to a new alert given a past pattern of true and false alerts

$p_i^*(\phi)$:   steady-state probability of each pattern of true and false alerts, given a sensitivity level

$V(\alpha_i)$:   speed of evacuation, given the number of people willing to respond. It is assumed that below a certain response rate there is a free flow, then a gradual engorgement (linear decrease of v with α), then total jam and v = 0 above a certain level of response

D_{MAX}:   critical threshold of the monitored variable (density of smoke above which people are hurt)

$D(\phi)$:   threshold of the monitored variable above which the signal is supposed to occur (decision variable)

$\bar{\nu}$:   mean rate at which the system deteriorates over time once the threshold $D(\phi)$ is exceeded (e.g., rate of increase of smoke density per time unit above the chosen threshold)

It is shown elsewhere (Paté, 1986) that the benefits of this warning system per time unit can then be written:

$$EV(\Delta L)=[\lambda_1(\phi)+\lambda_2(\phi)]N_E \times [\ \sum_{i\varepsilon\{k\}} \alpha_i(\alpha_i)p_i(\phi)] \times [D_{MAX}-D(\phi)]\bar{\nu}$$

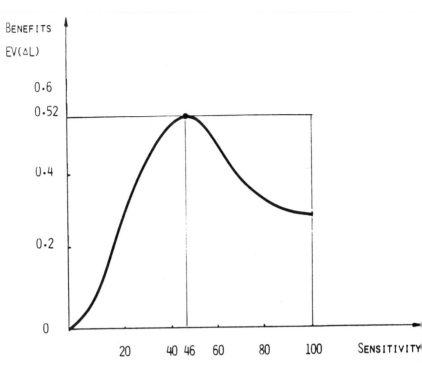

FIGURE 3. OPTIMUM SENSITIVITY FOR THE NON-PARAMETRIC
DESCRIPTIVE RESPONSE MODEL

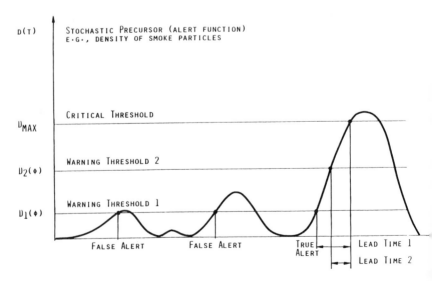

Figure 2. Stochastic Precursor; Monitoring an Underlying Phenomenon

Response Models

At least three models can be used to describe the effect of memory on the response to warnings (Paté, 1986):

- A non-parametric descriptive Markov model in which the response level is a function of the quality of the past alerts (true or false) that people remember
- A parametric descriptive model, in which the response level is determined by the impact of past true and false alerts on people's willingness to respond, given memory decay
- A normative model based on the assumption of rationality, in which each individual makes a decision to respond or not, given: (1) the probability that he assigns to the occurrence of the event following a warning; (2) his probability of being a victim of this event; (3) the costs to him of responding to the alert; and (4) the value that he places on his own life and his risk attitude.

The rate of response α can thus be modelled in the three different ways:

The non-parametric descriptive model requires the assessment of the number of past warnings n that are remembered and affect the response (e.g., n = 4), and the determination of the rate of response for each pattern. For example, α (TTFT) equals 0.7. The value of the rate of response (e.g., 0.7) can be estimated either on the basis of past statistical data, or by use of a questionnaire. This model has the advantage of simplicity: in the long run, what matters is the steady-state probability of each sequence of true and false alerts relative to true alerts. A large number of steps thus does not represent a computational problem. This model, however, has the clumsiness of such non-parametric descriptions: one must assess the rate of response of the population for each of the 2^{n+1} patterns of true/false alerts involved in an n-step memory model. It also presents one major drawback: it ignores the severity of past events (which may be critical to memory and response) and the time elapsed since they occurred (except indirectly through the rate λ of events E per time unit).

The parametric descriptive exponential model has the advantage of including the effects on memory and response of past events, their severity, and the elapsed time. It assumes exponential decay with time of the memory of past events, and requires adjustment of different parameters to reflect the respective influence of true alerts, false alerts, and missed predictions on future rates of response. For a given time t, the response factor α is of the following form:

$$Q(t)=\int_{-\infty}^{t} \{exp[-m(t-\tau)]\} \times [k_A M(\tau)-k_B M(\tau)-k_C(\tau)]d\tau$$

$$\alpha(t)=\Phi\left(\frac{Q(t)-\beta}{y}\right)$$

in which the parameters are the following:

$\alpha(t)$ is the proportion of people willing to respond to a warning at time t, given the past history of true alerts (A), Type I errors (B), and Type II errors (C) at all times τ preceding t.

$Q(t)$ is an intermediate function (before normalization) in the computation of α.

$\Phi(.)$ is a scaling function, for example, a Gaussian cumulative, in which β (≥ 0) and γ are scaling factors. Φ ensures that α remains between 0 and 1.

m is the memory-discounting factor.

$M(\tau)$ is the magnitude of the events that occurred at different times in the past.

k_A is the factor measuring the positive effect on future response rates of a true alert.

k_B is the factor measuring the negative effect on future response rates of a missed prediction (Type I error).

k_C is the factor measuring the negative effect on future response rates of a false alert (Type II error).

This model has the advantage of treating explicitly the length of the time intervals between warnings and between events and warnings. A disadvantage, however, is the assumed additivity of these effects. One may want to question in some cases the reality of exponential decay. A very long time after the last earthquake, for example, there is a popular perception that chances are increasing that another earthquake may occur, and there may be a greater willingness to respond to a warning. In other cases, one may argue that events that have occurred a long time ago may have left a stronger imprint in some people's minds than more recent ones. In general, however, memory decay seems a reasonable assumption.

The parametric normative model starts from an assumption of rationality from the individual point of view instead of the collective one. For each potential victim, the problem is to know whether he will respond or not to the warnings according to the decision analysis paradigm (Raiffa, 1968). This model includes (1) Bayesian updating of information every time a new event of type A, B, or C occurs, and (2) a study of individual decisions to take protective measures, given individual risk attitudes, costs, and values. A warning has been issued, the individual decision is modeled using a simple decision tree. This model is useful for collective decisions only if one assumes a spectrum of individual risk attitudes.

Illustration: Fire Alarm In A Student Dormitory

The following numerical application is presented for illustrative purposes (see, for a more complete description of the data, Paté, 1986). Assume that three fires occur per year on the average, exposing 100 students, that the number of people injured by the smoke in each event is 5 (mean value), and that the sensitivity of the fire alarm is defined on a scale of 0 (maximum sensitivity) to 100 (minimum sensitivity). The sensitivity is the decision variable. It determines the probability of an event conditional on the signal, which increases as sensitivity decreases, and the probability of a signal conditional on an event, which decreases as sensitivity increases. The non-parametric rate of response was assumed to vary from a low 20 percent for a system that tends towards 100 percent false alerts to a high 90 percent for a system that delivers 100 percent true alerts (but is prone to missing events). This very imperfect system is also assumed to issue irrelevant alerts at the rate of two per year. Figure 3 shows the variations of the expected number of avoided injuries as a function of (ϕ), the sensitivity of the device. A maximum of 5.5 avoided injuries (expected value) is reached for a high sensitivity of 30, at an average low cost of $3,000 per avoided casualty.

For the parametric descriptive model we assume for the values of the coefficients k_A=1.1, k_B=-1.1, and k_C=-1.2, therefore a slightly higher negative effect of false alerts than missed events. The memory decay factor is 10 percent per year and the factor is renormalized through a Gaussian cumulative curve. The variation of the expected number of avoided injuries as a function of ϕ show a maximum of 7.5 for a high sensitivity of 38. The discrepancy with previous results is not of major significance, given that there was no attempt to match the parameters of both models.

Finally, for the normative model of response, the example of a group of ten persons was considered. To each of them was attributed a different value reflecting the probability of being a victim if an event were to occur (from 0.1 to 0.99). The risk attitude, the cost of evacuation, and the value of avoiding injury were assumed to be uniform among individuals. The decision of each person to evacuate or not is analyzed assuming a rational decision model. The rate of response α is calculated in this manner for each considered value of ϕ. The variation of the expected number of avoided injuries as a function of ϕ shows a maximum of 10 for a high sensitivity of 24.

These three models illustrate clearly the memory effect and the variations of the benefits of a warning system between two extreme values of sensitivity. In all three cases one observes zero or very low benefit for extremely sensitive or extremely insensitive systems, and a maximum benefit at an intermediate point that depends on the physical characteristics of the system and the description of the memory effect assumed in the response model.

4.   RARE EVENTS: RESPONSE AND SOCIOLOGICAL CONTEXT

In the case of rare events or events for which even comparable occurrences are rare, the memory effect is no longer the key element. This is the situation encountered, for example, in emergency planning for nuclear power plants. In an accident sequence, following an initiating event, the plant state evolves in a manner that can be observed through the monitoring of different characteristics (e.g., temperature or pressure). A first choice has to be made as to which of these characteristics provides a signal on-site and above what threshold an alert should occur. Operator intervention following the in-plant signal can either accelerate, remain unchanged, slow down, or avoid the evolution of the accident toward radioactive release. A second threshold has to be set above which an off-site warning is issued and a recommendation is made to the public to either evacuate, shelter, take other protective measures, or do nothing. The lead time available for response, the public's willingness and capability to respond to the warning, and the probability of actual release of radioactive material are the three variables linking the dynamic evolution of an on-site accident and the public response to off-site warning.

The probabilistic analysis of alternative warning systems is done using a different stochastic model for each accident sequence (Claudio, 1985). Each of these models reflects the dynamics of the particular accident considered. An important element is the response model, which describes the public's willingness to respond and the capability to respond (for example, an evacuation model). A simplified behavioral response model has been developed to assess the public's willingness to respond to an emergency warning using previous work in the field of sociology and psychology (see Figure 4).

Two different theoretical approaches in the behavioral study of response to warnings can be found in the literature (Claudio, 1985). The first one focuses on the individual decisionmaking processes of warning respondents in a given psychological context. It gives importance to one or more of the following factors: cognitive abilities, conflict, learning and feedback, memory, and motivation. This includes, for example, Janis's conflict model of emergency decisionmaking (Janis, 1977) and models developed recently by engineering researchers such as the one presented above. Another approach is the sociological one. It emphasizes the role of social and cultural variables in shaping response and the actions of individuals and groups (see, for example, Mileti et al., 1975).

Indeed, many psychological and sociological variables influence the behavior of individuals in times of crisis. Which variables are most important is subject to contention. The qualitative model that we used in the analysis of warning systems for nuclear power plants (Claudio, 1985) is described by the influence diagram of Figure 4. It involves the following variables: (1) prior and situational risk perception; (2) social confirmation; (3) warning belief; (4) adaptive plan; and (5) distance from the plant.

Preliminary results of the study of warnings for nuclear power plants (Claudio, 1985) suggest that during the early stage of most accidents, when both the final plant state and the release of radioactivity are uncertain, it is better to wait for more information

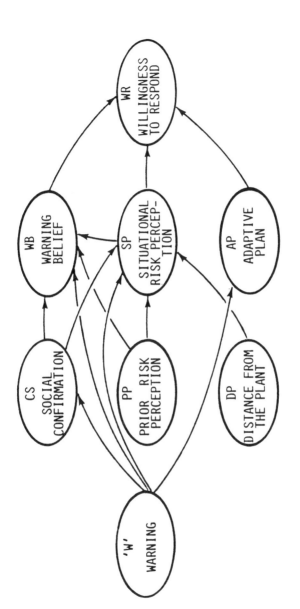

Figure 4. Behavioral Response Model for Nuclear Power Plant Emergency

before taking action. This is due in part to the negative effects of false alerts and to the low probability and generally slow evolution of events that could lead to off-site damage.

5.   OTHER APPLICATIONS AND FURTHER DEVELOPMENTS

Several other analyses of warning systems have been developed to emphasize the different aspects of the method presented in this paper. In an analysis of the benefits of camera monitoring in an oil refinery for the reduction of fire risks, the emphasis was put on the evolution of fires (ignition, development, and extinction), which was described using a Markov model (Paté, 1985b). In that study a model was designed to link the reduction of the detection time and the reduction of the final losses in fires of different types.

In another study, the method was used to provide information about the trade-offs between Type I and Type II errors in critical policy decisions. The problem was to link the reliability of the system of command and control of nuclear forces to the risks of different launch policies for intercontinental ballistic missiles (ICBMs) in the United States (Paté and Neu, 1985). Given the increasing vulnerability of the ICBMs, it may be tempting to shift from a policy of launch on impact to launch on warning. Yet there is a critical trade-off between the probability of not being able to respond (either because of failure of the control system or because of the destruction of the retaliatory forces) and the probability of a false alert, possibly followed by an accidental attack. In both cases, the model developed includes not only errors of information but also errors of action, such as an accidental strike. The first part of the study leads to the computation of the probabilities of errors of different types for different launch policies. The second part is an analysis of the consequences of these different launch policies in the first phase of a nuclear conflict. The results are the probability distributions for each element of the outcomes.

6.   CONCLUSIONS

Warning systems can provide powerful means of risk reduction. They may also create risks themselves. The best warning systems are not necessarily the most sensitive ones. The sensitivity that gives the optimum benefits (i.e., the maximum risk reduction) depends on the expected response to signals, the reliability of the warning system, and the preferences among the different elements involved in the trade-off between the consequences of Type I and Type II errors. In the simple case when the goal is clearly to protect the maximum number of lives, it was shown above how the memory effect can be taken into account to find this optimum.

Warning systems can be improved by more reliable signals, greater lead time, and better emergency plans. Our method allows the analyst to test the relative benefits of such improvements. A main feature of the model is the link between improvement of information and increase in the rate of response. Although one could sometimes consider forcing a response by coercive measures, this method may prove socially disturbing or simply ineffective and does not reduce the cost of false alerts. The paradigm proposed here for optimization of warning systems is a powerful complement to the traditional observation that "crying wolf" is hazardous in itself.

REFERENCES

Baecher, G. B, M. E. Paté, and R. de Neufville 1980. Risk of dam failure in benefit-cost analysis. _Water Resources Research_, 16(3): 449-456

Claudio, C. P. B. 1985. Design and optimization of warning systems: application to nuclear power plants. Doctoral thesis. Department of Engineering Economic Systems, Stanford University, Stanford, California.

Janis, I. L., and L. Mann. 1977. _Decision-Making: A Psychological Analysis of Conflict, Choice, and Committment_. New York: The Free Press.

Mileti, D. S., T. Drabek, and J. E. Haas. 1975. Human systems environments: a sociological perspective. Institute of Behavioral Science, University of Colorado, Boulder, Colorado.

Paté, M. E. 1984. Warning systems: application to the reduction of risk costs for new dams. In Proceedings of the International Conference on the Safety of Dams, Coimbra, Portugal. The Netherlands: Balkema Publishing. pp 73-83.

Paté, M. E. 1985a. Warning systems. _The Stanford Engineer_ 8(2): 17-24.

Paté, M. E. 1985b. Reduction of fire risks in oil refineries: economic analysis of camera monitoring. _Risk Analysis_ 5(4): 277-288.

Paté, M. E. 1986. Warning systems in risk management. _Risk Analysis_ 6(8): 223-234.

Paté, M. E., and G. Tagaras. Risk costs for new dams: economic analysis and effects of monitoring. _Water Resources Research_ 22(1):5-14.

Paté, M. E., and C. P. Benito-Claudio. 1987. Warning systems: response models and optimization. In Covello, Lave, Moghissi, and Uppuluri (eds.), _Uncertainty in Risk Assessment, Risk Management and Decision Making_. New York: Plenum Publishing. pp. 457-468.

Paté, M. E., and J. E., Neu. 1985. Warning systems and defense policy: a reliability model for the command and control of the U. S. nuclear forces. _Risk Analysis_ 5(2):121-138.

Raiffa, H. 1968. _Decision Analysis_. Reading, Massachusetts: Addison-Wesley Publishing.

SAFETY GOALS, UNCERTAINTIES, AND DEFENSE IN DEPTH

David Okrent

ABSTRACT

An abbreviated summary of the safety goals adopted by the U.S. Nuclear Regulatory Commission is first presented. This is followed by a display of some of the results reported in probabilistic risk assessments (PRAs) for specific reactors. Examples of the large uncertainties typically estimated for the PRA results are given. The difficulties of such uncertainties are illustrated by a particular risk management policy issue.

1. NRC SAFETY GOAL POLICY

The safety philosophy espoused by the U.S. Nuclear Regulatory Commission (NRC), and the U.S. Atomic Energy Commission (AEC) before it, has often been called "defense-in-depth." Measures were included on the design (and operation) of nuclear power plants to make very unlikely the chance that some transient event or accident would lead to damage or melting of the reactor core fuel, and a containment building was required to restrict the escape of radioactive fission products, although the containment design was not explicitly based on the possibility of coping with a molten core.

In judging that there was reasonable assurance that a plant could be operated without undue risk to the public, the NRC did not directly address the question "How safe is safe enough?" for light water reactors (LWRs), nor did it quantify the residual risk that was implicitly being accepted. In fact, until the Reactor Safety Study was completed (USNRC, 1975), a methodology did not exist.

In 1979 the NRC initiated studies toward the development of an approach to quantitative safety goals. In October 1980 the Advisory Committee on Reactor Safeguards (ACRS) of the NRC proposed a trial approach, which was intended to serve as a focus of discussion and, as such, was expected to be only a first step in an iterative process (USNRC, 1980).

The ACRS trial approach to risk management was a set of quantitative design objectives that were intended to reflect the following partially overlapping qualitative goals (UCNRC, 1980; Griesmeyer and Okrent, 1980; Okrent and Griesmeyer, 1981).

1. Future nuclear power plants should, if practical, present less risk to society than that from the principal competitor, coal-fired power plants.

David Okrent is Professor of Engineering and Applied Science, University of California, Los Angeles, California.

2. The risk arising from the presence of one or more LWRs at a particular site to those individuals at greatest risk (that is, those living or working close to the reactor site) should be small enough that (i) it does not significantly increase their risk of death from accidents or cancer and (ii) questions of equity are addressed (that is, any imbalance between the increment of individual risk and direct benefit should be small enough that a conscious effort to balance the disparity is not required).

3. The safety-related design should be required to place emphasis both on the prevention of accidents that could lead to severe core damage or large-scale core melt and on the mitigation of accidents involving large-scale fuel melt. The probability of such an accident should be very low, and there should be very low probability that an individual living near the plant would be killed, even if a large-scale core melt accident did occur.

4. Suitable additional design efforts should be made to reduce the risk below specified safety goals, with an as-low-as-reasonably-achievable (ALARA) cost-effectiveness criterion.

5. All LWR accident sources should be considered in evaluating the risks, and the safety goals should be compared against mean values with a prudent allowance for the presence of large uncertainties.

6. Incentives should be provided to reduce still further the likelihood of accidents involving large numbers of casualties, without unduly penalizing the technology or placing excessive costs on society.

After considering input from industry, the NRC staff, and others, the NRC commissioners adopted on March 14, 1983, a safety policy statement for trial use during a two-year evaluation period (USNRC, 1983). Also adopted was a staff implementation plan for evaluation of the quantitative guidance.

The safety policy included the following features:

Qualitative Goals

1. Individual members of the public should be provided a level of protection from the consequences of nuclear power plant accidents such that no individual bears a significant additional risk to life and health.

2. Societal risks to life and health from nuclear power plant accidents should be comparable to or less than the risks of generating electricity by viable competing technologies.

Quantitative Design Objectives (QDOs)

The risk to an average individual, in the vicinity of a nuclear power plant, of prompt fatalities that might result from reactor accidents should not exceed one-tenth of one percent (0.1%) of the sum of prompt fatality risks resulting from other accidents to which members of the U.S. population are generally exposed.

The risk to the population in the area near a nuclear power plant of cancer fatalities that might result from nuclear power plant operation should not exceed one-tenth of one percent (0.1%) of the sum of cancer fatality risks resulting from all other causes.

The latter risk was to be the average risk of cancer for the population living within 50 miles of the plant. The risk within one mile was a decade or more larger.

Benefit-Cost Guideline

The benefit of an incremental reduction of societal mortality risks should be compared with the associated costs on the basis of $1,000 per person-rem averted.

The benefit-cost guideline was only to be applied in judging whether it was worthwhile in bringing an existing plant up to safety policy objectives.

Plant Performance Design Objective

The likelihood of a nuclear reactor accident that results in a large-scale core melt should normally be less than one in 10,000 per year of reactor operation.

Implementation

The commission said that the basic impediment to the adoption of regulations that require that risks to the public be below certain quantitative limits, as exemplified by the quantitative design objective for large-scale core melt, is that the techniques for developing quantitative risk estimates are complex and have substantial associated uncertainties. Thus a serious question was whether, for a specific nuclear power plant, the achievement of a regulatory-imposed quantitative risk goal could be verified with a sufficient degree of confidence. For this reason, the commission decided that, during the evaluation period, implementation of the Policy Statement should be limited to such uses as examining proposed and existing regulatory requirements, establishing research priorities, resolving generic issues, and defining the relative importance of issues as they arose.

The safety goals were not to be used in the licensing process or to be interpreted as requiring the performance of PRAs by applicants or licensees.

The NRC had previously published for comment a draft Safety Policy Statement in February 1982 (USNRC, 1982) which was very much like the 1983 policy. Together with the draft Statement, they had listed a series of questions, including the following.

Some Questions Raised by the NRC in Draft Safety Policy

- Should the benefit side of the trade-offs include, in addition to the mortality risk reduction benefits, the economic benefit of reducing the risk of economic loss due to plant damage and contamination outside the plant?

- Should there be added a numerical guideline on availability of containment function, given a large-scale core melt?

- Should there be specific provisions for "risk aversion"? If so, what quantitative or other specific provision should be made?

- What further guidance, if any, should be given for decisions under uncertainty?

- What further guidance, if any, should be given on resolution of possible conflicts among quantitative aspects of some issues?

- What approach should be used with respect to accident initiators which are difficult to quantify, such as seismic events, sabotage, multiple human errors, and design errors?

- Should there be definition of the numerical guidelines in terms of median, mean, 90 percent confidence, etc.? If so, what should be the terms?

- Should the staff action plan include further specification of a process that will lend credibility to the use of quantitative guidelines and methodology? If so, what should be the principal bases and elements of such guidelines?

- On what basis should the numerical guidelines be applied to protection of individuals? Should they be applied to the individual at greatest risk, or should they be used in terms of an average risk limit over a region near the plant?

On August 4, 1986, the NRC published a final Policy Statement on safety goals (USNRC, 1986). The commission determined that the qualitative safety goals would remain unchanged from its March 1983 revised policy statement. Thus, the commission's two safety goals are: (i) Individual members of the public should be provided a level of protection from the consequences of nuclear power plant operation such that individuals bear no significant additional risk to life and health, and (ii) Societal risks to life and health from nuclear power plant operation should be comparable to or less than the risks of generating electricity by viable competing technologies and should not be a significant addition to other societal risks.

The safety goal Policy Statement also said the following (USNRC, 1986; p. 28045):

> Severe core damage accidents can lead to more serious accidents with the potential for life-threatening offsite release of radiation, for evacuation of members of the public, and for contamination of public property. Apart from their health and safety consequences, severe core damage accidents can erode public confidence in the safety of nuclear power and can lead to further instability and unpredictability for the industry. In order to avoid these adverse consequences, the Commission intends to continue to pursue a regulatory program that has as its objective providing reasonable assurance, while giving appropriate consideration to the uncertainties involved, that a severe core damage accident will not occur at a U.S. nuclear power plant.

This statement represented a major change from the 1982-83 draft Policy Statements. Furthermore, of the previous QDOs, the commission retained only the two which quantified early and latent cancer mortality risks, that is, that these risks should not exceed 0.1% of the normal background accident or cancer mortality risk. The cancer risk was to be averaged over the population within a ten-mile radius of the plant, as recommended by the NRC staff task force. The quantitative objective on core melt frequency and the quantitative cost-benefit criterion were omitted from the Policy Statement.

Guidelines for Regulatory Implementation

Regarding the regulatory implementation, the commission stated the following (USNRC, 1986; p. 28047):

> The Commission approves use of the qualitative safety goals, including use of the quantitative health effects objectives in the regulatory decisionmaking process. The Commission recognizes that the safety goals can provide a useful tool by which the adequacy of regulations or regulatory decision regarding changes to the regulations can be judged. Likewise, the safety goals could be of benefit in the much more difficult task of assessing whether existing plants, designed, constructed, and operated to comply with past and current regulations, conform adequately with the intent of the safety goal policy.

> However, in order to do this, the staff will require specific guidelines to use as a basis for determining whether a level of safety ascribed to a plant is consistent with the safety goal policy. As a separate matter, the Commission intends to review and approve guidance to the staff regarding such determinations. It is currently envisioned that this guidance would address matters such as plant performance guidelines, indicators for operational performance, and guidelines for conduct of cost-benefit analyses. This guidance would be derived from additional studies conducted by

the staff and result in recommendations to the Commission. The guidance would be based on the following general performance guideline which is proposed by the Commission for further staff examination.

Consistent with the traditional defense-in-depth approach and accident mitigation philosophy requiring reliable performance of containment systems, the overall mean frequency of a large release of radioactive materials to the environment from a reactor accident should be less than 1 in 1,000,000 per year of reactor operation.

This guideline, which was proposed for staff examination, was new and represented a major change from the 1982-83 Policy Statements.

2. SOME EXAMPLES OF UNCERTAINTY IN PRA

Table 1 gives a summary of results from existing published PRAs in the early 1980s, as summarized by the NRC staff (USNRC, 1984).

Table 1:   Results of Existing Probabilistic Risk Assessments

	INDIVIDUAL RISK WITHIN 1 MILE			
	Frequency Core Melt	Frequency Major Release	Early Fatality	Fatal Cancer
ANO-1	5×10^{-5}	2×10^{-5}	$\underline{6 \times 10^{-7}}$	2×10^{-7}
BIBLIS	4×10^{-5}	1×10^{-6}	3×10^{-8}	2×10^{-8}
BIG ROCK	1×10^{-3}	0	0	---
BROWNS FERRY	2×10^{-4}	4×10^{-5}	2×10^{-7}	1×10^{-6}
CALVERT CLIFFS	2×10^{-3}	1×10^{-3}	9×10^{-6}	2×10^{-5}
CRYSTAL RIVER	4×10^{-4}	2×10^{-4}	$\underline{3 \times 10^{-6}}$	2×10^{-6}
GRAND GULF	4×10^{-4}	4×10^{-5}	1×10^{-7}	1×10^{-7}
I.P. #2	$\underline{4 \times 10^{-4}}$	3×10^{-4}	3×10^{-8}	1×10^{-8}
I.P. #3	9×10^{-5}	3×10^{-5}	1×10^{-9}	3×10^{-10}
LIMERICK	2×10^{-5}	3×10^{-5}	1×10^{-8}	1×10^{-8}
MILLSTONE	3×10^{-5}		1×10^{-7}	6×10^{-7}
OCONEE	8×10^{-5}	4×10^{-5}	2×10^{-7}	1×10^{-7}
PEACH BOTTOM	3×10^{-5}	7×10^{-6}	4×10^{-8}	3×10^{-8}
SEQUOYAH	6×10^{-5}	4×10^{-5}	$\underline{1 \times 10^{-6}}$	5×10^{-7}
SURRY	6×10^{-5}	1×10^{-5}	2×10^{-7}	1×10^{-7}
ZION	4×10^{-5}	4×10^{-6}	2×10^{-8}	1×10^{-8}

These PRAs were performed using different assumptions and methodology, and represented various levels of completeness. Only a relatively few tried to evaluate the uncertainties in the reported results.

Figures 1-5 reproduce some of the kinds of uncertainties estimated for the PRAs on the Zion and Indian Point nuclear power plants (Commonwealth Edison, 1981; Consolidated Edison, 1982). (Figure 1 is Figure II.8-04, Figure 2 is Figure 11.8-14, Figure 3 is Figure 11-8-15, and Figure 4 is Figure II.8-58 of Commonwealth Edison [1981] and Figure 5 is Figure 8.3.9 of Consolidated Edison [1982]).

The PRAs performed for the NRC by its contractors had, for the most part, not included a detailed treatment of uncertainties. Some years ago the NRC decided to undertake a new set of PRAs, covering most major types of light water reactor designs in the USA, including improved PRA methodology and a detailed treatment of uncertainties. The result, draft NUREG 1150, became available for review and comment in early 1987 (USNRC, 1987). Some of the principal conclusions of the NRC staff are given below.

1. Plant-specific analyses are necessary to determine the likelihood of a severe accident in any of the current generation of U.S. reactors.

2. Radioactive source terms that could be released to the environment in a severe accident show a wide variability, with the upper bound values comparable to those predicted in the reactor safety study.

3. Early containment failure cannot be ruled out in certain severe accident sequences. Large, dry containments have a higher probability of withstanding the effects of severe accidents than do suppression-type containments. For pressure-suppression containments, however, if the pressure-suppression pool or ice bed is available after containment failure, the releases to the environment can be substantially reduced.

4. Application of the NRC's safety goal--i.e., risk of prompt and latent cancer fatalities--to the plants that were studied has shown that these goals have been met, considering the entire uncertainty range presented.

With regard to uncertainties, the NRC Staff said the following, in part:

• Uncertainties were not addressed adequately in Reactor Safety Study.

• Uncertainties are large because understanding of severe accident phenomena is not complete.

• Knowledge base has improved substantially since Reactor Safety Study, but still more to be learned.

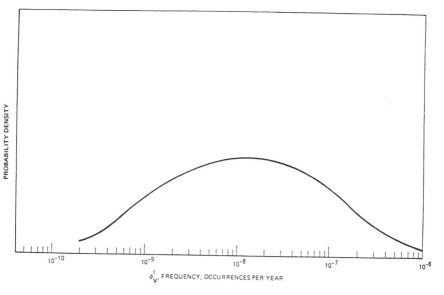

Figure 1. Probability Curve for Frequency of Initiating Event V.

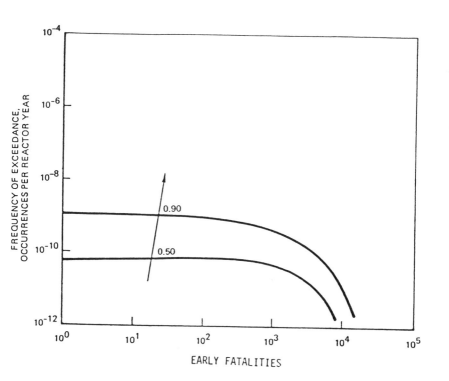

Figure 2. Risk With Uncertainty (Level 2) Diagram for Early Fatalities Risk from Internal Events.

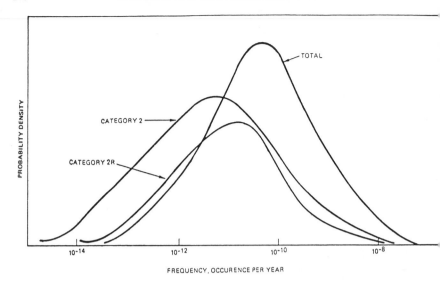

Figure 3. Frequency of ≥ 100 Fatalities Showing Contributions from
Release Categories 2 and 2R (Internal Events Only).

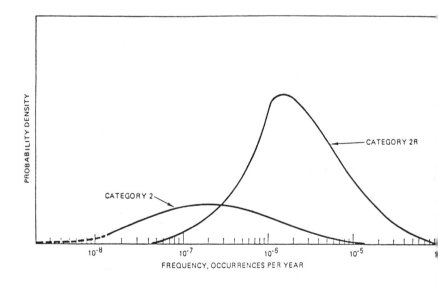

Figure 4. Frequency of Releases of Categories 2 and 2R as a Result of
Seismic Initiation (Base Case).

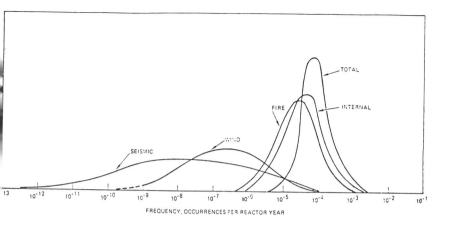

Figure 5. Core Melt Frequency from Various Sources--Indian Point 3.

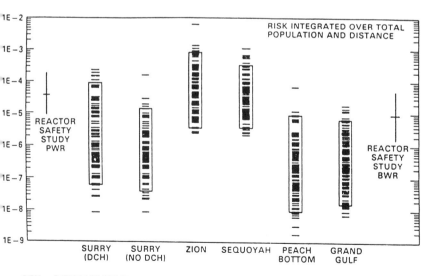

DCH = DIRECT CONTAINMENT HEATING

Figure 6. Comparison of Early Fatality Risks.

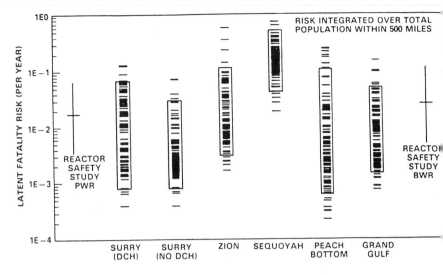

DCH = DIRECT CONTAINMENT HEATING

Figure 7. Comparison of Latent Cancer Fatality Risks.

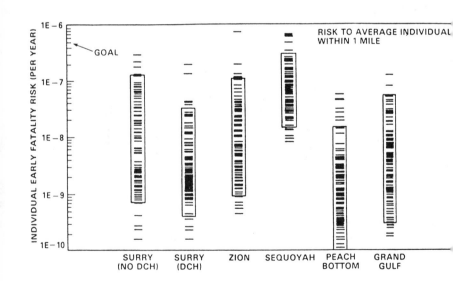

DCH = DIRECT CONTAINMENT HEATING

Figure 8. Comparison with NRC Safety Goal (Early Fatalities).

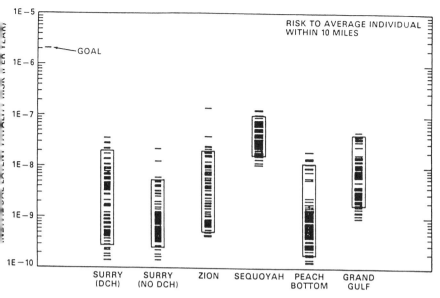

DCH = DIRECT CONTAINMENT HEATING

Figure 9. Comparison with NRC Safety Goals (Latent Cancer Fatalities).

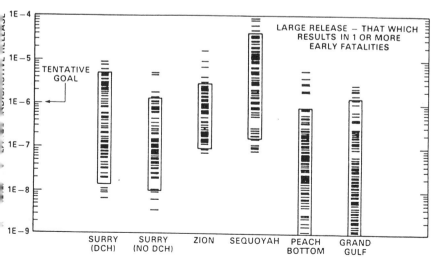

DCH = DIRECT CONTAINMENT HEATING

Figure 10. Probability of One of More Early Fatalities.

Therefore

- Subjective judgment of experts, based on extant knowledge, required to define reasonable ranges of potentially important issues.

- Subjective judgments by experts never substitute for good science; either experimental validation is needed or regulatory decisions must include margins to encompass uncertainties.

The conclusion in draft NUREG 1150 was that the uncertainties in risk were still very large.

Draft NUREG 1150 has been the subject of much peer review and has generated much controversy, particularly in its manner of assessing uncertainties. Nevertheless, it is useful to look at a few examples of the uncertainty spread as illustrated by Figures 6-10. (Figure 6 is Figure ES.3, Figure 7 is Figure ES.4, Figure 8 is Figure ES.5, Figure 9 is Figure ES.6, Figure 10 is Figure ES.7 of draft NUREG 1150.)

3.   HOW TO TREAT DEFENSE-IN-DEPTH

Liquid-metal-cooled fast breeder reactor design was generally aimed at fairly high overall reactor power levels to achieve improved economics. However, in the past few years, much of the design effort in the U.S. has been aimed at small reactors having special safety features which make them less vulnerable to certain accidents. These new "inherent" safety design features (frequently requiring no action on the part of the operator, or only much delayed action) were included in the SAFR (Rockwell International, 1985) and PRISM (General Electric, 1986) reactor designs. A brief summary of some features of SAFR is given below:

- Compact, pool-type, sodium-cooled reactor.

- Metal core design.

- Self-actuated shutdown system (SASS).

- Reactor air cooling system on exterior of vessel (RACS).

- Direct reactor auxiliary cooling system, heat exchanger in pool (DRACS).

- Should survive unprotected loss of flow or loss of heat sink for extended period without operator action.

Preliminary PRAs performed on the SAFR and PRISM designs suggest that the frequency of core melt or significant release of radioactive material is very low, as illustrated in Table 2 (Lee and Okrent, 1987).

Table 2:   SAFR/PRISM PRA Results

	SAFR	PRISM
Total Frequency of Significant Release	$\ll 10^{-6}$/R.Y.	$\ll 10^{-6}$/R.Y.
Acute Fatality per Reactor Year	6.3×10^{-17}	5.0×10^{-11}
Latent Cancer Risk per Reactor Year	very low	very low

Some of the observations on the SAFR/PRISM PRAs are as follows:

* Based on preliminary designs.

* Include only internal events in initial version.

* Structural and seismic contributions identified by designers as requiring study.

* How to include construction and design errors, poor management in PRA?

A key policy issued raised by the proposed SAFR/PRISM designs, one which must be addressed by the NRC, is the following:

SAFR and PRISM have certain containment capabilities but do not include a high-pressure-design containment structure around the entire plant. The designers argue that the PRA results are such that the current designs have residual risks far lower than the objectives in the Safety Goal Policy Statement. Hence, high-pressure containment is unnecessary.

In effect, this approach places great emphasis on reducing the likelihood of an accident leading to failure of the existing containment features.

Should the NRC be willing to accept this change in defense-in-depth? How should uncertainties (which are not estimated in the preliminary PRAs for SAFR and PRISM) affect this decisionmaking process?

REFERENCES

Commonwealth Edison. 1981. Zion probabilistic safety study.

Consolidated Edison. 1982. Indian Point probabilistic safety study.

General Electric. 1986. PRISM preliminary safety information document.

Griesmeyer, J. M., and D. Okrent. 1980. Safety Goals, Uncertainties, and Defense in Depth. Risk Analysis 1:121.

Lee, S. H., and D. Okrent. 1987. On the development of quantitative safety goals for inherently safe LMFBT design and licensing. In Proceedings of the Ninth SMIRT Conference, Lausanne, Switzerland, August 24-28, 1987.

Nuclear Regulatory Commission. 1975. Reactor safety study: an assessment of accident risks in U.S. commercial nuclear power plants. Report WASH-1400, NUREG-75/014. Washington, D.C.

Nuclear Regulatory Commission. 1982. Safety goals for nuclear power plant operation. Report NUREG-0880 for comment. Washington, D.C.

Nuclear Regulatory Commission. 1983. Safety goals for nuclear power plant operation. Report NUREG-0880, Revision 2 for comment. Washington, D.C.

Nuclear Regulatory Commission. 1984. Probabilistic risk assessment reference document. Report NUREG-1050. Washington, D.C.

Nuclear Regulatory Commission. 1986. In Federal Register 51:28044 (August 4, 1986). Republished with corrections in Federal Register 51:30028 (August 21, 1986). Washington, D.C.

Nuclear Regulatory Commission. 1987. Reactor risk reference document. Draft NUREG-1150. Washington, D.C.

Nuclear Regulatory Commission, Advisory Committee on Reactor Safeguards. 1980. An approach to quantitative safety goals for nuclear power plants. Report NUREG-0739. Washington, D.C.

Okrent, D., and J. M. Griesmeyer. 1981. Angew. Syst. 2:183.

Rockwell International. 1985. SAFR-sodium advanced fast reactor preliminary safety information document.

ANNULAR DISPOSAL OF OIL AND GAS BRINES: THE RISKS AND THE BENEFITS

Benjamin F. Hobbs, Carl Von Patterson, Jim S. Heslin, and
M. E. Maciejowski

ABSTRACT

Annular injection of oil and gas brines poses a danger of ground-water contamination. Predicting its impact is subject to many uncertainties. In this paper, we describe the risks associated with annular disposal of brine in Ohio and summarize the results of a risk-benefit analysis of the practice. The risks are quantified as the probability distributions of the number of rural home wells or municipal well fields which might be contaminated. The benefits are the avoided costs of safer (and more costly) disposal, and the employment and income benefits of a financially healthier oil and gas industry.

1. INTRODUCTION

Risk is ubiquitous. Risk is especially severe in groundwater quality problems because of the difficulty in observing and measuring leaks and transport of contaminants.

But the mere presence of risk is insufficient justification for doing a risk analysis. At least one of two other factors must be present. The first is that the performance of a system is a nonlinear function of the inputs. As a result, the expected performance cannot be calculated simply as the performance of the system under the mean inputs; the entire range of possible inputs must be considered. The second is that the expected performance does not provide enough information. In general, decisionmakers are not risk-neutral. They are concerned not only with the mean performance but also its distribution-- especially the risk of extreme events.

Both of these reasons for doing risk analysis motivate the present study. First, contamination of water supplies is a nonlinear process; for example, there may be concentration thresholds which, if exceeded, result in health impacts or financial costs. Second, public attention focuses on the risks of severe damage to groundwater supplies--on what might happen, rather than the expected value.

In this paper, a controversial waste disposal practice which poses risks to groundwater supplies is described in detail: the injection of oil and gas brines into the annuli of production wells. Because of

---

Benjamin F. Hobbs is Assistant Professor of Systems Engineering and Civil Engineering, Case Western Reserve University, Cleveland, Ohio.

Carl Von Patterson, James R. Heslin, and M. E. Maciejowski are graduate students in the Department of Systems Engineering, Case Western Reserve University.

these risks, this practice is used only in two states, and widely practiced only in one: Ohio. The results of a risk-benefit analysis of the practice in Ohio are presented. Details of methodology are omitted (the interested reader may refer to Patterson et al. [1987]; or Hobbs et al. [1987, 1988]). The purpose is to demonstrate one approach to quantifying and comparing the risks and the benefits of underground waste injection.

2. THE PROBLEM

It is an irony that oil production in the U.S. was first reported in 1814 as a by-product of salt brine wells (Templeton and Associates, 1980). Now, brine is considered a bothersome and environmentally hazardous by-product of the production of oil and natural gas. Oil production was not seriously undertaken until later in the 19th century and for a while, Ohio was the second largest oil-producing state. Since then, over 200,000 oil and gas wells have been sunk in Ohio (Ohio Legislative Service Commission, 1974), with approximately 50,000 still producing (Northeast Ohio Brine Disposal Task Force, 1984). The industry has expanded rapidly recently, having produced a record 15,271,000 barrels (bbls) of oil and 186,480,000 thousand cubic feet (mcf) of gas in 1984 (DeBrosse, 1985). These figures represent a two-thirds increase in production over that of a decade earlier and are a bright spot in an otherwise troubled Ohio economy.

With most oil and gas production comes brackish water or brine. Nationwide, ten billion bbls of associated brine are pumped annually. In Ohio, approximately 43,000 barrels are produced daily (Templeton and Associates, 1980; Ohio Oil and Gas Regulatory Review Commission, 1987).

The Composition of Brine

The largest source of brine in Ohio is the Clinton formation, which contains approximately 280,000 mg/l of total dissolved solids (TDS) and 170,000 mg/l of chlorides (Morth et al., 1979; Crist and Hudak, 1986). This level of chlorides is almost three orders of magnitude higher than the secondary national drinking water standard (U.S. Office of Technology Assessment, 1984). Sodium, present at about 72,000 ppm (Crist and Hudak, 1986), exceeds the recommended level for heart patients by about 3500-fold.

Chlorides and sodium are not the only problem, however. Heavy metals and organic chemicals are also of concern. In 1984, the Michigan Department of Natural Resources found significant concentrations of aromatic hydrocarbons, such as benzene (6.9 ppm:10,000 times the drinking water standard of 0.00067 ppm), ethyl benzene, and toluene, in Michigan brines. This discovery stimulated the Ohio Division of Oil and Gas (DOG) to analyze Ohio brines for organic chemicals. Crist and Hudak (1986) found that Clinton brines had a mean of 2.5 ppm of benzene, with a few samples approaching 10 ppm. This mean value is 3800 times the applicable drinking water standard.

These high concentrations of chlorides, sodium, and organic chemicals mean that brines should be handled carefully to avoid water supply contamination and other environmental impacts. Table 1 presents the maximum concentrations of selected contaminants in Ohio brines measured by Crist and Hudak (1986) and a comparison of them against drinking water standards. Clearly, benzene, sodium, and manganese, in that order, represent the worst threats to public health. Benzene is of particular concern, because it has been determined to be a carcinogen even at low levels; sodium only raises the risk of circulatory problems. On the other hand, sodium behaves conservatively and is not readily adsorbed, whereas benzene is often adsorbed quickly when released into groundwater environments.

Table 1

Maximum Concentrations of Selected Contaminants in Ohio Brines and Dilutions Necessary to Achieve Drinking Water Quality Standards

Constituent	Maximum Concentration (mg/l)	Standard (mg/l)	Dilution Factor
Benzene	10.	0.00067	14,925.
Alkylbenzene	2.19	1.36	1.6
Toluene	29.	0.343	84.5
Xylene	3.44	0.620	5.5
Chloride	180,000.	250.	720.
Chromium	2.2	0.050	44.
Manganese	69.	0.050	1,380.
Sodium	79,000.	20.	3,950.
Nickel	1.9	0.15	12.7
Lead	7.	0.050	140.
Zinc	2.2	5.	0.4

Source: Ohio Oil and Gas Regulatory Review Commission (1987), derived from data in Crist and Hudak (1986).

Brine Disposal Practices in Ohio

Unfortunately, careful brine disposal has been the exception, not the rule (Ohio Environmental Council, 1983). In the early days of the industry, most brine was dumped into nearby ditches and streams (and some of it in Ohio may still be disposed of in that manner). Later, permeable earthen pits were used for disposal, along with shallow wells (< 1000 ft deep). More recently, some producers have used earthen pits with clay or synthetic liners to store brine, disposing of it by deep well injection. Another disposal practice used by many local governments is spreading brine to control road dust and ice.

Deep injection of brines is recognized as the safest method of disposal (Thomas, 1973) and, in addition, is sometimes used to facilitate secondary recovery of oil and gas. Deep injection is done through wells designed or retrofitted for the purpose of disposal, usually under pressure. As Figure 1.1 shows, there are usually at least three major barriers between the injected brine and any freshwater acquifers the well may pass through: the tubing string; the production casing and associated cement grouting; and the surface casing and associated clay or cement grouting.

Another method is annular disposal, the focus of this paper. In annular disposal, the brine is injected into a disposal zone above the producing zone but beneath any fresh aquifers. The disposal zone used may be either (1) the annulus between the production and surface casings, in which case there is just one barrier, or (2) the annulus between the tubing string and the production casing, which results in two barriers. The first type of annular disposal is the most common type in Ohio. Annular injection is illegal in most other states because of the relative lack of protection compared to deep well injection, which has more casings between the brine and fresh water and injects the brine into deeper formations. Under Ohio law, annular injection takes place only under the force of gravity and in amounts no greater than 5 to 10 bbl/day, depending on the construction of the well. Further, annular wells must meet higher construction requirements than other normal producing wells (in particular, the outer casing must be cemented, not mudded, and the production casing must be new and not used). Finally, disposal of brine down any annular well can only come from wells on that particular lease or wells on leases directly bordering the lease of the annular well.

Table 2 sums up the amounts of brine disposed by various methods in Ohio in 1985. In Ohio, 183 deep injection wells have been granted state permits as of June 1, 1986 (Slutz, 1986), and most of the state's brine is disposed of in this manner (Northeast Ohio Brine Disposal Task Force, 1984). Annular disposal permits have been granted to 3521 operations (Slutz, 1986), although not all permitted wells are actually getting rid of their brine by that method (Ohio Environmental Council, 1983). Based on our analysis of annular permit data, we estimate that there are perhaps another 1750 other wells which depend on nearby annular wells for disposal of their brine.

The only other state in which there is a significant amount of annular disposal is Louisiana, where approximately 80 wells have annular disposal permits.

Table 2

1985 Estimated Brine Production and Disposal (in Barrels)

Disposal Method	Reported Disposal Amounts	Rate of Reporting Compliance
Deep Well Injection & Enhanced Recovery	$13,687,851	100%
Annular Disposal	1,192,485	80%
Surface Spreading	448,635	94%

Source: Ohio Oil and Gas Regulatory Review Commission (1987)

Water Supply Contamination by Brine

Haphazard disposal practices, not surprisingly, have ruined private and public water supplies and caused many other environmental impacts. The Ohio EPA has concluded that brine is one of the three most prevalent sources of groundwater contamination in the state (Stein, 1981). Improper disposal of this brine has resulted in many contamination incidents in Ohio. Between 1984 and 1986, DOG identified brine disposal practices as the cause of 71 water well contamination incidents (Table 3).

The U.S. EPA recently estimated that 0.1% of the nation's usable aquifers have been contaminated by petroleum and mining operations (National Research Council, 1984), and an estimate of 0.3% has been made for the amount of Ohio's fresh groundwater that has been contaminated by brine (Ohio Environmental Council, 1983). Although those amounts may not seem large, they translate into thousands of water well closings. For example, at least 17 states have reported brine-related contamination incidents in recent years (Miller, 1980) while 23,000 brine contamination incidents were reported in Texas in the 1960s alone (University of Oklahoma, 1983). Data on the number of well closings in other states due to brine are scarce, but there are many reported cases around the U.S. (e.g., Fryberger, 1975, and Baker and Brendecke, 1983; see the summaries in U.S. Congressional Research Service, 1980a and 1980b; U.S. Office of Technology Assessment, 1984; and National Research Council, 1984) and in Ohio (e.g., Pettyjohn, 1971; Gage, 1985; Lehr, n.d.; Roster, n.d.; Ohio Environmental Protection Agency, 1981; Shaw, 1966).

Table 3

Identified Causes of Oil-Field-Brine-Contaminated Water Wells
in Ohio, 1984-1986

Drilling Pits	39
Temporary Brine Storage Facilities	19
Annular Disposal	11
Brine Dumping	2

Source: Ohio Oil and Gas Regulatory Review Commission (1987)

In Ohio, 180 reports of brine spills or contaminations were made to the Ohio EPA in the late 1970s (Ohio Environmental Protection Agency, 1981). By 1983, the rate of reported incidents went up to 169 in northeast Ohio alone (Northeast Ohio Brine Disposal Task Force, 1984) (see also Table 3). This increase, of course, may reflect increased enforcement efforts and public awareness more than it does the actual number of incidents. Indeed, the vast majority of contamination cases are probably not reported, since most do not result in well foulings or other obvious impacts.

Most contamination incidents resulted from poorly designed storage pits, which allowed brine to percolate to the groundwater (Table 3). This is true even in the 1980s; for example, the contamination of 30 or more wells in Lake County, Ohio, in 1983-1985 has been blamed on pits

(Kell, 1985). In another instance, an earthen pit in Carroll County was sited on an outcrop of a sandstone aquifer; when it leaked, 150 cows located half a mile away were poisoned by lead and had to be slaughtered (Eckstein, 1984). Rigorous enforcement by regulatory agencies and prohibition of permanent pits should greatly decrease the number of such incidents. In addition, future drilling pits are required to have clay and/or plastic liners. Nonetheless, it is disturbing that even the most carefully constructed liners have a high probability of leaking within a few months (Montague, 1982); hence, additional contamination from pits can still be expected.

Yet leaky pits are not the only cause of groundwater contamination by brine, as Table 3 shows. Illegal surface dumping is also to blame. Injection wells, if poorly designed or installed or if built of poor materials, can also fail, leaking brine into fresh aquifers (e.g., Fryberger, 1975). Harrison (1985) argues that annulus overpressurization in production wells can result in contamination. Finally, improper road spreading for dust suppression and ice control can, in theory, foul groundwater supplies.

Annular disposal has also resulted in several cases of groundwater contamination in Ohio (Table 1). For example, a used casing was installed in a well by the Rocky Petroleum Company in Perry County in the mid-1970s; holes in the casing leaked brine, causing a spring to be created down-gradient with a salt concentration of approximately 100,000 ppm (Schultz, 1985). Of the annular well incidents in Table 1.3, most are caused by illegal annular disposal, in which the well is of the wrong construction or otherwise unsuitable and ineligible for a state annular disposal permit (Kell, 1987).

Recent Regulatory Developments

To help prevent annulus failures from causing contamination, DOG will begin testing wells used for annular disposal every five years starting in 1987-1988 (Slutz, 1986). Unfortunately, contamination still occasionally occurs and, as we show later, is likely to continue to happen. The consensus of most observers is that carefully sited, designed, and constructed deep injection wells are the safest way to dispose of brine.

But concerns about the shortage of deep injection capacity in Ohio and the financial burden it would place on producers has caused the state to hesitate to mandate this practice. H.B. 501, as passed into law by the Ohio legislature in 1984, permits continued use of annular injection and road-spreading of brine. This is in spite of recommendations by groups, such as the Northeast Ohio Brine Disposal Task Force (1984), which recommended elimination of those disposal methods. The compromise that H.B. 501 represented was reached because of concern in the Assembly about the negative economic impact that stricter disposal requirements would have upon the state's oil and gas industry. As Governor Celeste put it when signing the bill, H.B. 501 was balanced to protect both underground water supplies and an industry vital to the state's economy (Plain Dealer, 1985). Estimated deep well injection costs of $1.00 to $1.50/bbl (Templeton and Assoc., 1980; see also Section VI of this report) would be a significant burden upon small producers, many of whom pump three or more bbl of brine for every bbl of oil.

Nevertheless, DOG is enpowered to add restrictions to permits or to deny them if there is a substantial risk that the operation will present an imminent danger to public health or to the environment (Ohio General Assembly, 1983-1984). For example, the division has chosen to impose more exacting requirements on drilling pits constructed in Lake and Ashtabula Counties, where unconsolidated surface aquifers are an important source of water (Kell, 1985). Annular disposal is also forbidden in those counties. DOG also turned down several recent permit applications for annular disposal in Medina County because of local concerns about the risk to groundwater of that practice (Beeker, 1985). DOG has been giving out far fewer annular permits than in the past. For example, 1985 drilling data shows that only seven percent of new wells in Perry County are annular wells, whereas 72 percent were in 1980 (see Section VI, Table 6.3, below). It is the evident intention of DOG to phase out annular disposal.

Risks and Uncertainties Concerning Brine Disposal

Each of the steps in calculating the risks and benefits of annular disposal involves important uncertainties that must be recognized in the analysis. Under U.S. Water Resources Council definitions, "risk" refers to situations in which the probability distributions are known, whereas "uncertainty" concerns situations in which the distributions are unknown (Haimes, 1981; 1984). This section summarizes the risks and uncertainties to be encountered in an analysis of the benefits and costs of this brine disposal practice.

The failure of annular disposal is a random event whose probability and severity can only be roughly estimated. A framework for including random contamination events in a groundwater pollution assessment is presented by Kaunas and Haimes (1985), but a serious problem is the lack of data on failure rates. In the case of annular disposal, documentation of failures and resulting contamination is too sparse in Ohio to provide reliable estimates of failure probabilities. Underestimation of failure rates, due to undetected failures, is likely. A second uncertainty is that the chemical make-up of brines is not known with certainty; only very recently have samples of Clinton brine been analyzed for the presence of aromatic hydrocarbons.

The transport of solutes in groundwater, especially for chemical species not subject to sorption and transformation, is fairly well understood. Concludes a National Research Council panel (1984):

> Usually the analysis of flow and transport gives reasonably good results for contaminants in which no chemical reactions . . . occur. This provides some confidence in the general theory of both flow and transport. . . . Simple chemical reactions such as sorption are also simulated in our analysis reasonably well. . . . The current state of the art is such that predictions of conservative contaminant movement are possible with a reasonable degree of confidence, especially for the short term--periods of a decade or perhaps several decades.

This conclusion presumes that the models have been calibrated with field data.

Yet important uncertainties still confront anyone who wishes to predict contaminant concentrations in an aquifer. These include (Anderson, 1984; Cherry, 1984; Konikow, 1981; Harrison, 1984): geohydrologic uncertainties such as poorly known dispersivities and transmissivities; the difficulty of modeling solute transport in fractured consolidated media; inadequate knowledge about the general nature, much less actual rates, of transformation of organic chemicals; non-Fickian dispersion near the source of contamination to as far as several hundred meters; and insufficient data to allow accurate prediction of the fate of heavy metals. These problems are magnified when a hypothetical contamination event is to be modeled, as there are no observations of concentrations which would permit calibration of a solute transport model.

Even if concentration of contaminants could be accurately predicted, their impacts upon water supplies would still be subject to uncertainty. This is because impact depends on the number, location, depth, and size of water supply wells. Oil wells in sparsely populated areas are unlikely to affect water supplies greatly, while those near towns reliant upon groundwater (e.g., Chagrin Falls, Ohio) would have a much higher probability of imposing a severe impact. Further, impact depends on whether the contamination is detected; conceivably, low yet unsafe levels of heavy metals or aromatic hydrocarbons could persist in a water supply without being noticed. This is, however, unlikely to be the case with severe chloride contamination. Yet another uncertainty is that, given a level of contaminant in the water supply, the probability of health effects can, at best, only be roughly estimated, particularly if the concentration is very low. Finally, financial costs of contamination, such as loss of property value and the expense of alternative supplies, are subject to uncertainty.

The benefits of annular disposal--that is, the costs avoided by not requiring safer but more expensive alternatives--are also uncertain. A major uncertainty is the distance to an available injection well, which determines the cost of brine transport (the major cost component in Templeton's analysis [1980]). But the greatest uncertainty concerns the effect stricter regulations would have on the industry. Econometric analysis, which would relate drilling and production activity to the cost of inputs and the price of outputs, cannot yield definitive estimates of the effect of an increase in the cost of doing business. This is not so much due to data difficulties, which indeed can be considerable, but rather to the implicit assumption that past relationships among variables will continue into the future. Two reasons why this might not be true include: 1) future values of independent variables might be outside the range of past values; and 2) non-price variables which are omitted from the model, yet which have an important impact (such as public opinion and attitudes of regulatory agencies), might change.

Problem Statement

Ohio House Bill 501 continues to allow the controversial brine disposal practice of annular injection. This practice is permitted by only five other states, of which only Louisiana allows significant use. Because annular disposal provides but one or two barriers between the brine and fresh water aquifers, it is inherently more risky than deep well injection, which presents at least three. Groundwater wells have been contaminated by annular disposal wells with DOG permits; thus, this risk is a real one.

On the other hand, annular disposal is beneficial to Ohio. In particular, it is costless to the producer and allows well owners to avoid the cost of deep well injection, which can amount to $1 or more per barrel in some locations. If annular disposal is banned, the increased costs would likely lead to a decrease in production and drilling in areas currently dependent on annular disposal. The state would, in turn, lose employment and taxes. The importance of the oil and gas industry to the state is demonstrated by the approximately 60,000 jobs it provides.

The problem is that these risks and benefits are poorly understood. Better understanding of the risks and benefits is important because brine will continue to be a public issue. The Ohio General Assembly recognized this in passing H.B. 501, Section 1509.061, which established a "Brine Management Research Special Account." This account has funded research on several brine topics, including the effects of annular disposal upon public health and the environment. Continued controversy over brine handling is assured by the likelihood of future contamination incidents caused by (1) abandoned brine storage lagoons and wells, (2) continued illegal dumping, and (3) the relative unreliability of annular disposal and temporary drilling pits. Public outcry will continue to force the Ohio Department of Natural Resources to consider stricter regulation and permit conditions, and there is a strong possibility that the legislature will have to address the issue again. Indeed, both the U.S. Environmental Protection Agency, Region V, and DOG have been reconsidering their regulations which allow annular disposal (Houser, 1985). Local officials will also continue to press for stricter well permit conditions.

3.   SUMMARY OF THE METHODOLOGY

General Approach

Risk-benefit analysis, sometimes referred to as generalized benefit-cost analysis (Sharefkin et al., 1984), provides the framework of the analysis. Risk-benefit analysis consists of two phases: (a) the quantitative processing and evaluation of information, including the quantification of risk and the development of policy options, and (b) the introduction of value judgments concerning what risks are acceptable relative to their benefits and what policies are desirable (Haimes, 1981; 1984). Risk-benefit analysis has found wide application to problems involving risk and uncertainty within both government and industry.

A key feature of risk-benefit analysis is its recognition that decisionmakers are not risk-neutral, which implies that expectations alone do not provide sufficient information for a decision. Risk-benefit analysis can be used to quantify the probability of extreme events (Asbeck and Haimes, 1984), such as contamination of a large town's well field, which may be as or more important to decisionmakers than the expected level of contamination.

A second important feature of risk-benefit analysis is its recognition that public issues are multiobjective, multiparty problems, and that these multiple perspectives should not be collapsed into a single index of worth, such as dollars (Chankong and Haimes, 1983; Cohon, 1978; and Hobbs, 1985). Rather, the trade-offs that exist among objectives and interests should be made explicit. For example, the possible health risks of brine contamination and the financial benefits to the industry of annular disposal should not be lumped into one metric (e.g., dollars). Instead, the policy-making process is best served if the trade-offs among the objectives are displayed. Decisionmakers can then make any needed value judgments, rather than rely on analysts to impose their values through assumptions that are likely to be somewhat arbitrary.

The precise methodologies used in risk-benefit analysis depend on the nature of the problem and the information needs of the decisionmakers. Methods that are often found in the risk-benefit analyst's "tool box" include economic analysis, operations research, statistics, epidemiology, and process models of physical and engineering systems. In the context of groundwater contamination, economic analysis usually plays an important role in quantifying those costs and benefits that can be meaningfully expressed in dollar terms. Examples include the cost of spill and leak prevention (e.g., Kaunas and Haimes, 1985) and the expense of groundwater cleanup and other mitigation measures (e.g., Sharefkin et al., 1984). The methods of probability theory and statistics are of course very important in quantifying, to the extent possible, leakage frequency, duration (length of time), severity (rate of leakage), the probability distribution of water supply wells near the source of contamination, and the effect of input costs and oil and gas prices upon the level of drilling and production. Hydrologic models of flow and solute transport and transformation in aquifers are indispensable tools in any risk-benefit analysis of groundwater contamination, and can be used to translate probability distributions of leaks into probability distributions of contamination.

The framework for our analysis is shown in Figure 1. The risks of annular disposal would be assessed by estimating, in turn, the frequency and severities of: brine releases; concentrations of chemical species in freshwater aquifers; contamination of private and public water supply wells; and the health costs of contaminated supplies and financial expense of supply replacement.

The Risks of Annular Disposal of Brine:

| Sources of | \ | Solute Transport | \ | Water Supply |
| Contamination | / | & Transformation | / | Contamination |

The Benefits of Annular Disposal of Brine:

| Incremental Cost of | \ | Production & | \ | Tax, Employment |
| Deep Well Injection | / | Drilling Activity | / | Effects |

Figure 1. Framework of the Proposed Research

The methods we use to quantify the risks include those of probability theory, solute transport modeling, geography, and engineering economy. The benefits of annular disposal are (1) the avoided expense of safer and more costly disposal and (2) the secondary benefits of a financially healthier oil and gas industry, increased taxes, and employment. The tools of engineering economy, econometrics, and economic impact analysis are used to estimate the benefits.

Risk Assessment Methodology

The risk assessment consists of:

1. estimation of annulus reliability from radioactive tracer tests; and

2. quantification, for two important aquifers in Ohio, of the probability distributions for the number of community and rural home groundwater supplies that will be contaminated due to annular disposal.

Details can be found in Patterson et al., (1987) and Hobbs et al. (1988).

The first step consists of obtaining the probability distributions of the durations and magnitudes of leaks from annular disposal. To calculate the distribution of durations, we first obtained two probabilities:

p = the probability of immediate failure due to faulty construction

q = the probability per year of failure subsequent to well completion

These we estimated from radioactive tracer tests of annulus integrity, conducted by the Louisiana Division of Injection and Mining using maximum likelihood analysis. The results were $p = 0.0025$ and $q = 0.009$. Knowing these probabilities, the disposal well lifetime, and how often a well is inspected (currently, every five years) enables us to calculate the probability distribution of spill duration. The probability distribution of spill magnitudes is a subjective distribution based on interviews with experts on the topic.

The second step translated the leak probability distribution into probability distributions of the number of water wells contaminated. This was done by accomplishing the following tasks:

a.   Obtain parameters for a groundwater solute transport model of each aquifer of interest. We used the Konikow-Bredehoeft (1978) Method of Characteristics model of two-dimensional conservative transport. Parameters for two important water supply aquifers in areas where annular disposal takes place were taken from field studies. One is the Black Hand Sandstone, a consolidated aquifer with low flow velocities (Norris and Mayer, 1982); the other is the complex of unconsolidated buried valleys in the Muskingum River basin (Mayhew, 1973).

b.   Under the assumption that leakage rates are too low to significantly affect the flow field, obtain impulse response functions from each groundwater model. These functions describe the increase in concentration at each spatial point over time due to the pulse input of one unit of contaminant.

c.   Discretize the leak magnitude and duration distributions, and use the impulse response function to calculate the plume resulting from each combination of magnitude and duration. This yields a probability distribution of contaminant plume size. To obtain a plume from a particular leak, the aquifer's impulse response function is (i) multiplied by the actual size of the leak and (ii) convoluted over the duration of the leak. We trace plumes for 20 years after the start of the leak. A location is designated as being "contaminated" if the concentration at any time exceeds water quality standards.

d.   Use a probability distribution of the number of water supply wells, conditioned on the size of the area contaminated, together with the distribution of areas contaminated, to obtain the final distribution number of water wells contaminated. It is assumed that rural home water wells are located according to a spatial Poisson process; likewise with municipal well fields. The reason for treating water well location as a random variable is that there are several thousand annular wells and it is impossible to describe the location of water wells near each one.

An example of the output of this analysis is shown in Figure 1. It portrays the exceedence distribution of the number of rural homes contaminated in the Black Hand Sandstone as a function of the number of years between disposal well inspections. As inspections become more frequent, the risks decrease.

Benefits Assessment Methodology

The benefits assessment comprises:

1.   estimation of the short-run benefits of annular disposal as the avoided cost of deep well injection;

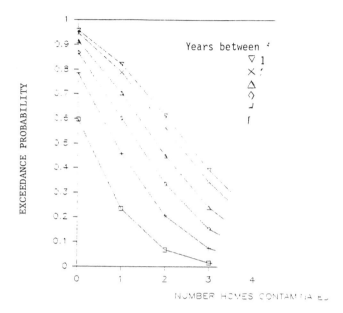

Figure 1. Exceedence Probabilities, Number of Homes Contaminated, Black
Hand Sandstone Aquifer

2. estimation of supply functions for the oil and gas industry in Ohio, including:

    a. drilling of oil, gas, and combination oil-gas wells in the six counties where the most annular disposal takes place, and

    b. oil and gas production statewide;

3. use of those functions to evaluate the impact upon drilling and production of the increased brine disposal costs that would result from a banning of annular disposal; and

4. input-output analysis of the effects of changes in drilling and production upon regional economic activity, employment, and tax revenues.

Details on the procedures used may be found in Hobbs et al. (1987).

Based on an estimate of $1.50/bbl for deep well injection, a ban upon annular disposal would impose an immediate direct cost of about $2,200,000/yr upon Ohio oil and gas producers. Because this assumes that no annular wells are shut down, this estimate is an upper bound on the short-run benefits of annular disposal.

The long-run benefits of annular disposal would depend on the reaction of the oil and gas industry to the expense of deep well injection. To estimate these, we created econometric models of drilling and production activity. The drilling models relate oil and gas drilling rates in a given year in the six most important annular disposal counties to past drilling rates and the anticipated present worth of drilling activity (based on prevailing prices and drilling costs). The production models regress production in a given year upon production in the previous year, new wells drilled in the previous year, and the price of oil.

These models are used to project to drilling and production over the next decade with and without a ban upon annular disposal. A ban affects drilling by lowering the present worth of new oil and gas wells and affects production by lowering the effective price received by well operators.

Decreases in drilling and production would affect regional economic activity and employment. Based on the estimated decreases in drilling and production, an input-output model previously constructed for southeastern Ohio (Young et al., 1985) is used to calculate the long-run direct and indirect economic impacts of a ban.

4. RESULTS

For several reasons, not all the risks and benefits of annular disposal could be estimated. On the risk side, only risks to water supply wells are considered; risks to natural ecosystems, human health, and domesticated plants and animals are not quantified, due to the absence of data on these effects. As an index of some of these unquantified risks, Tables 4 and 5 display the expected land area that would be contaminated due to annular disposal. A final limitation on the risk estimates is that they are obtained only for two regions in the state: the Buried Valleys of the Coshocton basin and the Black Hand Sandstone. These areas contain only about 25 percent of the annular disposal wells. Other areas were not investigated because of the absence of geohydrologic data. For this reason, the risks in Tables 4 and 5 are lower bounds on the statewide impacts. On the benefits side, our estimates of the effect of an annular disposal ban on income, employment, and economic activity are limited to the oil-producing regions of the state; impacts elsewhere in the state are not considered because no state-wide input-output model was available.

The risks are stated in terms of impacts resulting from 20 years of annular disposal (a typical disposal well lifetime). Releases of brine into the aquifers are traced for twenty years. In the Buried Valley aquifer, this period of time is sufficient to cleanse the aquifer, as the pollutants are ultimately discharged to the river. However, in the Black Hand Sandstone area, the pollution will remain and spread for centuries, because groundwater velocities are low. Hence, the risks presented in Tables 4 and 5 are lower bounds to the long-run risks to water supply in that region.

The analysis shows that the approximately 700 annular wells in the Black Hand Sandstone region are expected to contaminate 2.5 rural home wells due to leakages. There is also a 2 percent chance of contaminating a community well field, and a 4 percent chance of contaminating five or more rural home supplies. In the Buried Valley region, annular wells there are expected to contaminate one home, and have a probability of 1 percent of contaminating a town's well field. These calculations are based on an annular well inspection frequency of once every half-decade.

A ban would affect oil and gas drilling by lowering the present worth of new wells, and would have an impact on production by depressing the effective price received by drillers. The analysis shows that a ban would decrease drilling rates in Ohio by approximately 15 wells per year (less than 0.5% of the 1985 drilling rate). Production would diminish in the long run by about 90,000 bbl/yr of oil and 600,000 mcf/yr of natural gas. These impacts are based on an assumed price of $20/bbl for oil, $3.10/mcf for gas, and a 3 percent real interest rate.

Decreases in drilling and production would affect regional economic activity and employment. Based on the above assumptions and an input-output model previously constructed for southeastern Ohio, the long-run direct and indirect economic impacts of a ban would equal losses of about 80 jobs, $7,500,000 of regional output, and $1,100,000 of income (profit and royalties). This assumes a mean oil well cost of $100,000/well.

Table 4
Risks and Inspection Costs for Different Inspection Frequencies

| | Time Between Inspections | | | |
| | | | Current Law: | |
	1 Year	3 Years	5 Years	10 Years

BLACK HAND SANDSTONE STUDY AREA (728 Annular Disposal Wells)

Risks From 20 Years of Disposal[a]:

	1 Year	3 Years	5 Years	10 Years
Expected Number of Home Water Wells Contaminated	0.9	2.0	2.5	3.2
Probability of 5 or More Home Wells Contaminated	0.003	0.06	0.11	0.22
Probability of 1 or More Community Well Fields Contaminated	0.009	0.019	0.023	0.030
Expected Economic Damage Due to Contamination of Water Wells[b]	$57,000	$126,000	$153,000	$196,000
Expected Area Contaminated	0.09 mi^2	0.19 mi^2	0.23 mi^2	0.30 mi^2
Costs of Inspection:	$1,820,000/yr	$610,000/yr	$360,000/yr	$180,000/yr

BURIED VALLEY ACQUIFER (92 Annular Disposal Wells)

Risks from 20 Years of Disposal[a]:

	1 Year	3 Years	5 Years	10 Years
Expected Number of Home Water Wells Contaminated	0.28	0.69	0.87	1.05
Probability of 5 or More Home Wells Contaminated	0.0002	0.005	0.010	0.018
Probability of 1 or More Community Well Fields Contaminated	0.004	0.010	0.012	0.015
Expected Economic Damage Due to Contamination of Water Wells[b]	$17,000	$43,000	$55,000	$66,000
Expected Area Contaminated	0.04 mi^2	0.10 mi^2	0.12 mi^2	0.15 mi^2
Costs of Inspection:	$230,000/yr	$77,000/yr	$46,000/yr	$23,000/yr

---

[a] Calculations based on tracing contamination for 20 years, starting with the beginning of the leak.

[b] Assuming costs of $60,000/home and $200,000/town well field.

Table 5
Risks and Benefits of Annular Disposal

RISKS FROM TWENTY YEARS OF OPERATION[a]:

Black Hand Sandstone and Buried Valley Aquifers Only (820 Wells):

Expected Number of Home Water Wells Contaminated	3.3 wells
Probability of 5 or More Home Wells Contaminated	0.21
Probability of 1 or More Community Well Fields Contaminated	0.035
Expected Economic Damage Due to Water Well Contamination	$207,800
Expected Area Contaminated	0.36 mi^2

Entire State (3521 Wells):
Not estimated.

BENEFITS:

Short-Run Primary Benefits[b]:

Avoided Deep Well Disposal Costs, Black Hand Sandstone and Buried Valley Aquifers Only	$400,000/yr
Avoided Deep Well Disposal Costs, Entire State	$2,200,000/yr

Long-Run Secondary Benefits, Entire State[c]:

Effect on Oil and Gas Wells Drilled	16 wells/yr
Effect on Oil Production	94,000 bbl/yr
Effect on Gas Production	638,000 mcf/yr
Effect on Employment, Oil and Gas Industry	51 man-yrs/yr
Effect on Activity, Oil and Gas Industry	$5,600,000/yr
Effect on Total Employment, Regional Economy	82 man-yrs/yr
Effect on Total Activity	$7,500,000/yr

[a]Impacts of 20 years of operation of the annular wells in the stated study regions. Calculations based on tracing contamination for 20 years, starting with the beginning of the leak. Inspection frequency of once every five years assumed.

[b]Based on estimated amount of brine disposed by annular disposal, multiplied by a deep well injection cost of $1.50/bbl. This assumes that a ban on annular disposal would not result in any wells being shut down in the short run. If some operators shut down their wells rather than pay the additional cost of deep well injection, short-run primary benefits will be less than the amounts stated here. Under most price and interest rate scenarios of Section VI, the production equations show that little production is lost in the very short run (one to two years).

[c]Long-run calculations based on the following scenario: an oil price of $20/bbl; a gas price of $3.1/mcf; a three percent real interest rate; and a drilling cost of $30/ft. Employment and activity impacts based on the above prices and an assumed cost per oil or gas well of $100,000. To avoid overstating the benefits of annular disposal, the only production decreases included in the long-run calculations are those resulting from the long-run decrease in wells drilled. Production decreases resulting from possible shortening of well production life are not included; including those decreases would more than double the oil production impact and would increase the natural gas impact by ten percent.

These estimates of the benefits of annular disposal are not sensitive to oil and gas prices, but do vary somewhat with the assumed real interest rate.

Table 4 contrasts the impacts and inspection costs of different inspection frequencies. In neither region does the cost of more frequent inspections seem justified by the benefits. In the Black Hand Sandstone region, for example, increasing the frequency from once every five years to once every other year for the next 20 years would increase inspection costs by $550,000 per year. But this change would decrease the expected number of contaminated rural home wells by only one over the entire period.

In Table 5, the overall risks and benefits of annular disposal are summarized. A ban upon annular disposal would erase both. The direct benefits of annular disposal in just the Black Hand Sandstone and Buried Valley regions are roughly $400,000/year in avoided disposal costs. The loss of these benefits appears to be a high price to pay to lower the risks in Table 5 to zero. Evaluating home water supplies at $60,000/home and assuming that the replacement costs of municipal water supplies is $200,000 results in an expected annual financial cost of contamination one order of magnitude less than the direct benefits of annular disposal, assuming a real interest rate below ten percent per year. Therefore, if annular disposal were to be permitted and the industry required to compensate victims of contamination, both industry and water well owners would be at least as well off financially as under a ban.

Although we conclude that a ban would be undesirable on economic grounds, a ban may still be justifiable for public health or policy reasons. This study has not addressed the public health aspects of the problem; basically, we have assumed that any contaminated water wells are immediately detected and shut down. Another basic assumption has been that the only impact of groundwater contamination is the loss of water supply wells; this disregards the option and existence values of groundwater and possible impacts on fish and wildlife. As an index of these unquantified values, Table 5 shows the expected value of the area that would be contaminated due to annular disposal activities over the next two decades. We estimate this quantity to be approximately 200 acres in the Black Hand and Buried Valley acquifers.

This research has not addressed the desirability of annular disposal for particular wells or locations. Therefore, even if annular disposal should not be banned statewide, it might be desirable to disallow the practice in places where the risks to water supplies are particularly high and the costs of banning it are low. Based on our analysis, such areas include locations within half a mile or so of public water supply wells and where deep injection wells are nearby.

ACKNOWLEDGMENTS

This work was supported by the U.S. Geological Survey through the Ohio Water Resources Center. We thank Y. Y. Haimes for his collaboration on this project. Computational assistance was provided by D. T. Hoog, J. J. Kavlick, and C. Moy; J. M. Weber undertook the input-output analysis. Reviews and suggestions by J. T. deRoche of USGS and D. Calhoun of the Ohio Department of Natural Resources are gratefully acknowledged, as is the patient and helpful cooperation of the ODNR Division of Oil and Gas. We also thank the State of Louisiana, Bureau of Conservation, Division of Injection and Mining for making their tracer test results available to us.

REFERENCES

Anderson, M. T., 1984. Movement of contaminants in groundwater: groundwater transport -- advection and dispersion. In Groundwater Contamination. Washington, D.C.: National Academy Press.

Asbeck, E. L., and Y. Y. Haimes. 1984. The partitioned multiobjective risk method (PMRM). Large Scale Systems 6(1): 13-38.

Baker, F. G., and C. M. Brendecke. 1983. Seepage from oilfield brine disposal ponds in Utah. Ground Water 21(3): 317-324.

Beeker, J. 1985. Personal communication. Northeast Ohio Areawide Coordination Agency, Cleveland, Ohio.

Chankong, C., and Y. Y. Haimes. 1983. Multiobjective Decisionmaking: Theory and Methodology. New York: Elsevier-North Holland.

Cherry, J. A., R. W. Gillham, and J. F. Barker. 1984. Contaminants in groundwater: chemical processes. In Groundwater Contamination. Washington, D.C.: National Academy Press.

Cohon, J. 1978. Multiobjective Programming and Planning. New York: Academic Press.

Crist, R., and G. Hudak. 1986. Preliminary analysis of Ohio oil field produced brines for selected heavy metals and aromatic hydrocarbons. Ohio Department of Natural Resources, Columbus, Ohio.

DeBrosse, T. A. 1974-1986. Summary of Ohio oil and gas developments. Presented at winter meetings, Ohio Oil and Gas Association, Columbus, Ohio.

Eckstein, Y. 1984. Hydrogeological parameters and potential for groundwater contamination. Presented at annual meeting, Ohio Academy of Sciences, Cleveland, Ohio, April 28, 1984.

Fryberger, J. S. 1975. Investigation and rehabilitation of a brine-contaminated aquifer. Ground Water 13(2): 155-160.

Gage, C. 1985. Drilling turns water to brine. The Plain Dealer, Cleveland, Ohio, January 13, 1985.

Haimes, Y. Y. 1981. Risk/Benefit Analysis in Water Resources Planning and Management. New York: Plenum Press.

Haimes, Y. Y. 1984. Risk assessment for the prevention of groundwater contamination. In Groundwater Contamination. Washington, D.C.: National Academy Press.

Harrison, S. S. 1984. Evaluating system for ground-water contamination hazards due to gas-well drilling on the glaciated Appalachian plateau. Ground Water 21: 689-700.

Harrison, S. S. 1985. Contamination of aquifers by overpressuring the annulus of oil and gas wells. Ground Water 22: 317-324.

Hobbs, B. F. 1985. Choosing how to choose: comparing amalgamation methods for environmental impact analysis. In Environmental Impact Assessment Review, December.

Hobbs, B. F., C. V. Patterson, M. E. Maciejowski, and Y. Y. Haimes. 1988. Risk analysis of groundwater contamination by oil and gas brines. Submitted to Journal of Water Resources Planning and Management.

Hobbs, B. F., et al. 1987. Risk/benefit analysis of annular disposal of oil and gas brines. Final report. Department of Systems Engineering, Case Western Reserve University, Cleveland, Ohio.

Houser, R. 1985. Personal communication. Division of Oil and Gas, Ohio Department of Natural Resources.

Kaunas, J. R., and Y. Y. Haimes. 1985. Risk management of groundwater contamination in a multiobjective framework. Water Resources Research 21(11): 1721-1730.

Kell, S. 1985 and 1987. Personal communications. Division of Oil and Gas, Ohio Department of Natural Resources, Columbus, Ohio.

Konikow, L. F. 1981. Role of numerical simulation in analysis of ground-water quality problems. The Science of the Total Environment 21: 299-312.

Konikow, L. F., and J. D. Bredehoeft. 1978. Computer model of two-dimensional solute transport and dispersion in ground water. Chapter C2, Techniques of Water-Resources Investigations of the U.S. Geological Survey. Washington, D.C.: U.S. Government Printing Office.

Lehr, J. H. 19--. A study of ground water contamination due to saline water disposed in the Morrow County fields. Ohio State University, Columbus, Ohio. Submitted to Office of Water Resources Research, U.S. Department of Interior.

Mayhew, B. 1973. Groundwater investigation of the buried valley underlying the Muskingum River, Coshocton County and the Wakatomika Creek in Muskingum County, Ohio (Area 8). Prepared for the Ohio Department of Natural Resources. Ohio Drilling Company, Massillon, Ohio.

Michigan Department of Natural Resources. 1984. Analysis for aromatic hydrocarbons in oil field brines. Geological Survey Division.

Miller, D. W. (ed.). 1980. Waste Disposal Effects on Ground Water. Berkeley, California: Premier Press.

Montague, P. 1982. Hazardous waste landfills: some lessons from New Jersey. Civil Engineering 52(9): 53-56.

More, E. 1985. Personal communication. Division of Water Quality Monitoring and Assessment, Ohio Environmental Protection Agency, Twinsburg, Ohio.

Morth, A. H., J. E. Hatch, and D. A. Stith. 1979. Brine analyses, 1972-1974. Open file report 79-1. Ohio Department of Natural Resources, Columbus, Ohio.

National Research Council. 1984. Groundwater Contamination. Washington, D.C.: National Academy Press.

Norris, S. E., and G. C. Mayer. 1982. Water resources of the Black Hand Sandstone member of the Cuyahoga formation and associated aquifers of Mississippian Age in southeastern Ohio. Open file report 82-170. U.S. Geological Survey, Columbus, Ohio.

Northeast Ohio Brine Disposal Task Force. 1984. Brine disposal in northeast Ohio: a report on the problem and recommendations for action. Northeast Ohio Areawide Coordinating Agency, Cleveland, Ohio.

Ohio Environmental Council. 1983. Brine disposal in Ohio: an overview. Columbus, Ohio.

Ohio Environmental Protection Agency. 1981. Brine disposal from oil and gas production in Ohio. Columbus, Ohio.

Ohio General Assembly. 1983-1984. H.B. 501, 115th General Assembly, Regular Session, Columbus, Ohio.

Ohio Legislative Service Commission. 1974. Ground Water. Report #115, Columbus, Ohio.

Ohio Oil and Gas Regulatory Review Commission. 1987. Report of the Ohio Oil and Gas Regulatory Review Commission. Columbus, Ohio.

Patterson, C. V., B. F. Hobbs, M. E. Maciejowski, and Y. Y. Haimes, 1987. Annular disposal of oil and gas brines in Ohio: a risk analysis. In S. J. Nix and P. E. Black (eds.), Symposium on Monitoring, Modeling, and Mediating Water Quality. American Water Resources Association.

Pettyjohn, W. A. 1971. Water pollution by oil-field brines and related industrial wastes in Ohio. Ohio Journal of Science 71(5): 257-269.

The Plain Dealer. 1985. Ohio gets stiffer brine penalties. January 11, 1985, page 5A. Cleveland, Ohio.

Roster, R. J. 19--. A study of ground water contamination due to oil-field brines in Morrow and Delaware counties, Ohio. M.S. thesis, Ohio State University, Columbus, Ohio.

Schultz, K. 1985. Personal communication. Emergency Response Division, Ohio Environmental Protection Agency, Columbus, Ohio.

Sharefkin, M., M. Shechter, and A. Kneese. 1984. Impacts, costs, and techniques for mitigation of contaminated groundwater: a review. Water Resources Research 16(12): 1771-1784.

Shaw, J. E. 1966. An investigation of ground water contamination by oil field brine disposal in Morrow and Delaware counties, Ohio. M.S. thesis, Ohio State University, Columbus, Ohio.

Slutz, J. 1985-1987. Personal communications. Ohio Division of Oil and Gas, Columbus, Ohio.

Stein, R. B. 1981. Ohio's ground water resources and problem areas. In Proceedings and Recommendations, Workshop on Ground Water Problems in the Ohio River Basin, Cincinnati, Ohio, April 28-29, 1981.

Templeton, E. E., and Associates. 1980. Environmentally acceptable disposal of salt brines produced with oil and gas. Sponsored by Ohio Water Development Authority.

Thomas, W. E. 1973. The concentration of selected elements in brines of Perry County, Ohio. M.S. thesis, Ohio State University, Columbus, Ohio.

U.S. Congressional Research Service. 1980a. Six case studies of compensation for toxic substances pollution: Alabama, California, Michigan, Missouri, New Jersey, and Texas. Prepared for U.S. Senate Committee on Environment and Public Works.

U.S. Congressional Research Service. 1980b. Resource losses from surface water, ground water, and atmospheric contamination: a catalog. Serial no. 96-9, U.S. Senate Committee on Environment and Public Works.

U.S. Office of Technology Assessment. 1984. Protecting the nation's ground water from contamination, vols. 1 and 2. Washington, D.C.

University of Oklahoma. 1983. Groundwater contaminants and their sources. Draft report to Office of Technology Assessment, Norman, Oklahoma.

Young, K. R., D. L. Forster, L. J. Hushak, and G. W. Morse. 1985. Environmental regulation and regional economic growth: an input-output analysis of the Ohio coal mining region. Research bulletin 1170, Ohio Agricultural Research and Development Center, Ohio State University, Columbus, Ohio.

EVENT TREE SIMULATION ANALYSIS FOR DAM SAFETY PROBLEMS

Jery Stedinger, David Heath, and Neville Nagarwalla

ABSTRACT

Safety studies for existing dams have found that many do not
satisfy the latest estimates of the probable maximum flood, the current
design standard for high-hazard dams. This raises the question of how
safe is safe enough, and suggests that risk analyses may be required to
justify and evaluate the appropriateness of alternative retrofit and
operating policy changes. Risk-cost analysis for high-hazard dam safety
evaluation is an evolving methodology which has yet to be widely
adopted. The probability that a given flood level is surpassed, or a
given level of flood damages or loss of life is exceeded, depends upon
the probabilities of all combinations of factors which together would
cause those levels to be reached and exceeded. Here an event tree
describes the many random factors which contribute to major inflow
floods, determining reservoir operation and possible downstream damages;
this allows evaluation of the probability of dam failure and the distributions of damages and loss of life.

1.   INTRODUCTION

The design safety standard which should be imposed on high-hazard
dams in the United States has posed a special dilemma for the U.S.
water resources engineering community. The failure of large structures
built by the U.S. Army Corps of Engineers, the U.S. Bureau of
Reclamation, and the Tennessee Valley Authority, due to extraordinary
floods would mean the loss of an expensive structure and the services it
provides, and in many cases immense economic costs and loss of lives
associated with the flood as it passed downstream. The emergency
spillway of such structures and the flood storage zone should be sized
to pass or store some safety design flood which provides both a
reasonable level of safety and an acceptable compromise between
construction costs and the possible consequences of dam failure from the
occurrence of major floods.

Jerry Stedinger, David Heath, and Neville Nagarwalla are, respectively,
Associate Professor of Environmental Engineering, Associate Professor of
Operations Research, and a graduate student in the Department of
Operations Research, Cornell University, Ithaca, New York.

This work was supported in part by the Institute of Water Resources,
U.S. Army Corps of Engineers, through a grant through Battelle, Research
Triangle Park, North Carolina, and the National Science Foundation,
through grant CEE-8351819.

2.   BACKGROUND

At the turn and in the early part of this century, selection of design inflow floods using the floods-of-record and flood-frequency methods was found to be unsatisfactory. Available records could not be trusted to reveal the possible severity of the unusual and intense meteorological events that cause the extraordinary floods of concern (Myers, 1967), though flood-frequency techniques have been successful for traditional flood plain mapping and the design of levees, bridges, roads, and traditional flood control structures (Feldman, 1981).

In the 1930s the storm transposition approach was developed for estimating a spillway's design flood from the worst rainfall observed in a region, transposed to the reservoir's catchment. For high hazard dams, the safety standard has become the probable maximum flood (PMF), which is derived by conservative application (in terms of antecedent moisture, streamflow, and snowpack, and also storm placement and the time distribution of rainfall within the storm) from the probable maximum precipitation (PMP). The PMP is an estimate of the maximum possible precipitation depth over a given size catchment in a given length of time. It is calculated by maximizing the moisture in the worst storms of record, transposing those storms to the drainage (adjusting for maximum moisture and barrier and terrain effects), and finally using an envelope curve over duration and area to define a maximum possible precipitation for the duration and drainage size of interest (Myers, 1967; Stallings et al., 1986).

Budgetary concerns related to large dam retrofit decisions in the U.S. have created a demand for justification of hard-to-obtain appropriations; this has resulted in greater interest in risk-based analyses and possibly caused an easing of standards (Krouse, 1986). Recently, a National Research Council committee reviewed the standards and methods for dam safety evaluation (NRC, 1985). Risk-based procedures were encouraged for retrofit decisions when a structure does not pass the latest PMF estimate but still might be safe enough, or to see if the cost of upgrading to the full PMF is justified. Von Thun (1987) discusses the U.S. Bureau of Reclamation's dam safety risk evaluation methodology, which incorporates this idea, and illustrates the calculation of "risk costs." Two 1985 conferences addressed dam safety and risk assessment (McCann, 1986; and Haimes and Stakhiv, 1986; see also Bowles, 1987; and Moser and Stakhiv, 1987). Terry Coomes (McCann, 1986) has put the question clearly: "The debate is over money and the best allocation of the Nation's resources."

The general application of risk assessment to dam safety evaluation will require estimation of the frequency of very unusual and extreme events on the order of the PMF. One proposal is to assign the PMF a return period of from 10,000 up to a million years, and then to extend a flood flow frequency curve from the 100-year flood out to the PMF. That approach depends on those probabilities and the method of extension employed. Stedinger and Grygier (1985) show that those decisions can affect the ranking of retrofit alternatives. An alternative approach is to attempt to estimate the frequency of large storms and floods using regional storm data. A noble attempt at such an analysis is described by YAEC (1984): actually calculating the probability that the Harriman Dam in Vermont would be overtopped by a flood. Following the NRC Committee's charge, Stedinger and Grygier (1985), Von Thun (1987), and Resendiz-Carillo and Lave (1987) illustrate and discuss procedures and issues faced by such risk analyses. Bohnenblust and Vanmarcke (1982), McCann et al. (1984), Bury and Kreuzer (1986), and Bowles (1987) provide illustrations of more comprehensive analyses that include different possible causes of dam failures, including piping, structural, and earthquake-related damage.

3. A PROBABILISTIC DAM PERFORMANCE ANALYSIS

The probable maximum flood is based on conservative assumptions about the values of many factors (see Hansen et al., 1982, or UMO, 1986). As a result it is thought by many to be a very conservative design standard (Newton, 1983). In moving to risk-based analyses, the impact of the many factors contributing to the probability of dam failure and the magnitudes of damages should be considered (see Newton, 1983, Table 1). In particular, the probability that a flood might occur and exceed some threshold is the sum of the probabilities of all combinations of rainfall depths, rainfall distribution in space, rainfall distribution across time within a storm, and antecedent soil moisture levels that would yield a flood of that size or larger. Whereas the PMP is envisioned to be the worst combination of conditions possible, and therefore is defined by the single storm type that generates those conditions (WMO, 1986), a probability evaluation intended to also describe the distribution of less severe storms must consider the many storm types (winter or summer; hurricane, frontal, or thunder storms) that can lead to such major flood events.

Likewise, the probability that a dam fails, or a given level of damages is exceeded, depends upon all the ways a given outflow discharge rate may be achieved and thus depends upon the possible inflow floods, the reservoir operating policy, initial reservoir levels, if various outflow works and gates are operational, and if the turbines can be operated at capacity. Whether or not the structure actually fails depends on whether the emergency spillway can actually pass its rated discharge without failing. On the margin, it may also hinge on the direction of the wind across the reservoir's pool, the resultant wave runup on the dam, and the amount of overtopping the structure can withstand without failure.

Finally, the level of damages that results from the flood may depend on the time of year and time of day that the flood occurs. Other structures may or may not also fail and contribute to loss of life and downstream damages; likewise, flood warning systems have the potential to reduce loss of life from major floods, and from dam break floods in particular, but may not work perfectly (Pat[3]-Cornell, 1984; Pat[3]-Cornell et al., 1986).

Bury and Kreuzer (1986) show how the relationships between these events can be summarized by simple event trees. Figure 1 provides a more complex event tree describing how the various factors can

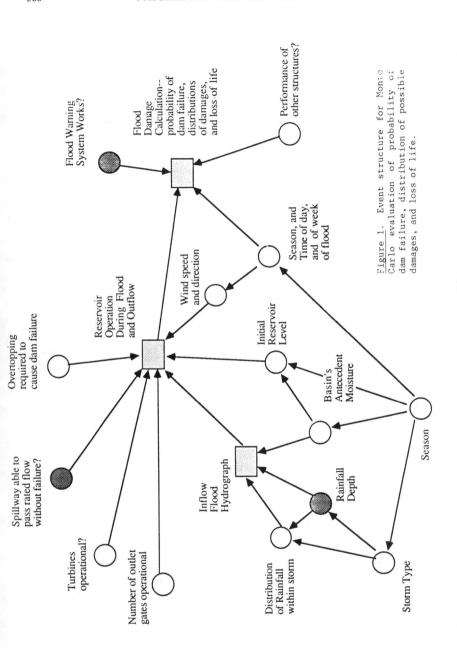

Figure 1. Event structure for Monte Carlo evaluation of probability of dam failure, distribution of possible damages, and loss of life.

contribute to a major flood which would cause downstream flooding and associated damages, and may even combine to result in dam failure. In the lower left corner, the circles describe seasonal and meteorological factors which combine to determine the inflow hydrograph. Such a flood needs to be stored by and passed downstream by a reservoir's turbines, outlet works, and emergency spillway, depending upon their ability to operate and the reservoir's operating guidelines.

Many of the factors described by circles in Figure 1 are best described by continuous random variables: rainfall depth, the possible temporal distribution of rain within a storm, and initial reservoir levels, for example. This precludes the simple event tree evaluation that is possible for simple binary trees in which each event can have only two values and the values realized at different nodes are independent.

The event tree in Figure 1 can be evaluated by discretizing the various random variables, being careful to capture the various interdependencies when evaluating the resultant probabilities of the various discrete outcomes. Alternatively, one could attempt to employ straightforward Monte Carlo procedures; but because of our interest in extremely rare events, which may occur with a probability of 1 in 10,000 or more, that approach is very inefficient. We have adopted a hybrid procedure which takes advantage of the strength of both approaches.

Table 1 describes our event tree simulation procedure. The procedure is based on the premise that the most important, perhaps dominant, random variable is the rainfall depth. Without sufficient rain, a flood won't occur. Therefore a reasonable hybrid Monte Carlo/analytical evaluation of the event tree in Figure 1 would randomly draw values of all the random variables in Figure 1 except rainfall depth. Then, conditional on those values, one can essentially analytically obtain the conditional probability of dam failure (which is the probability that the rainfall for that season, storm type, and so forth, would be sufficient to cause a spillway or overtopping dam failure) and the distributions of damages and loss of life. One then averages these conditional distributions across replicates to obtain the distributions of interest.

Table 1: Structure of Monte Carlo Simulation

---

1. Generate the values of all random variables in Figure 1, except rainfall depth.

2. For those values in 1, determine the rainfall depth that causes dam failure and calculate for this case:

   i)   Conditional probability of dam failure
   ii) Conditional distribution of damages, and
   iii) Conditional distribution of loss of life.

3. Repeat steps 1 and 2 the required number of times.

4. Average the conditional probabilities and conditional distributions obtained in step 2 across generated samples to obtain the average values.

In Step 1 the events were not generated randomly. Stratification is employed to increase the precision of the results by evaluating the distributions of some discrete random variables analytically. Assume that the spillway is able in actual practice to pass flows of q_1, q_2, and q_3 with probabilities of p_1, p_2, and p_3; then in a simulation with n replicates, in np_1 replicates the spillway would fail with a flow of q_1, in np_2 replicates it would fail at a flow of q_2, and so forth. Whether or not the flood warning system works, the numbers of gates and outlet working, the season, and the storm type within a season are all discrete random variables. The number of replicates for each combination can be made proportional to their joint probability; that would make the weights assigned to the resultant conditional rainfall and damage distributions proportional to their actual probabilities. It may also be necessary in the case of fractional values of np_i, and may sometimes be advantageous, to assign unequal weights to the various replicates so that the product of the weight times the number of replicates is still equal to the probability of each combination of discrete outcomes. In either case, for each such replicate, the values of the other discrete and continuous random variables can be drawn from their appropriate conditional distribution.

Figure 2 illustrates the kind of information generated by such an analysis. The figure displays the distribution of downstream damages for the original dam and a proposed modification to widen the spillway and raise the height of the dam. By widening the spillway, the damages caused by floods beyond the 100-year event increase because of the spillway's larger release. With the larger spillway and greater height, the dam fails with a probability of approximately 2×10^{-5}, instead of 4×10^{-4} without the modification. However, because the modified dam is higher than the original, if it fails, the downstream damages are larger than when the unmodified dam fails. The proposed modification has associated construction costs, losses associated with any decrease in the available active storage zone or other operating restrictions, and also a small increase in damages for floods with exceedance probabilities in the 1×10^{-2} to 4×10^{-4} range, and also for events with exceedance probabilities greater than 2×10^{-5}.

In Figure 2 the results of the evaluation of the event tree in Figure 1 are displayed using the distributions for damages, or similarly loss of life. These reveal in an uncondensed fashion the probabilities associated with various thresholds, possible trade-offs between increased damages for infrequent vs. very rare events (as illustrated in Figure 2), and discontinuities in the distributions. The analysis also yields the probabilities of dam failure from these large flood events and contributing factors. Traditional and simple summary statistics, which are also useful, include expected damages and loss of life; perhaps their variance, and the cost per expected life saved, are easily calculated. Sophisticated decision methodologies, such as multiattribute utility theory and the partitioned multiobjective risk method, could also be employed to help choose among alternatives.

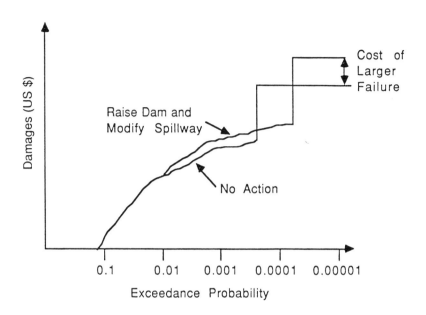

Figure 2. Impact on damage distribution of raising dam to allow large flood storage and widened spillway. The distribution of damages from major floods is obtained by combining all of the important factors including rainfall, rainfall timing, and initial reservoir level and soil moisture factors.

4.   CONCLUSIONS

Science and engineering tend to be evolutionary processes, with each model, set of theories, or method successively replaced by more sophisticated concepts and procedures. Simplistic flood-frequency analyses and flood-of-record planning have been rejected for determining flood safety evaluations for large high-hazard dams, and the PMP-PMF design standard has been adopted. It is now time to move from use of the implicit risk associated with such a deterministic or worst-case design standard to a standard determined by an explicit risk target, such as the ten-thousand or million-year floods or, even better, to quantify the costs of reducing the probability of large floods, dam failures, and the consequences of those events, and balancing those against the resulting benefits. This requires a thorough risk assessment in terms of the probabilities of various natural phenomena, the likely response of natural and human-engineered systems to those events, and the evaluation of the cost and consequences of the resulting disasters from various social, economic, and environmental perspectives. Water resources engineers have been extremely reluctant to assign a value to human life, which makes a quantitative risk/cost analysis of lives saved difficult.

Risk analyses have been readily adopted in many decisionmaking contexts, such as the design of modest flood control works and flood plain planning. It is profitable to reflect on why progress has been so slow in the safety evaluation of high hazard dams. Several factors may be relevant.

(1) Because we are dealing with such extraordinary extremes, far beyond what may actually have occurred in a river basin or a region, the analysis has a high "fantasy factor"; engineers face a challenge imagining and explaining the PMF, without having to develop the probability distribution of such events.

(2) Second, in moving to risk-cost analyses we may be trying to fine-tune a spillway design when only rough estimates of the possible magnitude of events with return periods in the hundred-to-thousand year range are possible; thus care should be exercised that the precision with which we can describe the natural phenomena of interest is commensurate with the sophistication of the analysis tools employed.

(3) Finally, and probably most important, the tradition in high-hazard dam spillway design and safety evaluations has for 50 years been to use a standard which with various titles has essentially corresponded to some approximation of a PMP-PMF analysis. That analysis is now widely accepted and agreed upon by the dam design community and is consistent with their desire to build dams that shouldn't fail. To many, there should be virtually no probability that large dams might fail and cause widespread loss of life and destruction. With such a mind set, risk analysis is not an attractive or even a reasonable activity. Still, when existing dams are found not to pass the latest estimate of the PMF, and the costs of a major retrofit are large, many ask: How safe is safe enough? and What cost for safety?

REFERENCES

Bohnenblust, H., and E. H. Vanmarcke. 1982. Decision analysis for prioritizing dams for remedial measures: a case study. Research report R82-12, Department of Civil Engineering, Massachusetts Institute of Technology, Cambridge, Massachusetts.

Bowles, D. S. 1987. A comparison of methods for risk assessment of dams. In L. Duckstein and E. Plate (eds.), Engineering Reliability and Risk in Water Resources. NATO Advanced Study Institute, Tucson, Arizona, May 1985. Dordrecht, The Netherlands: Martinus Nijhoff.

Bury, K. V., and H. Kreuzer. 1986. The assessment of risk for a gravity dam. Water Power and Dam Construction 38(12):36-39.

Collins, M. A. 1986. Workshop report: liability and accountability in risk-based decisionmaking in water resources. In Y. Y. Haimes and E. Z. Stakhiv (eds.), Risk-based Decisionmaking in Water Resources. New York: ASCE.

Duckstein, L., E. Plate, and M. Benedini. 1987. Water engineering reliability and risk: a system framework. In L. Duckstein and E. Plate (eds.), Engineering Reliability and Risk in Water Resources. NATO Advanced Study Institute, Tucson, Arizona, May 1985. Dordrecht, The Netherlands: Martinus Nijhoff.

Feldman, A. D. 1981. HEC models for water resources system simulation: theory and experience. Advances in Hydroscience 12:297-423.

Haimes, Y. Y., and E. Z. Stakhiv (eds.). 1986. Risk-based Decisionmaking in Water Resources. Proceedings of Engineering Foundation Conference, Santa Barbara, California, November 3-5, 1986. New York: ASCE.

Hansen, E. M., L. C. Schreiner, and J. F. Miller. 1982. Application of the probable maximum precipitation estimates -- United States east of the 105th meridian. Hydrometeorological Report no. 52. National Weather Service, National Oceanic and Atmospheric Administration, U.S. Department of Commerce. Washington, D.C.

Krouse, M. R. 1986. Workshop report: engineering standard versus risk analyses. In Y. Y. Haimes and E. Z. Stakhiv (eds.), Risk-based Decisionmaking in Water Resources. New York: ASCE.

McCann, M. W., Jr. 1986. Proceedings of Workshop on Current Developments in Dam Safety Management, Stanford University, Stanford, California, July 24-26, 1985.

Moser, D. A., and E. Z. Stakhiv. 1987. Risk analysis considerations for dam safety. In L. Duckstein and E. Plate (eds.), Engineering Reliability and Risk in Water Resources. NATO Advanced Study Institute, Tucson, Arizona, May 1985. Dordrecht, The Netherlands: Martinus Nijhoff.

Myers, V. 1967. Meteorological Estimation of Extreme Precipitation for Spillway Design Floods. Technical memo WBTM Hydro-5, Office of Hydrology, U.S. Weather Bureau, Washington, D.C.

National Research Council, Committee on Safety Criteria for Dams, Water Science and Technology Board. 1985. Safety of Dams: Flood and Earthquake Criteria. Washington, D.C.: National Academy Press.

Newton, N. W. 1983. Realistic assessment of maximum flood potentials. Journal of Hydraulic Engineering 109(6):905-917.

Pat³-Cornell, M. E. 1984. Warning systems: application to reduction of risk cost for new dams. In Proceedings of International Conference on Safety of Dams, Coimbra, Portugal.

Resendiz-Carrillo, D., and L. B. Lave. 1987. Optimizing spillway capacity with an estimated distribution of floods. Water Resources Research 23(11):2043-2049.

Revell, R. W. 1973. Reevaluating spillway adequacy of existing dams. Journal of Hydraulics Division, ASCE 99(HY2):337-372.

Stallings, E. A., A. G. Cudworth, E. M. Hansen, and W. A. Styner. 1986. Evolution of PMP cooperative studies. Journal of Water Resources Planning and Management 112(4):516-525.

Stedinger, J. R., and J. Grygier. 1985. Risk-cost analysis and spillway design. In H. C. Torno (ed.), Computer Applications in Water Resources. New York: ASCE.

Von Thun, J. L. 1987. Use of risk-based analysis in making decisions on dam safety. In L. Duckstein and E. Plate (eds.), Engineering Reliability and Risk in Water Resources. NATO Advanced Study Institute, Tucson, Arizona, May 1985. Dordrecht, The Netherlands: Martinus Nijhoff.

World Meteorological Organization. 1986. Manual for estimation of probable maximum precipitation. 2nd edition. Operational Hydrology Report no. 1, WMO 332. Geneva, Switzerland.

Yankee Atomic Electric Company. 1984. Probability of extreme rainfalls and the effect on the Harriman Dam. Report YAEC-1405. Framingham, Massachusetts.

THE CONSTRUCTION COST/RISK COST TRADE-OFF IN PUBLIC WORKS PROJECTS: NAVIGATION CHANNEL WIDTH DETERMINATION

Charles Yoe

ABSTRACT

Rising costs and financial strain at all levels of government have created pressures to cut the costs of public works projects. There is considerable opportunity for construction cost reductions through decrements to a project's margin of safety while remaining within good engineering practice. Absolute safety as an ideal is neither achievable nor economically rational. An example of the economic trade-offs between construction costs and the increased risk of project "failure" is illustrated for the determination of the channel width for Baltimore Harbor navigation improvements. The reduction of channel widths from 1000 feet to 800 feet saves $93 million in first costs of construction while imposing a $7 million increase in delay costs to avoid potentially dangerous passing situations within the narrower channel.

1. INTRODUCTION

The costs of risk-bearing are relevant in a broad class of economic problems, but particularly so for investments in the public infrastructure. Policies regarding construction, repair, maintenance, and rehabilitation of the public infrastructure are currently being formulated and revised in an environment of disinvestment in that infrastructure (Choate and Walter, 1981; Hulten and Peterson, 1984) and increasing competition for financial capital.

Earlier in this century an identified engineering need was often sufficient justification for a public works project. Because these projects were judged essential to the public's well-being, they were generally constructed according to the highest standards of good engineering practice. As a result, excess capacity and generous safety margins were routinely part of project design. In those bygone days, fewer demands were placed on the resources of the public sector and financing a project was rarely a problem.

Budget deficits at all levels of government, along with intense and increasing competition for limited financial resources to meet limitless demands placed on the public largess, have conspired to relegate public infrastructure expenditures to a relatively low priority on the public sector's agenda. Maintenance has been deferred, replacement and rehabilitation postponed, and new construction cancelled, delayed or stretched out. An engineering need is no longer sufficient to guarantee action.

Charles Yoe is a research economist at the U.S. Army Corps of Engineers Institute for Water Resources, Fort Belvoir, Virginia.

When action is taken, good engineering practice is still a top priority but it is no longer the only priority. Engineering practices have been forevermore altered by the intrusion of economic arguments into the domain of engineering science. Excess capacity is no longer automatic. It is widely recognized that no physical system can be made completely safe. We still want safety but realize "complete" safety is unattainable at any price, so we are satisfied with "enough" safety at an affordable price. Just how safe is "safe enough" is a definition that eludes a consensus. This paper considers the economic trade-offs inherent in determining the optimal safety designed in a public works project. The expansion of the argument to all manner of engineering projects is straightforward.

It should be clear from the outset that good, solid engineering practices, using the industry's engineering design standards, still rule the planning and design process and always should. Economic questions of trading off construction costs against risk costs become relevant only in those project design aspects where there is room for judgment in defining acceptable levels of public safety and residual risk.

This paper presents a simple model of the economic arguments relevant in the design of a project that meets an exogenous demand for the output of a public works project. The economic trade-off between construction and risk costs is illustrated in determining the width of the Baltimore Harbor deep draft navigation channel.

2.   CONSTRUCTION COSTS VS. COSTS OF RESIDUAL RISK

In the model developed to analyze the trade-off between reductions in public sector project costs and the residual risks imposed on the project's users, the underlying economic goal is to minimize construction costs and residual risk, subject to two constraints.

The first constraint is that demand for project output is exogenously given and is not sensitive to the risk assumed by the user of the project. For many projects, where users do not have the information or knowledge to assess the risks associated with different project designs, such an assumption seems reasonable. Demand for a museum will not be affected by the use of #4 reinforcing steel in a concrete slab rather than #5 steel. Use of a bridge is not likely to be affected by the decision to build it with one less pier. The addition of two rather than three feet of freeboard to a levee is not likely to affect land use patterns behind the levee.

There are other cases where an exogenous demand model is not as intuitively appealing. Airport runway length is a design issue of utmost importance to a small and very knowledgeable group of users. While the user of a museum may know little of and care less about the diameter of reinforcing steel in the concrete slabs, airline pilots care a great deal about runway freeboard lengths. Ship pilots likewise care a great deal about the width of navigation channels.

To make the model more acceptable for these types of projects, a second constraint is added. The project design is constrained to at least equal the minimum design safety standard necessary to realize the demand for project output that would be achieved under a consensus "safe" project design.

The cost-minimizing model is presented below:

$$\underset{Z}{\text{Min }} J = \int_{t_0}^{T} (C_p(Z) + C_r(Z)) e^{-rt} dt \tag{1}$$

s.t.

$$Z \geq K \tag{2}$$

$$Q \geq q_0 \tag{3}$$

C_p represents all project costs; C_d residual risk costs; Z is a vector of project design parameters that determine project safety or, conversely, residual risk costs; and $\partial C_p/\partial Z > 0$, $\partial C_r/\partial Z$. The discount rate is r; K is a vector of minimally acceptable safe design levels; and Q is project output, with q_0, the exogenously determined demand for project output, a suppressed argument of the cost functions. The problem is to determine the design parameters Z so as to minimize the sum of project costs plus residual risk costs over time.

Project costs are based on a life-cycle costing concept and include first costs of construction, operation and maintenance, major replacement, rehabilitation, and replacement that may be incurred over the life of the project. The residual cost function can be defined broadly so as to include all user costs associated with the project. To keep this paper more sharply focused, only the risk costs of varying project designs will be considered in the function, C_r.

The argument is that project costs, C_p, can be cut by shifting some of the burden of project failure to the users of the project through the function C_r. A decrease in C_p is offset to an unknown extent by an increase in C_r. Only the costs of the increased risk of project failure are considered in this analysis.

Defining project failure is a controversial issue that will not be addressed explicitly in this paper. Failure has been variously defined in the literature, from the project's failure to function consistently at the design level to the loss of the project. The definition of project failure in the example of navigation channel width determination that follows is simply the occurrence of a potentially unsafe passing situation for two vessels within the confines of the project.

Because project failure is a stochastic process, this can be reflected in the C_r function by defining it as follows:

$$C_r = \int_{t_0}^{T} C_f \left(\int f \ast g(f(Z)) df \right) dt \tag{4}$$

C_f is the cost of a failure; its arguments are suppressed but include the magnitude of the failure. Failures are designated by f (which can be a vector), a variable that captures the salient dimensions of failure for the given project: for example, severity or duration of the failure. The density function for failures is $g(f)$, which has Z as an argument.

This simple model will be used to illustrate the economic trade-offs between the costs of a narrower channel into the port of Baltimore and the costs of the increased risk associated with the narrower margin of safety attending the narrower project.

Application of the model in this analysis is limited by the availability of data. A true optimization was not possible because costs were developed for only two design alternatives. The example is sufficiently developed, however, to illustrate the significance of the main argument that construction cost reductions may have significant costs in terms of increased residual risk borne by project users.

3. BALTIMORE HARBOR

Deep draft navigation improvements are very expensive. They are routinely designed to serve the estimated future traffic that will demand the services of the port the improvements serve. The Coast Guard, pilots, Port Authorities, and shipping lines form a small and sophisticated users' group that effectively sets minimum acceptable design standards for a safe project that meets the economic needs of the port and its users. These standards influence the depth, width, and turning configurations of the channel improvements. As long as the minimum safety standards are met, demand is exogenous to the project's final configuration. Hypothetically, if economic or political analysis determines that a 50-foot-deep project is needed and users determine that the project has to be at least 700 feet wide to allow for safe passing situations under most conditions, it is reasonable to assume that a project width of 750, 817, or 1000 feet will have no effect on the demand for port services.

Preconstruction planning studies during the late 1960s determined that access channels to the Port of Baltimore should be deepened from 42 to 50 feet. From that point on, project costs became the overriding concern of all parties to the project.

There was significant opposition to the project from environmental interests. The plan for disposal of dredge material from the channel was challenged and the project delayed for a number of years. To say that costs were of utmost concern is not to denigrate the environmental concerns. In fact the environmental concerns were raised because of the project designers' efforts to find the least costly acceptable disposal site. The site ultimately met those criteria, but not without considerable controversy.

Figure 1 presents project cost estimates* prepared at various times during the planning process by the Baltimore District for the Baltimore Harbor and Channels, Maryland and Virginia Project. Estimates of project costs in current dollars increase steadily from $99.7 million in 1969 to $379 million in 1985. The range in constant dollar cost estimates still approaches $200 million after they are adjusted to an October 1985 price level.

As the quality of the cost estimates improved over time with the addition of more and better information, the issue of cost-sharing for such projects became a bigger issue. Because federal cost-sharing was in transition from essentially 100 percent federal costs to some new and unknown local cost share at the time the Baltimore Harbor Project was being planned, the State of Maryland had an unprecedented incentive to hold down the costs of this project.

When P.L. 99-662, the Water Resources Development Act of 1986, set cost-sharing for projects like the one in Baltimore at 50-50 for the state and federal shares, this represented a huge increase in the price of the project to the state. The state, faced with paying half of the $379 million project cost, was motivated to seek innovative ways to pay for the project.

Perhaps the most straightforward innovation was to cut the cost of the project. This was to be done by constructing the project with a narrower channel. The rationale was simple: when the federal government was picking up the tab a 1000-foot-wide channel was great, even necessary some said. (Throughout the remainder of this paper I will refer to the reduction in channel width as going from 1000 to 800 feet. The actual channel widths varied over the 53-mile length of the project.) Once it was clear the state must pay half of project costs, the choice was essentially a narrower project or no project. The 1000-foot channel still seemed great, but neither the Corps of Engineers, the Coast Guard, Maryland Pilots, nor the state thought it was necessary. An 800-foot channel would do quite nicely.

---

* Cost estimates presented in this paper are total cost estimates readily available from project documents. They are used for example purposes. An item-by-item comparison of project costs was not undertaken, and it is likely that the project elements which comprise project costs over the life of the study differ from estimate to estimate. Comparison of cost estimates over time is intended to imply a general comparability of projects over time. To the extent that they do not, the basic arguments of the paper are in no way affected.

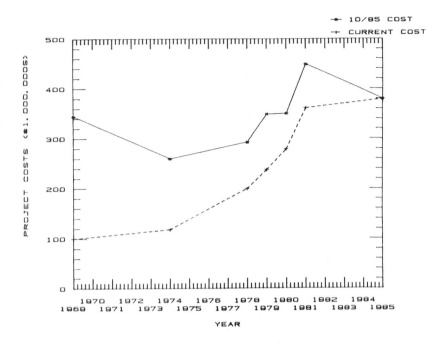

Figure 1. Comparison of Project Costs

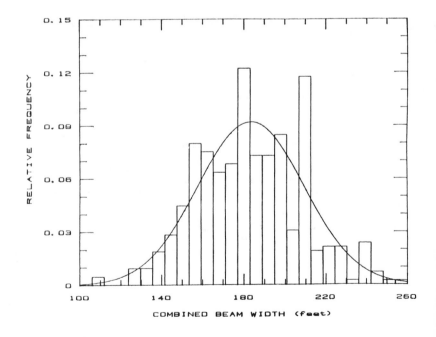

Figure 2. Observed and Hypothetical Relative Frequency Distribution

The political pragmatism of this decision is flawless. The project is now under construction, more than 17 years after the first study was completed. There were two significant flaws in this solution from an economic viewpoint, however. First, no one considered what the costs of going from a 1000-foot channel to an 800-foot channel would be. Second, once there was a decision to deviate from the original 1000-foot channel, no one looked at the economically optimal channel width. Nonetheless, the interested parties determined that an 800-foot channel was the minimal acceptable width for a safe channel, and this reduced project costs from $379 to $286 million.

The implicit trade-off in this solution was that the narrower channel saved the state nearly $50 million, while forcing the pilots and shippers to assume a greater risk of collision, grounding, or delay in using the project. Clearly the subjective judgment of the pilots and shippers was that the increased risk was, if not negligible, at least not unreasonable, because they readily agreed to the trade-off.

In this instance the users' instincts were probably good, as I will shortly demonstrate. Nonetheless, reliance on such instincts in trading off construction cost reductions for the imposition of increased risks to project users in any engineering project cannot be allowed to substitute for analysis.

4. ECONOMICS VS. SAFETY IN CHANNEL WIDTH DETERMINATION

Channel width determinations are made primarily on the basis of economics and safety. The channel must be wide enough for the ships that will call at the port, but is must also be safe for ships to use during passing situations within the channel. The cost savings from building a narrower channel must be weighed against the residual risk of collisions, groundings, or delays that can result from the provision of a narrower channel. Because of the changes in the cost-sharing formula, project sponsors in many cases are now willing to assume those residual risks in order to save millions of dollars in project costs. The analyst's challenge in this circumstance is to quantify this risk to ensure that its assumption is a rational and defensible alternative in achieving the planning objectives.

House Document 94-181, the 1969 original project feasibility study and the authorizing document for Baltimore Harbor and Channels, Maryland and Virginia, based its determination on channel widths on a design vessel* defined as a 90,000 DWT bulk cargo ship with a beam of 122 feet and engineering judgment current during the 1960s. The 1985 Supplemental Information Report (SIR) bases channel widths on a 150,000 DWT bulk carrier with a 145-foot beam and improved engineering judgment. Table 1 presents a comparison of some of the channel width computations for these two reports and the intermediate 1981 General Design Memorandum (GDM). In general, channel element widths are obtained by multiplying vessel beam width by some percentage of beam width.

---

* The design vessel is not the largest vessel that will use the project but the largest class of vessels that will use the project in significant numbers.

During initial study of the project, the design vessel was based on the fleet in operation at that time (late 1960s) and then-current plans for new vessels. The fleet which prevailed at the time of the 1981 GDM, from which the 1985 SIR derives its fleet forecasts, reflected the changes brought on by the energy crisis, the grain embargo, and other market trends of the 1970s, as well as by technological change. These events were not foreseen by the 1969 report, providing a perfect example of the uncertainty that is inherent in any description of the future.

The Corps of Engineers' criteria for determining safe channel widths come primarily from the proposed Sea Level Panama Canal Studies and subsequent modifications of these findings. Corps guidance provides for channel widths that provide safe and efficient movement of the vessels expected to use the channel.

Engineering Manual 1110-2-1613 provides general criteria for channel widths. The factors considered in deriving the criteria and their subsequent uses include whether a passing situation can occur, vessel controllability, vessel speed relative to channel bottom, current velocities and direction, depth of water under the keel of the vessel, whether the channel occupies the entire waterway (i.e., is the channel in open or confined waters), and the characteristics of the channel banks. The channel width design criteria used by the Corps continued to evolve over the period of study, as is evident in Table 1.

The table shows that determination of channel width evolved over time in response to a number of factors, some economic, some engineering. One of the changes in the underlying assumptions was that the passing of two design vessels within the confines of the channel was so unlikely as to not be worth worrying about. Nonetheless, there is a distinct possibility that two large vessels could pass within the channel or that one would delay its transit through the project area to avoid passing within the channel. The risk of either of these occurrences has associated costs. The trade-off between lower project costs and the risks of unsafe passing situations is the subject of this section.

In Baltimore Harbor it is common practice for vessels traversing a channel alone to navigate along the channel center line. In passing situations, the vessels move to the outside of the channel and return to the center of the channel upon completing the passing maneuvers. The Maryland pilots coordinate meeting and overtaking situations so they occur in either open areas of the Bay or in channel sections where vessel speeds are reduced and interactive forces minimized.

Table 1

A Comparison of Channel Width Determinations

VIRGINIA CHANNELS

HD 94-181

Bank clearance	150% x 122'=183'
Maneuvering lane	180% x 122'=220'
Ship clearance	100% x 122'=122'
Maneuvering lane	180% x 122'=220'
Bank clearance	150% x 122'=183'

Total width:928' rounded to 1,000'

1981 GDM

Bank clearance	120% x 145'=167'
Maneuvering lane	180% x 145'=261'
Ship clearance	100% x 145'=145'
Maneuvering lane	180% x 145'=261'
Bank clearance	120% x 145'=167'

Total width:1001' rounded to 1,000'

1985 SIR

Bank clearance	80% x 145'=116'
Maneuvering lane	180% x 145'=261'
Ship clearance	100% x 145'=145'
Maneuvering lane	180% x 105'=189'
Bank clearance	80% x 105'= 84'

Total width:795' rounded to 800'

MARYLAND CHANNELS

HD 94-181

Bank clearance	100% x 122'=122'
Maneuvering lane	180% x 122'=220'
Ship clearance	100% x 122'=122'
Maneuvering lane	180% x 122'=220'
Bank clearance	100% x 122'=122'

Total width:806' rounded to 800'

Table 1 (cont.)

1981 GDM

Bank clearance	60% x 145'= 87'
Maneuvering lane	170% x 145'=247'
Ship clearance	100% x 145'=145'
Maneuvering lane	170% x 145'=247'
Bank clearance	60% x 145'= 87'

　　　　Total width:813' rounded to 800'

1985 SIR

Bank clearance	60% x 145'= 87'
Maneuvering lane	170% x 145'=247'
Ship clearance	80% x 145'=116'
Maneuvering lane	170% x 105'=179'
Bank clearance	60% x 105'= 63'

　　　　Total width:692' rounded to 700'

Source: Planning documents of the Baltimore District, U.S. Army Corps of Engineers

The first task in analyzing the costs of the risk from a potentially unsafe passing situation is to define an unsafe situation. The most appropriate measure, based on the Corps' engineering design criteria, is the combined beam width of the two vessels passing. If the combined beam width of the two vessels exceeds the design maximum safe combined beam width for a particular channel width, then the situation is considered potentially unsafe. Ideally the analysis will also take weather, speed, and handling characteristics into account.

Based on the Corps' channel width criteria contained in Engineering Manual 1110-2-1613 and the engineering judgment of the Baltimore District, the Virginia channel displayed in Table 1 is 620 percent of beam width. In the Maryland channel it is 540 percent. Because a passing situation involves two vessels that may not have the same beam width, we halve the percentages that apply to two vessels and use 310 and 270 percent of the single combined beam width measure. Thus 310 percent of the combined beam width in a passing situation must be less than or equal to the channel width to be defined as safe. For the Virginia channel the maximum is 258 feet, for the Maryland channel 259 feet. Thus, for example, any vessel with a beam greater than 113 feet should not pass the design vessel (145-foot beam) in the Virginia channel.

Having defined the threshold measure of a safe or unsafe passing situation, the next step is to estimate how often potentially unsafe passing situations will occur. Existing passing situations were analyzed by estimating the combined beam widths of 426 passing situations that occurred over 21 randomly selected days between May 1984 and May 1985.

Passing situations were identified from Association of Maryland Pilots' logs, which record vessel beam, direction, and the time a vessel entered and left the project area for billing purposes. A pass was assumed to have occurred when there is an overlap of transit time for two vessels headed in opposite directions. It was not possible to determine conclusively whether the pass took place in open waters or the confined waters of the channel. Each vessel made an average of 3.8 passes per round trip, and the effective probability that any vessel will pass another somewhere within the project area, not necessarily the channel, is 1.0.

Figure 2 presents the observed frequency and a hypothetical density function for a normally distributed random variable with the mean and standard deviation of the observed passing situation for combined beam widths. Descriptive statistics for the combined beam widths of the passes are shown in Table 2. The Chi-Square Statistic for the hypothesis that the data are distributed normally was 72.7 with 17 degrees of freedom, significant at a 7.2E-9 level. A more appropriate test, the Kolmogorov-Smirnoff statistic, indicates a less comfortable 0.14 significance level.*

Table 2
Descriptive Statistics for Passing Situations

Sample size	426
Mean	183.913
Median	182
Mode	212
Variance	661.379
Standard deviation	25.717
Minimum	109
Maximum	260
95% Confidence interval	180.978 185.877

Figure 3 shows a normal cumulative distribution with a mean of 183.4 and a standard deviation of 25.7 superimposed on the observed distribution. Based on the observed data, the probability of a potentially unsafe passing situation (combined beam in excess of 259 feet) is 0.0016 per pass. Approximately two passes in 1000 will be potentially unsafe.** The probability of a potentially unsafe passing

*Numerous alternative distributions were tested, including the Wiebull, Gamma, Chi-Square, Erlang, and others. The Weibull distribution, a three-parameter distribution, provided the best fit for the data. The Weibull distribution was not used because it is intuitively less appealing than the normal distribution in a heuristic example. An analysis was done of the probabilities of interest derived from the various distributions tested, and there was no significant difference in these values; thus, little is sacrificed by using a normal distribution.

**Passes with a combined beam width in excess of 259 feet are designated "potentially unsafe" rather than "unsafe" because the criteria for establishing this threshold are very subjective and embody a great deal of uncertainty. To the extent that the criteria illustrated in Table 1 account for the factors which contribute to the potential hazard of a passing situation, the potential for an unsafe pass will increase.

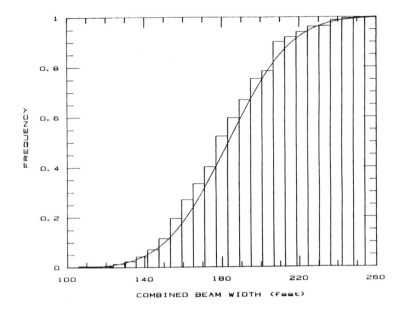

Figure 3. Cumulative Distribution Observed and Hypothetical

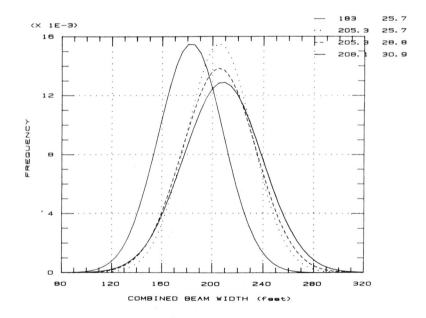

Figure 4. Density Functions for Passing Situations Present and Future

situation obtained from a stochastically simulated, normally distributed sample of size 1500, with a mean and standard deviation identical to that used above, is estimated to be 0.0015 per pass.

Because the project is being built to handle navigation needs for the next 50 years, it would be improper to base the safety/cost trade-off entirely on existing passing situations. The average beam for the current Baltimore bulk commodity fleet is estimated to be about 94.1 feet (U.S. Army Corps of Engineers, 1985). The average beam of the future fleet is estimated to be 105.3 feet, an increase of 12 percent. Future passing situations are expected to involve larger vessels and, consequently, there should be more potentially unsafe passing situations.

To simulate the distribution of future passing situations, a normal distribution with a mean combined beam width of 205.3 feet (12 percent more than the current mean) was generated. The variance was initially assumed to remain unchanged. A 1500 observation sample was used in all simulations. The resultant probability of a potentially unsafe passing situation with the larger fleet was 0.0198 per pass. To investigate the possibility of a more dispersed distribution in the future, the larger mean was used, with a 12 percent larger standard deviation to estimate the probability of an unsafe pass. This stochastic simulation yielded a probability of 0.0304 that the combined beam would exceed 259 feet.

A final sensitivity analysis was conducted using the upper confidence limit value for the mean shown in Table 2. The 185.9-foot mean was increased by 12 percent (to 207.8 feet) and the upper confidence limit value for the variance was increased by 12 percent (to 30.6 feet). This distribution represents the largest possible fleet that can be reasonably anticipated from the data in the 1981 GDM and 1985 SIR. The probability of an unsafe pass estimated in this manner was 0.0470. These probabilities are summarized in Table 3 and the density functions are presented in Figure 4.

Table 3
Selected Probabilities of Unsafe Passing Situations

Distribution	
Observed 183' mean & 25.7' standard deviation	0.0016
Simulated 183' mean & 25.7' standard deviation	0.0015
Simulated 205.3' mean & 25.7' standard deviation	0.0198
Simulated 205.3' mean & 28.8' standard deviation	0.0304
Simulated 208.1' mean & 30.9' standard deviation	0.0470

The 1985 SIR estimates that 7404 passes will occur each year; thus based on the probabilities in Table 4, we can expect from 12 to 348 potentially unsafe passings by the time the ultimate fleet size is operating in the project area if all the passes take place in the channels.

Not all passing situations will occur in the confined waters of the channel. Many will occur in the safety of the unconfined natural depths of the Chesapeake Bay. To estimate the number of passes that occur in the channels, passing is assumed equally likely to take place anywhere throughout the 175-mile length of the bay. The probability of passing within the channel is the ratio of channel length to the distance from the mouth of Chesapeake Bay to Baltimore Harbor, or 53 miles/175 miles = 0.34.

The probability of an unsafe passing situation, $g(f)$ of Equation 4, is most accurately expressed as a conditional probability density where

$$P(A|B|C) = P(A)P(B)P(C) \tag{5}$$

and A is the event that the combined beam is greater than 259 feet, B the event that a pass occurs in the channel, and C the event that a pass occurs. All events are assumed to be independent. $P(C)$ is effectively estimated to be one. The available data do not permit a more careful analysis, but such an analysis is straightforward. The use of $P(C)=1$ in no way limits the applicability of this example. $P(B)$ is estimated to be 0.34 and $P(A)$ is shown in Table 7 above. The conditional probability in Equation 1 above ranges from 0.0005 to 0.0160. These probabilities represent from four to 118 passing situation per year.

If 118 potentially unsafe passing situations occur annually, procedures for avoiding them are needed. The simplest solution is to simply restrict channel traffic to one way for those passing situations. This means one ship would have to delay entrance to the channel until the other ship has cleared the channel. To determine the economic efficiency of a narrower channel, the accumulated costs of these delays must be weighed against the cost savings that result from dredging and maintaining a narrower channel.

To estimate the delay, an average two-hour delay per vessel is assumed. A detailed analysis could determine the procedures for regulating traffic and the characteristics of the average delayed vessel. For example purposes, delay costs are assumed at \$2,886 per hour.* The annual cost of delay at this rate for 118 vessels is \$681,000. The current cost of delay for four vessels is \$23,000 per year.

The objective function for this example can be expressed as:

$$\text{Min}_{w} J = \int_{t_0}^{T} (C_p(w) + C_d(w)) e^{-rt} dt \tag{6}$$

s.t.

$$w \geq k \tag{7}$$

---

*This is the estimated hourly at-sea cost for a U.S. flag tanker in the 175,000 DWT range. Costs were obtained from U.S. Army Corps of Engineers Engineering Circular 1105-2-177, dated 24 July 1987. These costs were chosen simply because they represent an upper limit on the delay costs and not because they are considered representative of the vessels that would be delayed.

C_d are risk costs measured by the costs of delay, w is channel width, and k is the minimum acceptable design channel width.

Data are not readily available for the Baltimore Harbor project to allow an optimization analysis based on Equation 2. The most appropriate analysis for the Baltimore Harbor project for which data are available is a comparison of the accumulated present worth of delay costs from the narrower channel to the construction cost savings of the narrower channel. Reliable estimates of annual project operation and maintenance costs are not available for the original project and the revised state version. Thus it is impossible to compare all costs, as Equation 2 suggests.

It is likely that the increasing occurrence of potentially unsafe passing situations will take place gradually over time. In the absence of better data, one option for analyzing this time trend is to assume that these situations grow from four per year to 118 per year at a constant percentage. Such an analysis is a straightforward application of discounting future costs. In order to establish an upper bound on the delay costs, however, I assumed that the estimated maximum of 118 unsafe passing situations occurs in each of the 50 years of the project's life.

Building the narrower channel incurs residual risk or delay costs equal to the accumulated present worth of $681,000 per year for 50 years, evaluated at a 9 percent interest rate or $6.8 million. Construction cost savings from the narrower channel are $93 million. The cost savings from the narrower channel are $86.2 million.

The heuristic example presented above was limited by the availability of data. The analysis would ideally estimate the costs of collisions and groundings which could be expected to occur due to violations of the one-way traffic rule. (These costs can be expected to be several orders of magnitude greater than the costs of delay. Unfortunately, estimating the probability of such events, a conditioned subset of the probabilities in Equation 5, must rely on subjective methods.) Likewise the residual risk costs of all channel width alternatives would be estimated in a more complete analysis.

5. SUMMARY

Minimizing the costs of public works infrastructure investments entails a trade-off between project costs and risk costs borne by users of the project.

The uncertainty inherent in the economic and engineering analyses and judgments which form the basis for channel width determination provides considerable leeway in the determination of channel design and project costs.

While the cost savings from a narrower project are obvious and perhaps compelling for political reasons, the hidden costs of these savings--i.e., residual risk of collision, groundings, or delay--can be substantial. Estimated residual risk costs based solely on delays amount to a fraction of the construction cost savings for Baltimore Harbor. Residual risk costs must be evaluated on a case-by-case basis. Analysts formulating a comprehensive plan cannot afford to overlook these costs.

REFERENCES

Choate, Pat, and Susan Walter. 1981. America in Ruins. Durham, North Carolina: Duke Press Paperbacks.

Hulten, Charles R., and George E. Peterson. 1984. The public capital stock: needs, trends, and performance. American Economic Review: 166-173.

Kray, C. J. 1973. Design of ship channels and maneuvering area. Journal of Waterways, Harbors and Coastal Engineering, Proceedings ASCE: 89-110.

Minorsky, V. U. 1977. Grounds probability studies. Final report on task IV, February 1976--January 1977. Prepared for U.S. Maritime Administration by George C. Sharp, New York.

National Research Council, Marine Board Commission on Engineering and Technical Systems. 1983. Criteria for the Depths of Dredged Navigational Channels. Washington, D.C.: National Academy Press.

U.S. Army Corps of Engineers, Baltimore District. 1979. Preliminary review draft, Baltimore Harbor and Channels Maryland and Virginia. 1979 combined phase I and II general design memorandum.

U.S. Army Corps of Engineers, Baltimore District. 1981. Final report, Baltimore Harbor and Channels Maryland and Virginia. 1981 combined phase I and II general design memorandum.

U.S. Army Corps of Engineers, Baltimore District. 1985. Supplemental information report, Baltimore Harbor and Channels Maryland and Virginia.

U.S. Army Corps of Engineers, Office of the Chief of Engineers. 1965. Tidal hydraulics. Engineering manual 1110-2-1607. Washington, D.C.

U.S. Army Corps of Engineers, Office of the Chief of Engineers. 1981. Engineering and design: deep draft navigation project design. Engineering regulation 1110-2-1404. Washington, D.C.

RISK ANALYSIS IN PUBLIC WORKS FACILITIES PLANNING, DESIGN, AND CONSTRUC-
TION

Walter Diewald

ABSTRACT

The National Council on Public Works Improvement was created by
Congress to examine the state of the infrastructure. The Council's
final report, entitled Fragile Foundations: A Report on America's
Public Works, specifies the broad range of problems extant because of
decades of neglect. The Council recognized early in its investigations
that there is a need to examine risk more thoroughly in public works
facility planning, design, and construction. Within this context risk
is a factor in many activities, ranging from planning and forecasting
future demands to the testing and demonstration of new technology. This
paper examines how risk analysis can be used within the public works
decisionmaking process to assist decisionmakers in evaluating alternatives. Specific examples illustrate how risk managers, risk analysis
specialists, safety experts, and public works professionals interact
within the process.

1. BACKGROUND

The U.S. Congress created the National Council on Public Works
Improvement in late 1984. The Council consists of five public citizens
supported by a staff of six professionals charged to examine the state
of the nation's infrastructure. Of the many topics which the Council
and staff have addressed, we have given considerable attention to the
issue of needs. The Council decided early in its deliberations that a
single-value needs estimate does not adequately address the problem of
adequate public services. Needs assessments rarely provide sensitivity
analyses to show which standards are key determinants of the total cost
and how greatly costs are affected by the standards and, further, how
changes in the standards affect risks and by how much. Moreover,
evidence suggests that needs estimates may vary by as much as 30 or 40
percent, depending on the standards selected.

As we researched this subject it became apparent that the adequacy
of public works is a relative, not an absolute, concept. Physical
assets comprise only one component of adequacy. Other components such
as product or service delivery, quality of service to users, and investment efficiency are also important. Thus the capacity of the nation's
public works actually depends on how the facilities and equipment are
used and on alternative ways to supply services, not simply the miles of
highways or sewers we have or the number of airports, etc.

Walter Diewald is Senior Engineering Policy Analyst, National Council on
Public Works Improvement.

We sought to explore this approach to needs and capacity--based on performance rather than simply physical assets--and its ramifications. It brings into question many of the traditional ways of providing and evaluating public works. In fact, we found that decisionmakers, budget directors, program managers, and others were seeking more information about standards and about the risks and costs associated with them. We found them asking some very important questions, such as:

- What are we paying for?

- How do standards reflect a particular level of service or safety or reliability?

- What risks are associated with specified levels of service and/or safety?

- What are the risk/cost implications of alternative forms of service delivery?

- Do engineering standards preempt alternative solutions that would provide equivalent service?

- Do standards limit the flexibility of the decisionmaker in identifying potential solutions?

These questions are driven by a number of factors, including an expanding engineering knowledge base, changing federal cost-sharing rules, and limited fiscal resources.

2.   WHICH RISKS ARE WE ADDRESSING?

The term "risk" can have a number of meanings, depending upon the subject and the speaker. In discussions of public works infrastructure, for example, we can consider risk in a number of project-level and program-level tasks:

- planning and forecasting future needs: risk stems from the uncertainty in estimates of future demands and supplies

- evaluating project alternatives: the design, construction, and operation of the facilities involve risk

- comparing capital-intensive alternatives with low-cost options: the use of low-cost options as alternatives to capital improvements requires the assessment of future performance, introducing uncertainty and risk

- preparing benefit-cost assessments

- evaluating the financial and technical risk of new technology

- planning research and development programs

- testing and demonstrating new technology.

Each of these tasks includes elements of risk which may be considered implicitly or explicitly depending on the methodology in use. The following discussion examines certain risk-related aspects of project-level decisionmaking in the field of public works.

3.   BASIS FOR STANDARDS

The body of traditional design standards, criteria, safety factors, and computational procedures represents the codification of early engineering practice and an evolutionary process which includes considerable scientific research. Water resource professionals meeting to examine risk in decisionmaking have reported that engineering standards have some significant advantages to the engineer and to the public; these advantages include but are not limited to the following (Haimes and Stakhiv, 1985):

- Equity is achieved because similar engineering works are designed and operated in a similar fashion so that everyone is subjected to similar risks.

- Safety and reliability are "built in" at a level which is uniformly accepted and applied.

- Future uncertainties are handled in a practical and replicable manner.

- Considerable experience in design and operations is represented in the standards.

- Legal liability of the engineer and organizational integrity of the agency is protected in case of failure.

Thus, standards have become a very important part of the engineering planning and design process. There is reluctance on the part of practicing engineers to move away from traditional, standards-based designs. This reluctance is understandable when alternative procedures such as risk analysis include large uncertainties regarding loss of life, economic consequences of failure, and ultimate liability.

Nevertheless there are limitations associated with standards-based engineering. It obscures many uncertainties by preempting them from the planning and evaluation process because they are implicit in the standards and thus not subjected to any explicit analyses. For example, traditional dam design treats all the elements needed for selecting the design standards in a generalized way; the appropriateness of the design standards as applied to individual dams is generally unknown (NRC, 1985). Highway design standards are based on a given number of locational characteristics; the importance of other factors at a specific location cannot be directly evaluated and is left to "engineering judgment."

Some of these uncertainties could, and perhaps should, be dealt with in a probabilistic manner so as not to eliminate potential solutions from consideration in the public policy area. For example, standards should not necessarily lead to a specific project design which is more expensive than necessary simply to meet a given level of performance.

Other limitations include the following:

- Application of new knowledge/data or new technology may be hampered or prevented by the standard.

- Design flexibility for a specific situation may be limited unnecessarily.

- Information about costs, risks, and benefits is not fully utilized.

4.   THE PROCESS OF RISK ANALYSIS

In view of these limitations, the Council examined some of the broader considerations related to risk, cost, and service implications inherent in standards and specifications, particularly with regard to the impact of standards upon design flexibility and the availability of alternative solutions to the decisionmaker. In this context it is appropriate that risk analysis techniques be examined more thoroughly to determine how they can be used more effectively in public works planning and the decisionmaking process.

Risk analysis is the process of establishing the relationship between an action or a situation and its impact. In most cases, risk analysis deals with adverse impacts. Risk analysis can be characterized by two distinct components: risk assessment and risk management. In risk assessment, one attempts to quantify in a statistical or probabilistic manner what is certain and uncertain about the natural hazards and project performance reliability in order to select an optimal alternative. Risk assessment involves hazard identification, source assessment, exposure assessment, effects assessment, and risk assessment policy (NRC, 1983).

Risk management, on the other hand, focuses on public policies and institutional response mechanisms as a complement to technical solutions designed to reduce economic and public health and safety risks from potential hazards. A risk manager bases his decision on the characterized risk and several other parameters. Risk management includes characterized risk, cost assessment, benefit-cost assessment, and consideration of the subjective values inherent in risk perception, social and legal constraints, and political reality.

5.   EXAMPLES OF ALTERNATIVE APPROACHES

The Council organized a workshop to examine some of these questions and to put risk analysis into the context of public works decision-making. The workshop participants included risk managers, risk analysis specialists, highway safety experts, and public works professionals. The presentations and discussions of the workshop provided specific examples of potential applications of risk-based decisionmaking. These are discussed below in terms of specific infrastructure categories but are applicable to the entire range of categories.

In-Service Inspections

Maintenance and rehabilitation are becoming extremely critical components of public works service delivery, particularly as our facilities age and deteriorate. Maintenance actions, however, do not receive the attention that new construction does and thus they are predominantly secret activities.

Maintenance management is getting increasing attention in many areas and it appears that reliable information obtained from in-service inspections, coupled with effective decision support tools and models, can lead to more cost-effective maintenance management. This offers the potential of reducing the cost of providing public works facilities.

Ted Mayer (of Westinghouse) suggested that an inspection program can play a significant role in reducing in-service failures and their associated costs (Meyer, 1987). As facilities wear and age this role becomes more important because the predictability of equipment behavior decreases. Typically, inspection guidelines or requirements are based on engineering judgment, along with a knowledge of the structure and its potential failure modes. However, consideration of the probability of failure or the consequences of failure is generally not a factor.

Risk-based methods, on the other hand, can be used to develop a prioritized inspection program where the goal is the explicit reduction of risk. A risk-based process can be used to prioritize the system components which contribute most to the probability of failure. An example of such a process was discussed at the Council's workshop, and it involved the following tasks:

(1)   Define components eligible for inspection

(2)   Examine and evaluate facility records

(3a)   Estimate the failure consequences

(3b)   Identify components within zone of interest

(3c)   Estimate the probability of failure

(4)   Rank results using weighting factors.

Inspection programs can employ a variety of inspection techniques with different impacts on the resulting probability of component failure.

Inspection can mean visual inspection, physical testing, failure testing, etc. An inspection program usually implies a sampling of components and, as a result, a probabilistic statement of results. Further, there is the need to address uncertainties that inherently exist in defect characterization. (In another portion of the Council's research effort, the National Research Council examined the potential role of nondestructive evaluation [NDE] in infrastructure maintenance programs (NRC, 1987). Proposals were set forth for the application of NDE inspection programs for highway systems, building systems, and water piping systems.)

Fritz Seilor (of Lovelace Research Institute) introduced to the workshop a rationale for selecting specific highway components, such as road surfaces, roadsides, bridges, and support structures, for more frequent inspection and/or maintenance. This approach is based on a risk-weighted random sampling plan, with the risk expression obtained on the basis of management concerns such as cost, risk reduction, and risk management (Seilor, 1987). Seilor also suggested the concept of "fuzzy sets" as a means of introducing the less quantifiable risk considerations into the decision process.

Risk-Based Standards

Chris Whipple (of EPRI) examined the issue of risk-based standards for the workshop (Whipple, 1987). He pointed out that risk-based standards are a reaction to a more complicated set of objectives for risk management and to the increased licensing of activities that were previously regulated by market forces, likely to continue to increase as the private sector continues to provide more public works services. Although privatization may provide more cost-effective delivery of services, public solutions provide greater political accountability. The ultimate success of privatization will depend upon the ability of the public sector to retain the safeguards that are required of public works.

As was noted earlier, standards are well-based in engineering experience, often bad experience. Boiler standards were initiated or tightened when boilers exploded. Today's emerging technologies and solution options are not necessarily well-suited for trial-and-error management; the cumulative operating experience has not lasted long enough to provide a good data base of risks.

Standards also have the characteristic of reflecting professional judgment and not much public participation. In today's environment the public is taking a more active role in public works decisionmaking; questions as to how safe a facility or situation is, how it can be made safer, and what the engineering details are may not be of general interest, but questions as to how safe a facility is and what we are willing to pay for safety are public value questions.

The value of risk analysis in this case is not to learn whether one critical number has been exceeded. The purpose is to find out what the sources of risk are and where the biggest uncertainties are. Risk analysis seeks to identify the major contributors to risk so efforts more effectively fix things that contribute to safety; in effect, it indicates where you should be worrying. One of the big benefits of risk assessment has occurred when it showed that safety work was directed at trivial contributors to risk. In nuclear power plant design it turns out that the factors that contribute heavily to risk are the things that people did not pay much attention to initially: things such as loss of electrical power to the site, loss of in-site emergency power, and earthquakes.

Quite clearly shifting to risk-based standards from traditional professional consensus standards is not a switch from one point of view to another. It is more appropriate to overlay the risk point of view onto the standards or codes and to sort through them to find which ones do not make sense. Thus, we redefine and re-evaluate codes from a more visible, more analytical framework that explicitly includes risk analysis. This approach also enables broader consideration of alternative solutions, including low-cost alternatives and innovative technology.

Highway Design

In the field of highway design it is well known that the cost-effectiveness of roadside design improvements can vary widely between highway sections because of accident rates, traffic volumes, terrain, required construction quantities, unit construction costs, and right-of-way requirements (Graham and Harwood, 1982). Jorgensen developed a methodology that avoids the specification of a rigid solution for improvement design based on standards. Instead it provides the designer with information on cost-effectiveness to make an objective decision from amongst a set of alternatives. This approach works well for low-volume roads. The authors noted the sensitivity of cost-effectiveness to traffic volumes (Jorgensen Associates, 1978).

Problems can arise when design improvements act principally to reduce the severity rather than the number of accidents, or when design improvements result in higher speeds and increased accident severity.

Other aspects of highway engineering are also important because the legal liability of governments (the taxpayers) for deaths and injuries arising from street and highway defects has been growing at an increasing rate--$6.8 billion in claims in 1983, $2.1 billion in California alone (Anderson, 1987). Highway maintenance activities in general and traffic control devices in particular, constitute the major portion of alleged deficiencies in negligence claims against governments. These two aspects are quite dissimilar in many ways, and yet they are the most important regarding claims and decisions.

Traffic control devices are based on fairly detailed and understandable criteria from a national set of standards and generally adopted by state governments. Since the criteria for location and placement are so well prescribed, deviations or damage or lack of maintenance can be easily cited by aggressive plaintiffs.

On the other hand, highway maintenance does not have definitive standards and criteria for individual work elements. Although some public officials believe that vague standards will protect them in court, in fact, the courts are left to set standards of care through their decision and awards.

Thus the call for a risk management program for local authorities is increasingly necessary. The risk management program should include the following (Anderson, 1987): (1) identification of the risks; (2) determination of an appropriate risk management method; (3) implementation of the method; and (4) monitoring of the method. Risk management at the local level will involve improving safety, handling claims, training personnel, updating standards for design, operations, and maintenance, and reducing exposure from others' liability.

5.   LIMITATIONS OF RISK ANALYSIS

Risk analysis has proven helpful in quantifying uncertainty in public decisionmaking. It does have limitations, however, including the following:

- Estimates of the probability of extreme (low-probability) events is imprecise, whereas the total costs associated with different alternatives may be sensitive to these estimates.

- Those factors that cannot be measured in economic terms, such as social issues, environmental impacts, and loss of life, are more difficult to reflect in the risk analysis process but may be the most important considerations in making decisions.

- Results of a quantitative, risk-based analysis reflect probable annual cost, but a system failure may result in a single catastrophic loss from which the owner and others may never recover; thus the relevancy of the analysis to the interests of the parties involved may be questionable.

- Other developments, actors, and/or factors which enter the project's sphere of influence in the future are not entirely predictable and may invalidate the risk-cost determination.

- Risk analysis depends on the accuracy or reasonableness of the use of present techniques of system modeling.

- Not all combinations of consequences which can result in system failure are known.

- Risk-based analysis is not inexpensive, although such analysis could result in savings to the system operator.

These limitations should not deter engineers from using risk-based analysis because the procedures are not intended to replace appropriate design standards. It is a tool for decisionmaking with a basis in logic, unlike arbitrarily developed tables of hazards or lists of specifications.

6. CHALLENGE OF RISK-BASED DECISIONMAKING FOR PUBLIC WORKS

The challenge at this time is to examine the cost, risk, and performance implications inherent in engineering standards and specifications for public works facilities. It would be helpful to examine how a risk-based methodology might be applied to the engineering planning, design, and construction phases of project decisionmaking so that a broader range of alternative solutions is available for public works decisionmakers. Consideration can be given to how risk assessment and risk management can be applied to decisions regarding public works facilities prior to their design and construction. Future work should address such questions as:

- What are the component parts of a risk-based methodology for public works planning and design?

- How does a risk-based decisionmaking methodology compare with one which is standards-based, and what are its advantages and disadvantages?

- What are the most favorable aspects of the two approaches, and are they sufficiently compatible that a more effective process can be developed?

- How can risk-based decisionmaking be integrated into the existing public works decisionmaking process?

REFERENCES

Anderson, Roy W. 1987. Factors commonly found in negligent highway design and maintenance cases and methods to reduce risks. Presentation to National Institute on Municipal Liability, American Bar Association.

Graham, J. L., and D. W. Harwood. Effectiveness of clear recovery zones. NCHRP Report 247. Transportation Research Board, National Research Council. Washington, D.C.

Haimes, Yacov Y., and Eugene Z. Stakhiv (eds.). 1985. Risk-Based Decision Making in Water Resources. American Society of Civil Engineers.

Jorgensen Associates. 1978. Cost and safety effectiveness of highway design elements. NCHRP 197. Transportation Research Board, National Research Council. Washington, D.C.

Meyer, T. A. 1987. Risk based decision making with potential application to public works improvements. Report prepared for the National Council on Public Works Improvement.

National Research Council. 1987. Opportunities for improving reliability of public works using nondestructive evaluation. In Infrastructure for the 21st Century: Framework for a Research Agenda. Report prepared for the National Council on Public Works Improvement.

National Research Council. 1983. <u>National Risk Assessment in the Federal Government: Managing the Process</u>. Washington, D.C.: National Academy Press.

National Research Council, Committee on Safety Criteria for Dams. 1985. <u>Safety of Dams: Flood and Earthquake Criteria</u>. Washington, D.C.: National Academy Press.

Seilor, Fritz A. 1987. Risk assessment, weighted random sampling, and highway improvement. Report prepared for the National Council on Public Works Improvement.

Whipple, Chris. 1987. Risk-based standards in engineering. Report prepared for the National Council on Public Works Improvement.

SUMMARY OF RESPONSES TO PARTICIPANT QUESTIONNAIRE

On the last evening of the conference, a questionnaire was distributed to all participants, requesting answers to three key questions about the conference: 1) important issues/aspects/elements raised, 2) new ideas/concepts predicted to be helpful on the job, and 3) issues needing further study. An edited list of participant responses follows.

1) List the three most important issues/aspects/elements that were raised during the conference related to risk-based decisionmaking.

   • Risk analysis (RA) is an additional tool, knowledge-based, to help decisionmakers.

   • We still have considerable differences in opinion on the applicability of RA. We range from assuming all risk is covered in the event probability to thinking that RA is the actual, final decision.

   • A standard without RA might be "just" but very "unequal."

   • Improving risk-based decisionmaking.

   • Measuring WTP for risk reductions.

   • Processes for decisions are important.

   • Definition of risk analysis.

   • Differentiate cause and effects. Look at sources of uncertainty problems instead of how to get by it for decisionmaking.

   • Understanding and characterization of health hazards, man-made engineering hazards, and natural hazards.

   • Non-homogeneity among members of a class of decisionmakers.

   • Time sequencing of decisions incorporated in risk analysis.

   • Tendency to postpone risky decisions may be rational.

   • Need for risk analysis in making policies.

   • Methods of incorporating people's willingness to pay for risk reduction.

   • Importance of perceptions in dealing with risk.

   • Brought Paté-Cornell and Lester Lave together with the engineers.

   • Engineers' liability.

   • Determining the objectives of the risk analysis prior to determining methodology.

   • Work with variance ranges as well as data point means when determining risk-based decisions.

   • Communicating risk issues and risk analysis results to lay audiences, i.e., decisionmakers and general public. Obvious in difficulty some had communicating results to assembled experts.

- Need to develop analytical techniques which do not rely on expected values.
- Need to reconcile our analytical models/tools with way in which public regards risks.
- Risk analysis has a wide range of applications (regulation, management, public education, etc.).
- There are no generally accepted definitions or unique methods available.
- Risk analysis can be used to establish standards.
- Few decisions are based on realistic assessments.
- Disconnect between security of risks raised by a problem and research fund attention to the problem.
- Split between focuses of technical impacts people and people who deal with popular attitudes and behavior.
- Risk communication.
- Health risk management.
- Risk analysis is simply a "Decisionmaking Process" function.
- The issues surrounding the integration of risk assessment into decisionmaking (i.e., where and how does it fit?).
- The whole area of risk perception/risk communication.
- Raised (but not adequately addressed): the fact/value issue in risky decisionmaking.
- Standards versus risk analysis.
- Risk communication.
- Unified approach for risk analysis.
- Risk analysis is basically a case-by-case technique or approach.
- There is little agreement in definition of risk analysis concepts. (There seems to be more consistency in practice, ironically.)
- Risk analysis versus standards use -- discussion of liability and how risk might be used in lieu of standards.
- Use of risk analysis rather than extreme value to determine design criteria.
- Is risk analysis new, and what is it in the methodology that sets risk analysis apart from traditional engineering analyses? It is difficult to understand what it is and how it is different from some existing DM techniques.
- Difficulty in reconciling scientific/engineering risk assessments with social (normative) decision theories.
- There are a significant number of seemingly adequate techniques for scientific/engineering risk assessments.
- New methods do encode uncertainty in risk analysis.

- Risk perception studies.

- Liability versus risk analysis.

- The diversity of important natural and man-made hazards which are candidates for application of risk assessment.

- The importance of risk communication to enable decisionmakers to be made aware of risk assessment information.

- Importance of sharing risk among larger population to enable hazardous activities to be accepted in a community.

2) List the three most important new ideas/concepts that you learned during this conference which would be helpful in your job.

- Adhering to standards only may not only not give the best solutions but may not be professionally defensible.

- We've been neglecting the final step of RA. It might well give a decision other than the economic or strict efficiency choice.

- There is much more good work being done in risk analysis than I imagined.

- Minimax regret versus expected value decisionmaking.

- Dam management principles.

- Irrigation management issues.

- More about Bayesian techniques.

- New flood forecasting techniques.

- Risk modeling techniques.

- Irrigation management issues.

- Successes and failures of getting risk assessment adopted by large agencies.

- Caught up with what others were doing.

- Risk communication techniques studied by Bill Desvousges.

- Smaller noxious facility siting discussed by Sam Ratick.

- Conflict between expert opinion and public perception presents an interesting dynamic optimization problem.

- Interesting/novel concepts: Micromorts, fuzzy set theory, nucle-arization, etc.

- When the issue is complex we have difficulty communicating the impacts to each other, much less to laymen.

- The idea that belief systems have an impact on risk analysis.

- Almost any kind of analysis of natural and man-made hazards is considered as being risk analysis.

- Delli Priscoli's notion of "risk taking" as a shared endeavor rather than a cost to be imposed.

- Ratick's noxious facility siting material that related facility size, number, spatial factors, and acceptability.
- Desvousges' use of colors to present risk information regarding radar.
- Risk communication.
- Public perception of risk is influenced/established by the media (TV).
- We as scientists must help to focus public attention on the most immediate scientific problem rather than the sexy, media-hyped problem.
- Micromorts and other buzzwords for communication.
- Importance of being a Bayesian.
- When to do or when not to do risk analysis.
- Research on warning systems.
- Warning system concepts.
- There is less than anticipated transfer of risk analysis technologies from other fields to water resources planning.
- Pate-Cornell Warning Theory.
- Added dimension of risk versus standard use.
- Possible use of many fewer years of data for design studies (Lester Lave).
- What risk analysis is and is not.
- Application of fuzzy set theory to risk-based decisionmaking.
- Sensitivity to the limits of risk analysis and management.
- Network of experts to aid in guidance and review of work and issues.
- Use of fuzzy set theory.
- Combination of engineering and health risk.
- Use of available risk assessment.
- Risk assessment--not a cookbook. Tailor the analysis to the problem.
- Importance of risk assessment to institutions as well as natural/-man-made systems.
- Key role of warning systems in risk assessment in certain situations.

3) List the three most important issues needing further study in risk-based decisionmaking in water resources.

- Try to develop a uniform definition or description of risk analysis.

- Try to determine if RA is more effective if it's applied throughout the investigation or as a final "check."

- Look for examples where risk analysis gave a "better" or at least different decision.

- Techniques to identify sources of uncertainties.

- Methods to estimate risk quantitatively, accounting for various uncertainties and randomness.

- Avenues to use the above information in decisionmaking.

- Involving multiple parties in risk-based decisions.

- Relative efficiency (overall, and in terms of opportunities for poor professional judgment) of high analysis or standards.

- How to make risk analysis relevant in actual policy decisions.

- How to include social, economic, legal, and institutional parameters into risk assessment.

- How to make interdisciplinary studies successful in the policy arena.

- Need to improve calculation of risk.

- Need to learn to evaluate cost of loss of life, etc.

- Need to keep pushing to get techniques accepted.

- Risk communication.

- Risk mediation.

- Risk management.

- Interface between the complex models of the researchers and the pragmatic situation of limited data, expertise, time, and money among practitioners.

- Reconciling our models to way people really make decisions/choices -- e.g., minimizing the maximum effects, extreme value models, etc. Some analytics needed to show effects of alternative preference/decision models.

- Communicating risks to various lay publics, particularly general public, to deal with the nuclearization of issues.

- Peripheral impacts of climatic change (vegetation, water seasonal distribution, supply).

- Develop a better definition of risk analysis procedures so they can be communicated to others.

- The probability of remote events.

- Relation between world views of public agency decisionmakers and scientists/engineers.

- California water pollution resulting from agricultural investigation.

- Match between public expectations and technical possibilities.

- Risk communication to public.

- Change/modify public perception of risk.

- Policy and priority setting based on risk analysis.

- The adequate development of water resource decisionmaking institutions to accommodate risk-based information.

- The precise mechanisms of identifying "acceptable risk" in an appropriate manner (not as scientific judgment, but as social choice).

- Identifying a mechanism to match resource allocation to "highest priority" risk (however defined).

- Clear definitions for risk and a unified approach for risk analysis (I believe this is possible).

- Ways to educate decisionmakers of elements of risk analysis.

- Techniques of risk analysis.

- Need for better means of communications of risk to decisionmaker and public.

- Need to get decisionmakers to sit in on sessions like this (of course, somewhat less technical content).

- Need more work on identification of benefit risks.

- Whether Lester Lave's study is valid with respect to data set used for the analysis.

- In the area of climate analysis: would like to see some attempt made at use of probabilities in GCMs and analysis of climate change effects.

- Issue of whether probabilities are essential to risk analysis.

- Development of probabilistic forecasting schemes for extreme natural hazard events.

- Modeling and estimation of uncertainties in sequential forecasts of natural hazards.

- Modeling and development of warning-response systems for natural hazards.

- Attention to demand side risk-based decisionmaking.

- More work on integrating risk assessment, risk management, and decision theories.

- A focus on the ethics of using "risk" in public sector decisionmaking.

- Use of fuzzy set theory.

- Communication of risk analysis results to decisionmakers.

- Risk management.

Engineering Foundation Conference

RISK ANALYSIS AND MANAGEMENT OF NATURAL AND MAN-MADE HAZARDS

Sheraton Santa Barbara Hotel & Spa
Santa Barbara, California

November 8-13, 1987

Sunday, November 8, 1987

3:00 p.m. - 9:00 p.m.   REGISTRATION

6:00 p.m.   DINNER

9:00 p.m. - 11:00 p.m.   SOCIAL HOUR

Monday, November 9, 1987

7:30 a.m.   BREAKFAST

9:00 a.m. - 12:00 Noon   RISK ANALYSIS PERSPECTIVES

                             Chairman: Y. Haimes, University of Virginia

                             "Alternative Risk Evaluation Paradigms"

                                  W. D. Rowe, Rowe Research

                             "Research Issues in Natural and Man-Made
                                  Hazards"

                                  N. Sabadell, NSF

                             "Engineering Approaches to Risk and
                                  Reliability Analysis"

                                  B. Yen, University of Illinois

                             "Public Engineering, Ethics and Risk
                                  Analysis"

                                  J. Delli Priscoli, IWR

                             "An Overview of Risk Analysis Approaches at
                                  EPA"

                                  T. Barry, EPA

12:00 Noon   LUNCH

2:00 p.m. - 5:00 p.m.   AD HOC SESSION

Monday, November 9, 1987 (continued)

5:00 p.m.   SOCIAL HOUR

6:00 p.m.   DINNER

7:30 p.m. - 10:00 p.m.   RISK COMMUNICATION DECISIONMAKING
                             Chairman: Curtis Brown

                             "Eliciting Risk Values and Measuring
                                 Preferences"
                                 R. Howard, Stanford University

                             "Risk Communication"
                                 D. von Winterfeldt, U.S.C., and
                                 V. Covello, NSF

                             "Risk Evaluation of Extreme Events"
                                 Y. Haimes, University of Virginia

                             "Measuring Risk Perceptions: Two Case
                                 Studies"
                                 W. Desvouges, RTI

                             "A Case Study in Water Supply Risk
                                 Communication"
                                 J. Boland, Johns Hopkins University

                             "Issues in Climate Uncertainty"
                                 S. Schneider, NCAR

10:00 p.m.   SOCIAL HOUR

Tuesday, November 10, 1987

7:30 a.m.   BREAKFAST

9:00 a.m. - 12:00 Noon   ENVIRONMENTAL RISK ANALYSIS
                             Chairman: T. Barry, EPA

                             "Managing Environmental Hazards to Health"
                                 P. Portney, RFF

                             "Dredge Material and Ocean Sludge Disposal"
                                 T. Leschine, University of Washington

Tuesday, November 10, 1987 (continued)

"Ecological Risk Analysis"

S. Bartell, Oak Ridge Laboratory

"Marine Pollution: Perception of Risks"

H. Levenson, OTA

"Uncertainty in Environmental Risk Analysis"

I. Bogardi, University of Nebraska,
L. Duckstein, University of Arizona, and
A. Bardossy, University of Karlsruhe

12:00 Noon	LUNCH
2:00 p.m. - 5:00 p.m.	AD HOC SESSION
5:00 p.m.	SOCIAL HOUR
6:00 p.m.	DINNER
7:30 p.m. - 10:00 p.m.	ENVIRONMENTAL HEALTH HAZARDS

Chairman: Jonathan Bulkley

"Risk Analysis of Groundwater Pollution"

V. Changkong, Case Western University

"Risk Sharing in the Location of Noxious Facilities"

S. Ratick, Clark University

"Managing Environmental Health Hazards for Agricultural Runoff"

E. Lichtenberg, University of California, San Francisco, and
D. Zilberman, University of California, Berkeley

"Risk Assessment of Heavy Metals in Irrigation Return Flows"

M. Caswell, University of California, and L. D., James, Utah State University

10:00 p.m.	SOCIAL HOUR

Wednesday, November 11, 1987

7:30 a.m.	BREAKFAST

Wednesday, November 11, 1987 (continued)

9:00 a.m. - 12:00 Noon   GLOBAL WARMING AND CLIMATE CHANGE

Chairman: E. Stakhiv, IWR

"Modeling the Hydrologic Effects of Climate Change -- A Case Study"

D. Sheer and D. Lettenmeier, University of Washington

"Hydrologic Modeling of Climate Change -- A Case Study"

P. Gleick, University of California

"Risk-Cost Analysis of Sea Level Rise Solutions"

D. Moser and E. Stakhiv, IWR

"Application of Extreme Value Theory to Great Lakes Rise"

J. Bulkley, University of Michigan

12:00 Noon   LUNCH

2:00 p.m. - 5:00 p.m.   AD HOC SESSION

6:00 p.m.   DINNER

7:30 p.m. - 10:00 p.m.   WORKSHOP DISCUSSION OF KEY ISSUES

Chairman: J. Delli Priscoli, IWR

Workshop 1.   Risk Evaluation Paradigms

Chairman: D. Bauman, Southern Illinois University

Workshop 2.   Analytical Approaches to Risk Analysis

Chairman: R. Krzysztofowicz, University of Virginia

Workshop 3.   Standards Setting Versus Risk Analysis

Chairman: B. Paul, Bureau of Reclamation

10:00 p.m.   SOCIAL HOUR

Thursday, November 12, 1987

7:30 a.m.	BREAKFAST

9:00 a.m. - 12:00 Noon   RISK MANAGEMENT STRATEGIES FOR NATURAL HAZARDS

Chairman: R. Krzysztofowicz, University of Virginia

"Possibilities for Implementing Probabilistic Flash Flood Forecasts"

S. F. Zevin, NWS

"Warning Systems in Risk Management"

M. E. Paté-Cornell, Stanford University

"The Costs of Not Being a Bayesian"

R. Krzysztofowicz, University of Virginia

"Alternative Risk Management Strategies"

D. Okrent, UCLA

12:00 Noon	LUNCH
2:00 p.m. - 5:00 p.m.	AD HOC SESSION
5:00 p.m.	SOCIAL HOUR
6:00 p.m.	DINNER

7:30 p.m. - 10:00 p.m.   RISK MANAGEMENT STRATEGIES FOR TECHNOLOGICAL HAZARDS

Chairman: A. Slay, Slay Enterprises, Inc.

"Oil Drilling Wastes and Risk of Groundwater Contamination"

B. Hobbs, Case Western Reserve University

"Dam Safety Risk Analysis, Rio Grande River"

L. Lave, University of Pittsburgh

"Fault Tree Simulation Analysis of a Dam Safety Problem"

J. Stedinger, Cornell University

Thursday, November 12, 1987 (continued)

"Risk Analysis in Navigation Planning"
C. Yoe, IWR

"Risk Assessment of Dams in Series"
D. Bowles, Utah State University

10:00 p.m.   SOCIAL HOUR

Friday, November 13, 1987

7:30 a.m.   BREAKFAST

9:00 a.m. - 12:00 Noon   WRAP-UP, PLENARY SESSION, PANEL REPORTS,
                          DISCUSSION

                          Panel 1. Risk Evaluation Paradigms

                          Panel 2. Analytical Approaches to Risk
                                    Analysis

                          Panel 3. Standards Setting vs. Risk Analysis

SUBJECT INDEX
Page number refers to first page of paper.

AUTHOR INDEX
Page number refers to first page of paper.